W9-BGI-386

Handbook
for
Space
Pioneers

Prepared by L. Stephen Wolfe and Roy L. Wysack

Publishers • GROSSET & DUNLAP • New York
A FILMWAYS COMPANY

Acknowledgments

The Pioneer Information Division of the Galactic Association gratefully recognizes the special contribution of the following pioneers and staff members in the preparation of this bulletin:

Aristotle Chianakis
James Colligan
Jeannette M. De Wyze
Kirk Dupuis
Indira S. Hodara
Earle G. Horne
Megan Kearin
Alex R. X. Kotorovich
Ezra P. Lilly
Rose McGillicuddy
Francoise Patreau
Diana Price
George Soonge
Valerie Wysack

Published simultaneously in Canada.
Library of Congress catalog card number: 78-55235
First printing.
ISBN: 0-448-16178-8 (hardcover edition)
ISBN: 0-448-16185-0 (paperback edition)
Printed in the United States of America.

Inforetrieval Computer Abstract

DATA FILE CODE: INFOPIONEARTH

SCOPE:
This bulletin contains introductory information about the program of planetary colonization sponsored by Earth Branch of the Galactic Association of Intelligent Life (GAILE).

PURPOSE:
This bulletin is intended to introduce the GAILE program to Earth residents. It provides general information about the administrative aspects of the program and a brief description of each of the eight planets to which Earth citizens may emigrate.

APPLICABILITY:
For more detailed information about other aspects of the GAILE colonization program for Humans, refer to the bulletins listed in appendix B. Members of other Association species should refer to the applicable general bulletins listed in appendix C. This bulletin supercedes previous bulletin GAI-GB-502. Relevant material has been excerpted therefrom.

PRECEDENCE:
None higher within the following reference files: GAIL, GAILE, PIONEERING EFFECTIVE DATE: 4 July 2376 SPACE MIGRATION, COLONIES, EMIGRATION.

Contents

TABLES

ILLUSTRATIONS

"Just as a star's light shines across the galactic void, the life of its planets transcends individual worlds. Earth's life has outgrown it, sucked dry its resources, polluted its air and water, and crowded its lands from shore to shore.

"Today begins our migration to new worlds and to a new, fresh, clean life for Humankind. As we spread across the galaxy, we will grow in wisdom, knowledge and ability. This wisdom will in turn revitalize the Earth that nurtured us. You who depart today comprise the first droplets of a giant flood that soon will follow, for nature dictates that all people will become PIONEERS FROM EARTH!"

From remarks by Lee Alan,
Director of Space Flight Operations,
International Council for Space Exploration,
at the departure ceremonies for the first
Wyzdom pioneers, 23 April 2114 adtc

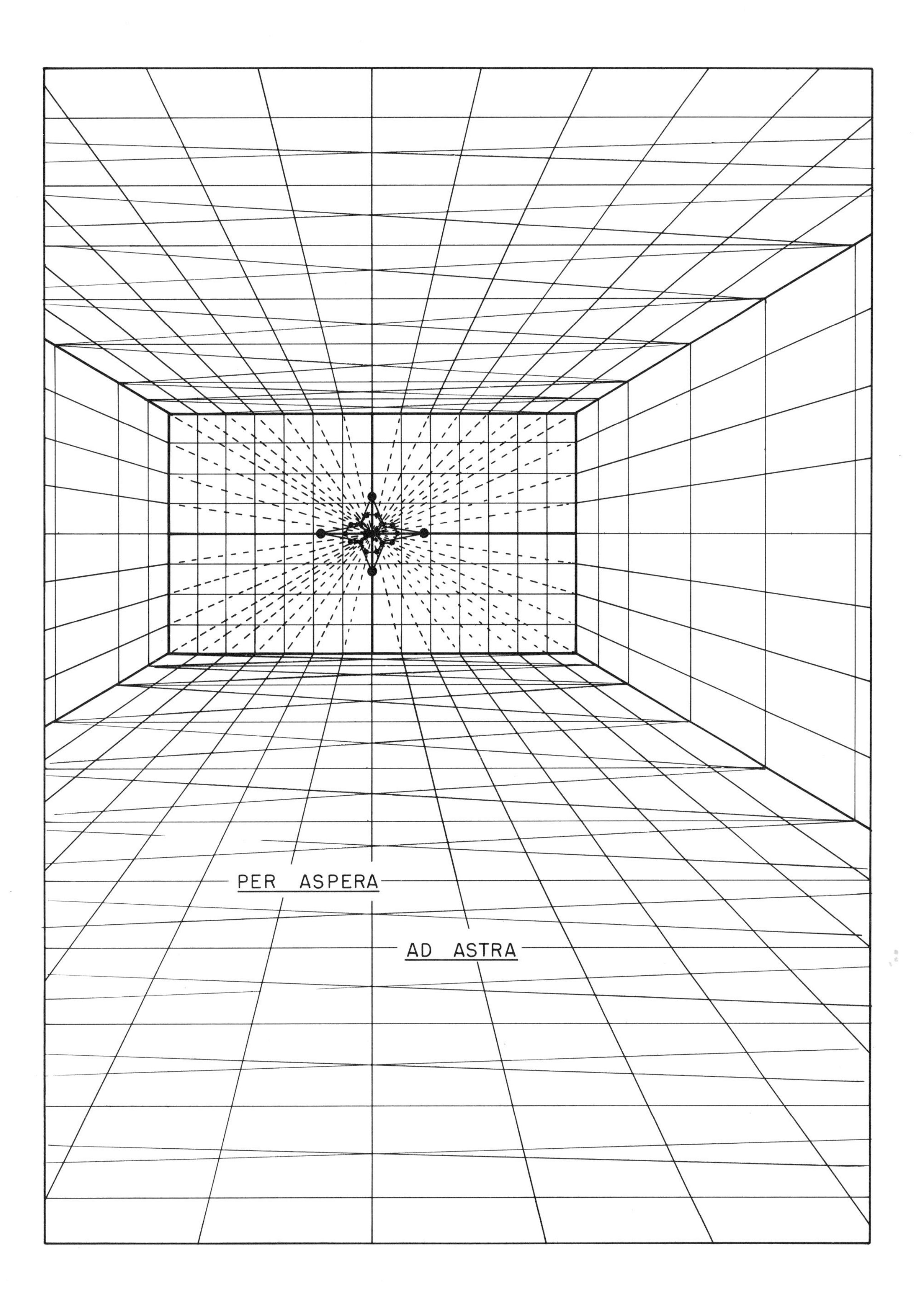
PER ASPERA
AD ASTRA

Introduction

Pioneering isn't easy. It wasn't easy when Polynesians crossed the storm-swept Pacific Ocean to Hawaii. It wasn't easy when Europeans hacked their farms out of the cold and dangerous forests of colonial America. Today, it still takes courage and determination for 24th century Humans to forsake all Earthly possessions and travel light years to unfamiliar worlds filled with alien life.

Yet contemporary pioneers have advantages their historic counterparts lacked. Modern starships transport passengers across interstellar distances in greater safety and comfort than the leaky sailing vessels that carried ancient emigrants. Modern science provides us with more thorough portraits of the new worlds than past pioneers had of new continents. That information, combined with modern technology, enables today's pioneers to confront their new environment comfortably. Few people today would want to live in log cabins, eat spoiled food, or forgo the conveniences of artificial light, climate control, and powered appliances. On today's frontiers, they need not.

Yet creating in new worlds the advantages of modern life requires many people with many skills. It takes more than space pilots and scientists to build a new society. Farmers, factory workers, cooks, computer programmers, accountants, and garbage collectors all play important roles. You don't have to be rich, brilliant, beautiful, or politically well-connected to stake your claim on the frontiers of the galaxy. All you need is a desire for a better life and a willingness to work for it.

The Galactic Association of Intelligent Life (GAIL) is the organization which makes it possible for anyone to become a pioneer. GAIL's Earth Branch (GAILE) is committed to the development of eight planets suitable for Human beings. This current bulletin has a threefold purpose: to acquaint readers with the GAILE program, to tell them how they can become a part of it, and to help prospective pioneers make the most important decision of their lives—the choice of their future world.

Figures 1.1 and 1.2 show how you can see the eight planets displayed in the night sky. These suns of the colonial worlds aren't the brightest stars, but they are no doubt the most important for Humankind. Each wears about it a unique jewel of the heavens, a planet as big, as varied, and as beautiful as Earth once was.

The planets range in size from giant Wyzdom, with 1.3 times Earth's surface area to tiny Mammon with just 60 percent of it. Their oceans vary between the all-encompassing ocean of Poseidous to the isolated seas that cover just half of Mammon's surface. Some planets boast extremely complex plants and animals such as the giant domesauris of Brobdingnag or the dexterous trup of Yom. One planet, Genesis, has no land life to call its own.

More important than the planets' natural phenomena are the states of their Human development. Wyzdom, halfway through its third century, has a self-sufficient industrial economy offering all the products and services available on Earth. The rest of the colonies display comparatively less complex industry, in proportion to their ages. Romulus, just sixteen years old, still depends heavily on Earth's imports, while Athena has yet to receive her first permanent settler!

This bulletin cannot hope to explain in detail every aspect of life on all eight colonies. It is meant merely as a sampler of the opportunities GAILE offers. While you are reading, remember that each planet is far vaster and more varied than any single continent of Earth. Each offers every climate from tropical forests to arid deserts and frozen tundra, and every landscape from seacoast bays to mountain valleys and great plains. No doubt some readers will find physical features or forms of life that make one planet particularly attractive, but most will find the Human lifestyles of each world more important. The older worlds offer the comforts of Earth and the chance to employ your practiced trade or profession while still partaking of the benefits of a young and growing society. The younger colonies offer more ambitious pioneers the chance to build a new culture from scratch, to participate in forming its institutions and shaping its values. The risks of the newer colonies are higher, but the rewards may be infinitely greater.

After you have read this handbook, you will undoubtedly want to learn more about specific planets. The bulletins listed in appendix C describe each world in greater detail. GAILE wants every pioneer to choose the planet that best suits his or her desires, ambitions, and abilities, for GAILE and the entire Human species benefit from the success, happiness, and prosperity of the pioneers from Earth.

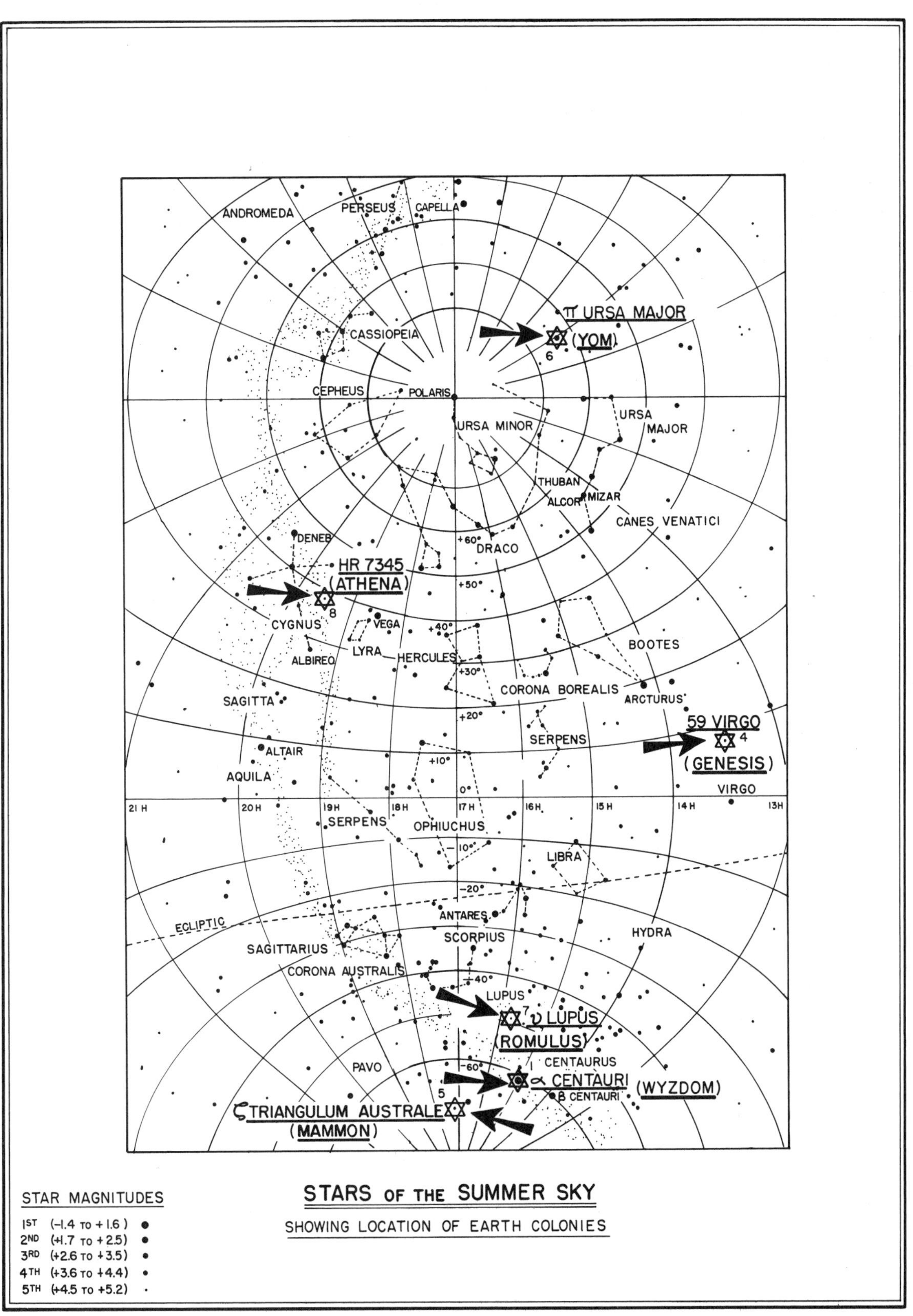

ANDROMEDA
PERSEUS
CAPELLA
π URSA MAJOR
(YOM)
6
CASSIOPEIA
CEPHEUS
POLARIS
URSA MINOR
URSA
MAJOR
THUBAN
MIZAR
ALCOR
CANES VENATICI
DENEB
+60°
DRACO
HR 7345
(ATHENA)
+50°
8
CYGNUS
VEGA
+40°
LYRA
HERCULES
BOOTES
ALBIREO
+30°
CORONA BOREALIS
SAGITTA
ARCTURUS
+20°
59 VIRGO
4
(GENESIS)
SERPENS
ALTAIR
+10°
AQUILA
0°
VIRGO
21 H
20 H
19H
18H
17H
16H
15 H
14H
13H
SERPENS
OPHIUCHUS
-10°
LIBRA
-20°
ECLIPTIC
ANTARES
SCORPIUS
HYDRA
SAGITTARIUS
CORONA AUSTRALIS
-40°
LUPUS
7
υ LUPUS
(ROMULUS)
CENTAURUS
PAVO
-60°
1
α CENTAURI
(WYZDOM)
5
β CENTAURI
ζ TRIANGULUM AUSTRALE
(MAMMON)
STAR MAGNITUDES
1ST (-1.4 TO +1.6)
2ND (+1.7 TO +2.5)
3RD (+2.6 TO +3.5)
4TH (+3.6 TO +4.4)
5TH (+4.5 TO +5.2)
STARS OF THE SUMMER SKY
SHOWING LOCATION OF EARTH COLONIES

FIG. 1.1

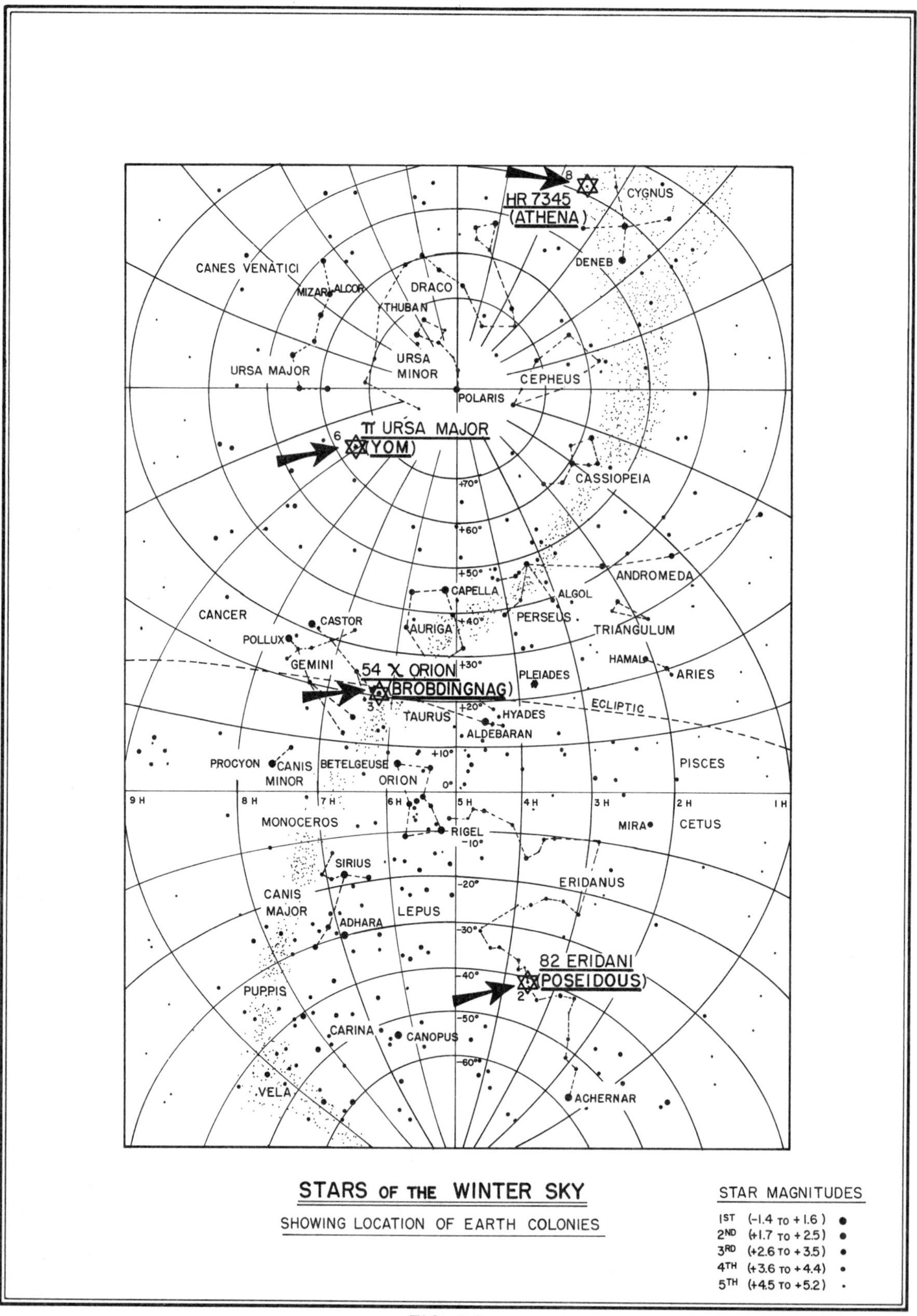
HR 7345
(ATHENA)
CYGNUS
DENEB
CANES VENATICI
MIZAR
ALCOR
DRACO
THUBAN
URSA MAJOR
URSA MINOR
CEPHEUS
POLARIS
π URSA MAJOR
(YOM)
CASSIOPEIA
ANDROMEDA
CAPELLA
ALGOL
CANCER
CASTOR
AURIGA
PERSEUS
TRIANGULUM
POLLUX
HAMAL
ARIES
GEMINI
54 χ ORION
(BROBDINGNAG)
PLEIADES
ECLIPTIC
TAURUS
HYADES
ALDEBARAN
PROCYON
CANIS MINOR
BETELGEUSE
ORION
PISCES
9 H
8 H
7 H
6 H
5 H
4 H
3 H
2 H
1 H
MONOCEROS
RIGEL
MIRA
CETUS
SIRIUS
ERIDANUS
CANIS MAJOR
ADHARA
LEPUS
82 ERIDANI
(POSEIDOUS)
PUPPIS
CARINA
CANOPUS
VELA
ACHERNAR
+70°
+60°
+50°
+40°
+30°
+20°
+10°
0°
-10°
-20°
-30°
-40°
-50°
-60°
STARS OF THE WINTER SKY
SHOWING LOCATION OF EARTH COLONIES
STAR MAGNITUDES
1ST (-1.4 TO +1.6)
2ND (+1.7 TO +2.5)
3RD (+2.6 TO +3.5)
4TH (+3.6 TO +4.4)
5TH (+4.5 TO +5.2)

FIG. 1.2

The Pioneering Program

GAIL is a voluntary association of intelligent and scientifically advanced species formed to promote the exchange of information between their civilizations and the orderly and peaceful exploration of space. It includes four member species—Ardotians, Chlorzi, Ergints, and Humans—and negotiations are under way to include a potential fifth species, the Minutae. GAIL is not an interplanetary government. It has no power to tax and must rely on the resources provided by each member for that member's own exploration and development activities.

GAIL maintains a permanent headquarters orbiting a small class M star about 221 light years from Earth. There, a small staff composed of members of each GAIL species coordinates efforts between members, transmits information, and serves as diplomatic liaison. It is GAIL Earth Branch, however, which undertakes the space effort on behalf of Earth's people. The Earth Branch performs four basic functions:

1. Exploration of the galaxy for new worlds.
2. Transportation of pioneers to the colonial planets.
3. Provision of critical equipment and supplies during the early years of a colony's development.
4. Exchange of information between the colony planets, Earth, and the rest of GAIL.

A corresponding organization exists for each of the other Association species, but Earth's program is currently the most active.

The vigor with which we Humans now pursue space exploration may result from our coming to it fairly recently. Little more than four centuries have passed since people made their "first" feeble excursions into space. The discovery of Wyzdom, the first habitable planet, just three centuries ago was as much a matter of good fortune as careful design. The development of space warp (see Interstellar Transportation and Communication) greatly extended people's reach into space and allowed us to discover two more habitable planets before encountering the Ardots in 2217. After the initial shock of that meeting wore off, the exchange of ideas that followed enhanced the knowledge of both civilizations tremendously, and did more to advance Humanity's exploratory capabilities than all previous space research. Ardots and Humans agreed to assist each other in the exploration of space to avoid duplicating each other's efforts. The Ardot's discovery of a third intelligent species, the Chlorzi of Ignati, necessitated a more formal system of cooperation which developed into GAIL.

Though GAIL has grown to encompass four members and a potential fifth, there will be no shortage of new worlds for all of them. Several members have more worlds than they need at present, for they have been exploring longer. Humans, too, may one day exceed their capacity to populate new planets, but today our supply of new worlds is limited only by our ability to find them.

The Milky Way galaxy contains about one hundred billion stars, but GAILE starships need not visit each one to determine if it possesses planets suitable for Humans. Astronomical observations made from Earth and the established colonies can eliminate 88 percent of all stars from the list of potential candidates. The power levels of very large or very hot stars fluctuate too rapidly to support life on any planets they might have. Though small or cool stars possess stable power levels, a habitable planet would have to orbit so close to them to receive enough energy that it would be torn apart by tidal forces. The remaining 12 percent of all stars must be explored one by one. To date, the entire space known to be explored by GAIL's members encompasses but a tiny fraction of our great galaxy as shown in figure 2.1. Although most of the galaxy lies beyond our current techology, our reach grows each year. As it grows, Humankind's potential for finding new worlds will cease to have limits. If our ratio of exploratory successes to date can be extrapolated, then our galaxy must hold at least six hundred million worlds habitable by Human beings!

Each year Earth's starships contact up to a dozen star systems. Of those, perhaps one may contain a planet suitable for colonization by any of GAIL's members. The discovering starship remains in orbit about its find for several years, gathering data and transmitting it simultaneously to Earth and to GAIL central. To avoid political bias and to insure equitable treatment for all GAIL members, a computer selects which member will colonize based upon both objective criteria and each member's desire for additional planets. The computer's program, approved unanimously by all Association members, considers such factors as the population of each member species, the number of exploratory starships it supports, and its economic ability to support additional colonies. The proximity of the new planet to each member's home planet and, most important, the suitability of the new environment to the needs of each member species have great influence on the decision.

One factor that rules out colonization of an otherwise suitable planet is the presence of a native species with potential for developing an intelligent civilization. The colonization of Earth by advanced aliens just 10,000 years ago, when people lived in caves and manifested no civilized behavior, would have precluded the development of Human civilization as we know it. Today GAIL's members unanimously concur that all beings capable of conceptualization, self-awareness, rational behavior, and the intelligence and ability to alter their environment should be allowed to develop their inevitable civilizations without interference from others. When such civilizations have matured to the point that they leave their home planets, they will be encouraged to join the Galactic Association.

In recent years GAIL's exploratory program has dramatically increased the number of colonizable planets available to people. As that number grows, and as the cost of interstellar transportation drops, Humanity may someday be able to reduce the population of Earth!

Figure 2.2 shows the relative positions of the Human colonies and the home planets of the other GAIL members as viewed from above the galactic plane. The actual distances to the colony worlds are shown as dashed lines for comparison. Contour lines depict the three-dimensional region of space explored by GAILE starships. It is a highly irregular space because exploratory missions are designed along paths which maximize the probability of finding habitable planets. The other intelligent species have similar explored regions around their home planets (which are not shown for clarity), and these regions now overlap those of the Human species. The extent to which the Ergints have explored the galaxy remains a mystery. They currently evince little interest in exploration but are interested in monitoring the activities of other GAIL members. Being creatures of "pure energy," the Ergints cross the interstellar voids, in some sense, without spaceships. Many scientists suspect that long ago these advanced beings mastered the entire galaxy and contacted civilizations that Humans will not know for centuries.

GAILE provides one-way transportation for pioneers from Earth to the colonial planets. The trip across the vast interstellar distances is made in Planck's torch-driven starships of modular design. The nature of the ships and their accommodations are described in detail in Interstellar Transportation and Communication. The ships employ space warp to shorten travel time, but nevertheless, voyages appear to travellers to take several months.

The huge cost of interstellar transport dictates that pioneers keep personal possessions to a minimum. Clothing, furniture, appliances, robots, and vehicles are generally not transported; instead they will be created from the basic materials of the colony planet. Credit, available to all new ar-

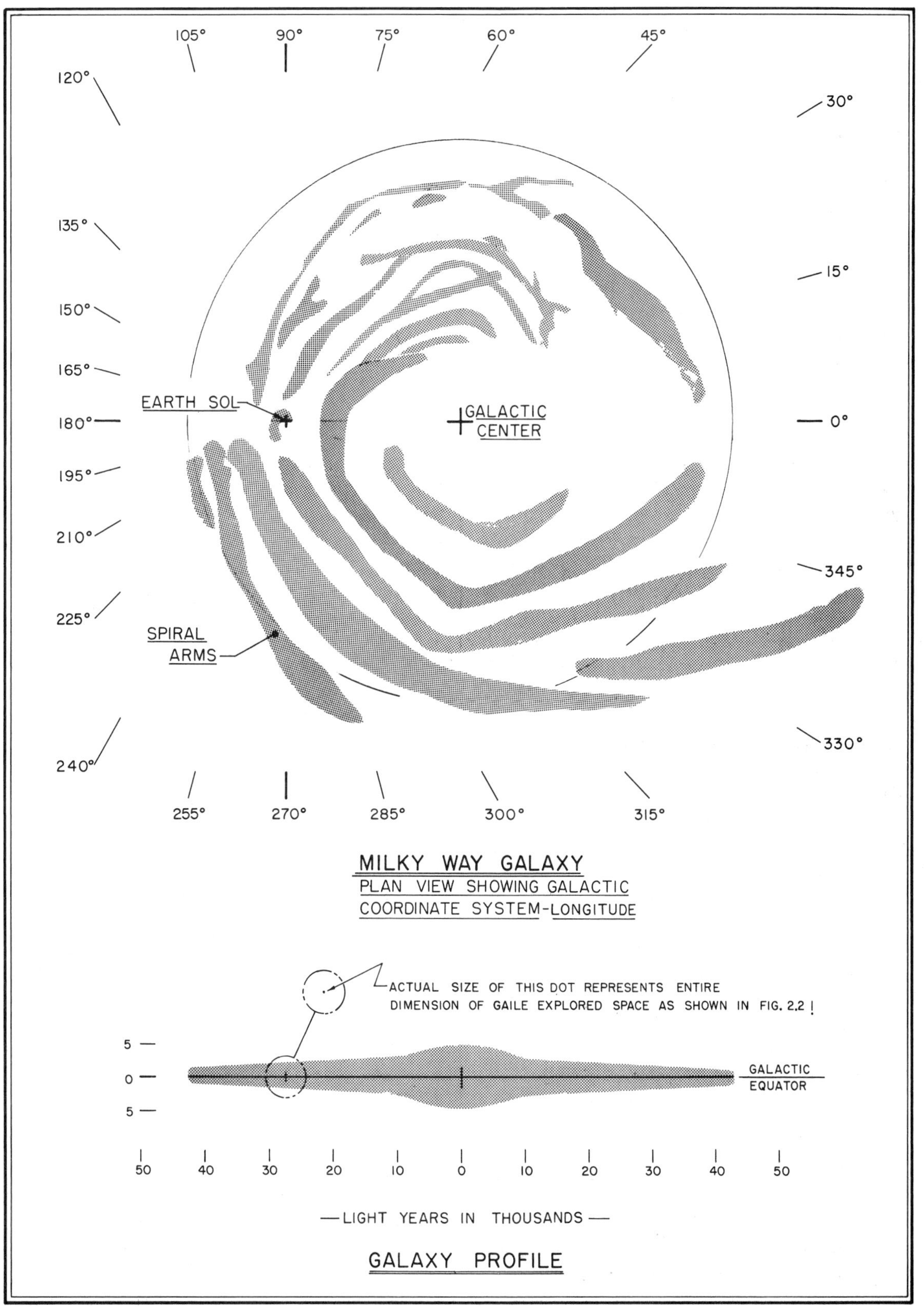
105°
90°
75°
60°
45°
120°
30°
135°
15°
150°
165°
EARTH SOL
180°
GALACTIC CENTER
0°
195°
210°
345°
225°
SPIRAL ARMS
240°
330°
255°
270°
285°
300°
315°
MILKY WAY GALAXY
PLAN VIEW SHOWING GALACTIC COORDINATE SYSTEM-LONGITUDE
ACTUAL SIZE OF THIS DOT REPRESENTS ENTIRE DIMENSION OF GAILE EXPLORED SPACE AS SHOWN IN FIG. 2.2 !
5
0
5
GALACTIC EQUATOR
50
40
30
20
10
0
10
20
30
40
50
—LIGHT YEARS IN THOUSANDS—
GALAXY PROFILE

FIG. 2.1

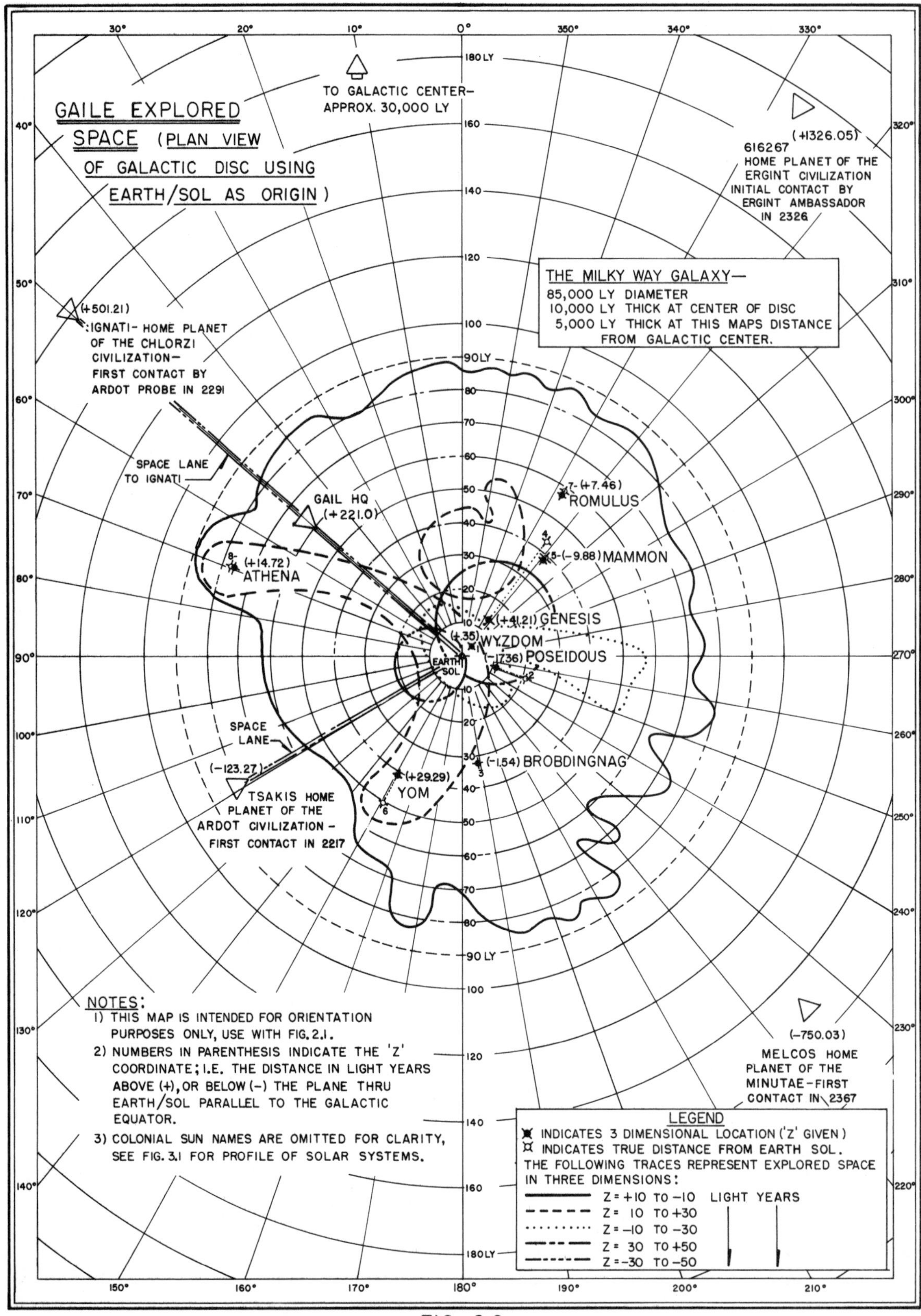
GAILE EXPLORED SPACE (PLAN VIEW OF GALACTIC DISC USING EARTH/SOL AS ORIGIN)
TO GALACTIC CENTER- APPROX. 30,000 LY
(+1326.05)
616267
HOME PLANET OF THE ERGINT CIVILIZATION INITIAL CONTACT BY ERGINT AMBASSADOR IN 2326
THE MILKY WAY GALAXY—
85,000 LY DIAMETER
10,000 LY THICK AT CENTER OF DISC
5,000 LY THICK AT THIS MAPS DISTANCE FROM GALACTIC CENTER.
(+501.21)
IGNATI- HOME PLANET OF THE CHLORZI CIVILIZATION- FIRST CONTACT BY ARDOT PROBE IN 2291
SPACE LANE TO IGNATI
GAIL HQ (+221.0)
7-(+7.46) ROMULUS
5-(-9.88) MAMMON
8-(+14.72) ATHENA
(+4.21) GENESIS
(+.35) WYZDOM
(-17.36) POSEIDOUS
EARTH SOL
SPACE LANE
(-123.27)
TSAKIS HOME PLANET OF THE ARDOT CIVILIZATION - FIRST CONTACT IN 2217
(+29.29) YOM
(-1.54) BROBDINGNAG
(-750.03)
MELCOS HOME PLANET OF THE MINUTAE-FIRST CONTACT IN 2367
NOTES:
1) THIS MAP IS INTENDED FOR ORIENTATION PURPOSES ONLY, USE WITH FIG. 2.1.
2) NUMBERS IN PARENTHESIS INDICATE THE 'Z' COORDINATE; I.E. THE DISTANCE IN LIGHT YEARS ABOVE (+), OR BELOW (-) THE PLANE THRU EARTH/SOL PARALLEL TO THE GALACTIC EQUATOR.
3) COLONIAL SUN NAMES ARE OMITTED FOR CLARITY, SEE FIG. 3.1 FOR PROFILE OF SOLAR SYSTEMS.
LEGEND
INDICATES 3 DIMENSIONAL LOCATION ('Z' GIVEN)
INDICATES TRUE DISTANCE FROM EARTH SOL.
THE FOLLOWING TRACES REPRESENT EXPLORED SPACE IN THREE DIMENSIONS:
Z = +10 TO -10 LIGHT YEARS
Z = 10 TO +30
Z = -10 TO -30
Z = 30 TO +50
Z = -30 TO -50

FIG. 2.2

rivals, enables them to set up housekeeping and to pay expenses until they have found work. Personal mementoes, special medical apparatus, and art objects may be brought, but a five kilogram per person limit is strictly enforced.

During the early years of a colony's development, GAILE provides the basic industrial equipment necessary to the survival of a modern society. This equipment varies from one colony to the next depending on such factors as climate, landing sites, terrain, mineral and food sources available, and the general development plan formulated by the pioneers themselves. The following is a typical list of the items GAILE might provide:

1. A gigawatt torroidal fusion core power plant
2. A computer complex including data files containing all of Human knowledge
3. Basic seed crops for up to ten years food supply
4. Cell banks containing fertilized eggs of domestic animals
5. A biolaboratory for analysis of native life, control of disease and the development of food sources better adapted to the new environment
6. Automining and refining equipment for the production of industrial materials
7. Factory equipment capable of producing, on a limited basis, all forms of mechanical and electronic equipment up to and including computer master switching centers and class one robots
8. A few antigravity vehicles for local transportation of people and goods
9. Modular sections which also can serve as temporary housing for new immigrants are carried on GAILE's starships.

With this basic equipment and the knowledge stored in their central computer, pioneers have been able to construct modern industrial societies from wilderness within 100 years. During the first century of a colony's development, colonists must do without many of the conveniences of 24th century Earth, such as personal robots, food autoselectors, somafields, and computer-guided air cars, but the pioneers of all the colonies to date have prospered and enjoyed healthy, happy lives building their new worlds. They have looked upon virgin plains and forests, drunk from unpolluted streams and tasted natural foods, fresh from the land. These things cannot be had on Earth at any price.

As the industries of a colony take root and grow, the supply of industrial goods from GAILE decreases and the flow of new colonists to help with the work at hand increases. Eventually the new worlds become self-sustaining, and even the temporary housing for new immigrants is no longer required. From this point onward, GAILE continues to transport new pioneers and to update the colony's computer libraries with new information from Earth and the other GAILE members.

Interstellar travel requires immense resources. A typical starship capable of carrying 10,000 passengers costs more than 11 billion Unicredits to build, equip, and stock with supplies and fuel. One round trip to a colony world consumes fully 30 percent of this value. As the mean personal income of Earth citizens is about UCr 12,000 per year, it would take 92 years of wages from an average group of 10,000 pioneers to build one ship, and 28 years of wages just to pay their passage! And these astronomical costs don't even include the costs of exploration or the expensive industrial equipment necessary for the new colonies. Clearly the cost of colonization cannot be borne by the pioneers themselves.

Earth's early space exploration offered little tangible benefit in exchange for great expense. Proponents of space travel justified early explorations of the moon and the other planets of the Solar system as quests for scientific knowledge. Arguments were advanced about the benefits of "technological fallout" from space exploration, but in fact, military and political considerations provided the principal motivation for early programs. Space did soon yield some real economic benefits, first from communication and weather satellites, then later from great orbiting solar power stations and factories. Space proponents pointed to these as justification for further space exploration despite the fact that such exploration had little relevance to the current space industrialization.

The proponents won out, however, and throughout most of the first century of the space age, governments funded their own space programs to greater or lesser degrees in the name of science. By the late 2040's, however, cost of programs for each country became so high that governments pooled their resources to form the International Council for Space Exploration (ICSE).

ICSE began a space program for all people, launching the first exploratory mission to Alpha Centauri, the star system nearest Earth.

The discovery of Wyzdom in 2074 added unprecedented momentum to the space movement. Despite the planet's lack of tangible economic benefits, it seemed somehow unthinkable to discover a fresh, new world and then ignore it. The people of crowded Earth, eager for someplace to go, launched the first Wyzdom colony with the vague hope that somehow, some good would come out of it.

Some good did. Examination of the life of Wyzdom added much to Human understanding of the general laws which govern all life by providing a comparison with Earth. Study of the stars of the Alpha Centauri system gave Humans a better understanding of Earth's own sun and allowed them to predict its future more accurately. The opening of this new frontier provided a relief valve for the ambitious but stifled people of Earth, and offered a chance for alternative societies to develop in contrast with the densely populated and increasingly homogeneous Earth.

Yet the ICSE realized that a massive exploration and colonization program would not be financed voluntarily by the people of a crowded Earth who would not be able to go, and who would derive little benefit from it. They knew that to realize the potential of Wyzdom and worlds yet undiscovered, some way would have to be found to make space exploration pay for itself. On Earth, new territories of the past supported themselves by exporting minerals or foodstuffs, but few material items are as valuable as the cost of transporting them across interstellar distances. The principal export of the colonies had to be knowledge. And the colonies excelled in producing it.

In new worlds where needs are great and the restrictions of a crowded Earth absent, human beings demonstrate their greatest creative ability. The pioneers from Earth produce a variety of new products and processes, as well as methods for performing old tasks more efficiently. Many of these ideas can be patented, although the rights to them have little value to the inventor outside of his home planet. It is natural that first ICSE and now GAILE, which provide the communications link between the new worlds and the old, should have, as compensation for their services, the rights on Earth and the other Association planets to all patentable ideas originated in any colony. Copyrights, too, provide important sources of revenue, for people on the new planets do not confine their ideas solely to inventions, and the people of Earth have great interest in reading what the pioneers write. Finally, the vast body of knowledge collected on the colony worlds serves as an important data base for scientists in research institutions and industry. Though this data is technically public knowledge, many institutions and individuals pay handsome fees to GAILE for the privilege of accessing their computer data banks.

Patent royalties and sale of information have not generated enough revenue to pay all the costs of GAILE's program to date, but the flow of money from ideas developed on the older colonies is substantial and meets about 50 percent of current expenses. As in all long-term investments, the bulk of the profits come late in the cycle. The earliest three colonies are just entering their most productive period, and GAILE economists predict that during the next three centuries they will return 10,000 times the amount GAILE has invested in them. By then, Earth's development probably will have expanded to several dozen planets, and many of the older colonies may be shouldering the burden of space exploration by building and operating their own starships.

Until that time, however, GAILE must continue to borrow part of its funds from the people of Earth and therefore must exercise great care in conserving resources expended by the planet settlement program. GAILE attempts to conserve in three basic ways:

1. Transportation of goods is limited to the barest essentials, and the colonial development plan is carefully built around reducing the bulk and mass of the items transported.
2. GAILE chooses pioneers who will best benefit the colonies and the people of Earth.
3. All Earthly possessions of pioneers become the property of GAILE to partially compensate the cost of passage. In cases of extreme need, however, GAILE may waive its claim to property and allow it to be passed on to immediate family members.

GAILE does not like to rely heavily on the contributions of departing pioneers, for often young people and people without great wealth on Earth will work hardest to make the colony a success. With one exception (Mammon), GAILE makes no economic claims upon existing colonies, preferring to allow the colonies to conserve their resources for their development, though all colonies may be called upon to repair or resupply starships when needed.

Earth's colonies are self-governing. Earth's government makes no attempt to appoint a governor or to establish a constitution, and Earth levies no taxes on the pioneers. GAILE hopes that all colonies will respect the inalienable rights of intelligent individuals, and that their method of government, if any, will be democratically constituted. To date, all colonies have lived up to these expectations. If a dictatorial tyrant did seize control of a planet, the government of Earth would probably take measures to preserve Human rights. The great potential for the people of Earth lies in the diversity and growth of Human knowledge that the colonies ultimately will supply. Humanity holds a direct interest in the free and unrestricted development of the new worlds.

Selection of Pioneers

How can *you* go to the new worlds? Few Earth people can afford the enormous cost of transporting not only themselves, much less the equipment and supplies they require to survive on an unsettled planet. Those few that have such wealth may make the trip with no questions asked; yet for the great multitude, there is another way.

Earth's branch of the Galactic Association of Intelligent Life sponsors a program of emigration to the colonial planets. Association employees arrange for purchase and operation of starships and provide critical equipment for the developing planets. The people of Earth finance colonial development because they expect the value of the knowledge and ideas gained on the new worlds to repay their investment with interest. GAILE is willing to bear this cost only if assured that the colonies will succeed.

"Success" of a colony implies that it will be permanent, that it will grow in population and per capita income and that a fertile environment for ideas will exist. It is pointless to promote colonies if all the inhabitants starve to death, or die of exposure or disease. Similarly, Earth's people can not hope to benefit from planets where life exists at bare subsistence levels. Such societies do no research and don't develop new ideas. Growing modern, industrial societies on totally primitive worlds requires above average populations, and colonies need people with a wide variety of skills. Craftsmen as well as scientists, farmers as well as computer programmers, teachers, writers, home makers, lawyers, physicians, pilots, and even pastry chefs, all contribute to make a whole society.

To assure that the colonies have pioneer populations sufficiently hardy and diverse, GAILE selects people to go to the new worlds from volunteers and any Earth citizen can apply. A complex computer program written by the GAILE staff aids in choosing pioneers and guarantees that the selection process will not be biased by personal preferences of individual staff interviewers. The program analyzes hundreds of factors for each applicant, comparing them with the current needs of each planet. Design of the program and the weighting factors used in it have been approved by the GAILE Policy Board. Colonial governments in existence more than 100 years also may approve the factors used for their colony.

The selection process begins with the application. Nobody is asked to emigrate and no one is coerced to go. Each prospective pioneer must take the initiative by applying in person. This is the most important test of the screening program, because by taking the initiative to apply, an individual demonstrates his or her desire to make a new life. The will to succeed is more important to the new planets than any other factor. Only three other characteristics are *essential* for any person to become a pioner: *honesty, resourcefulness,* and the willingness, to *work hard.*

No one guilty of crimes of violence, fraud, embezzlement or of persistent failure to pay his financial obligations may be selected for the pioneering program. GAILE has made a few exceptions to this policy, but, in these rare circumstances, the individuals considered had redeemed themselves by performing outstanding services to their communities and had proven they would not repeat their transgressions in the future.

Pioneers must also be resourceful to succeed on their new worlds. This does not imply that they be unusually brilliant or creative, but they must demonstrate the capability and willingness to learn new tasks and take on new challenges. These qualities are as important for an unskilled laborer as they are for a scientist. The colonies cannot afford people so set in their specialized ways that they are unable or unwilling to perform new jobs that may not even exist on Earth.

GAILE considers a variety of other attributes of each candidate in making its selections, but none of these are prerequisite. Good health and physical fitness, for example, are important. Life on the colonies is physically more demanding than life on Earth. Pioneers have machines, but young industries require more manual labor than the highly automated factories of Earth. In addition, the higher gravitational attraction of some planets makes simply standing up and walking a much greater effort than it is on Earth. Most people adapt readily enough to this new condition, but those who are grossly overweight might not adjust before succumbing to heart or respiratory failure. GAILE often makes exceptions to the physical fitness requirement, especially in the case of a person with highly desirable skills or abilities. Even people with severe handicaps, such as the loss of both natural eyes or several limbs, have become pioneers by demonstrating exceptionally strong motivation to overcome their affliction.

Youth is not a requirement for pioneers, but few people over 60 are inclined to make the trip. It takes a middle-aged person many years to reacquire the wealth given up upon leaving Earth. Despite this fact every shipload of colonists holds a few determined individuals who have made such a sacrifice to start a new life. Their maturity and experience greatly benefit the younger pioneers. In fact, the oldest person ever admitted to the pioneering program was more than 110!

Children usually accompany their parents to the new worlds, though families with two or three very young children are not encouraged on the more primitive planets. Children over fourteen who have no legal guardians or who secure written permission of their guardians may apply for the pioneering program alone, and each year many are accepted.

At any given time, colonies need some skills more urgently than others. The particular skills needed vary from year to year and from planet to planet. Though GAILE does not discourage applicants without special skills, they should recognize that GAILE gives preference to people whose particular abilities fill the needs of their first choice planet.

Each applicant and his or her entire family must apply in person for the pioneering program at one of GAILE's Regional Processing Centers. The process takes two or three days and includes a complete bioscan, the administering of several computer-controlled tests, and a computer-aided personal interview with a GAILE selection counselor. Biotypes are checked to guard against people with false identities.

After completing this process, a Human interviewer, aided by computer, makes a recommendation. If the recommendation is negative, the applicant may request a second interview and may change his or her first choice planet prior to this. If the second interviewer rejects the applicant, the applicant may appeal the decision to the Selection Appeals Board. The Board uses the computer's data in assigning priorities to requests for appeal so that those with the greatest chance of having their decisions reversed will be heard first. Though the Board's decision is final on Earth (except for applicants to Mammon who may appeal directly to the MMC corporate staff), applicants can still appeal by letter to the government of the colony to which they have applied. For reasons that are not always clear to the GAILE staff, colonial governments occasionally overturn decisions of the SAB, but such instances are rare. In less than one appeal in 10,000 has a colonial government overturned a selection decision made on Earth.

Applicants may be placed on a provisional rejection list and reconsidered for up to three years after the date of application. Conditions on the colonies do change, and these changes affect the selection process. An individual's skills may come into demand, and this may make the difference required to include him in the program. After three years, applicants will receive notification that they have been finally accepted or rejected. They may begin the application process again if they wish and receive consideration on an equal basis with new applicants. Although it is extremely rare, applicants who have been rejected three or four times previously are sometimes accepted to the program.

Filing the formal application for the pioneering program is both time-consuming and expensive, particularly when the cost of travel to a Regional Center are figured. Before applying in person, most prospective pioneers like to get an idea of their chances of being accepted for the planet of their choice. The Preliminary Application Form in the back of this bulletin is designed specifically for this purpose. (See page 195). It contains a series of questions designed to measure what a prospective applicants' chances of being selected will be when he or she undertakes the formal application process. Throughout the years, the Preliminary Application has been refined so that today it predicts most applicants' chances of selection with an accuracy of ± 5 percent.

To save time and money, the Preliminary Application is evaluated by computer only using the same weighting factors approved by the GAILE Policy Board and the colonial governments. The computer returns to each applicant its estimate of the probabilities that the applicant will be selected by GAILE to emigrate to his or her first or second choice planets. For example, if an applicant listed Wyzdom and Genesis as his or her first and second choices, the computer might reply, "Probability of selection for Wyzdom = 75%, probability of selection for Genesis = 81%."

Data from the Preliminary Application is not used in the formal selection process and in no way influences the decision on any application. It is therefore in the applicant's best interests to answer questions on the Preliminary Application as accurately and as honestly as possible. Failure to do so will only distort the probabilities calculated by the computer and give the applicant a false assessment of his or her chances.

If you want to apply for the pioneering program, file the Preliminary Application in the back of this bulletin. If your probability of selection is greater than 75 percent for either of your first two choices, the computer will make an appointment for you at the nearest Regional Processing Center. If your probability of selection is less than 75 percent, you are encouraged to consider other planets.

Potential applicants should find out more about the colonies of their choice by reading the detailed bulletins about each planet listed in Appendix C. These may be accessed through your computer library service. Hardcopies may be purchased from certain commercial book outlets, or from GAILE Regional Offices. The decision to emigrate is the most important one a person can make in his life. Consider it carefully, and may you find your dream in the stars.

Life on the Planets

Why would anyone leave the comforts and security of Earth to struggle for survival on an untamed world? In 9,000 years of civilization, life on Earth has grown soft. Humanity has tamed the ravages of nature; the average lifespan has lengthened to 120 years; the work week has shrunk to 30 hours, and hundreds of diversions compete to fill the leisure time. Civilization has brought losses too, both of beauty and opportunity. Few people on Earth know how blue the sky or sea can be, and few have seen the undistorted splendor of the night stars. Just as few know what it is like to create and achieve without securing permission from dozens of their fellows.

Many people feel stifled by this state of affairs, but Earth is too small to afford them any relief. They dream of a world in which they can own land, explore uncharted wilderness, discover new species, look upon untamed vistas, and breathe clean, fresh air. Colonial worlds offer all this and much more to those who are willing to work hard and who crave freedom and opportunity. They offer a clean slate on which to inscribe a new society; one boasting the advantages of the knowledge gained during the last 9,000 years while escaping the social prejudices and political controls unavoidable in modern life.

GENERAL CHARACTERISTICS

It is hard to generalize about the Association colonies, although a few generalizations can be made before beginning the descriptions of each. As table 3.1 readily illustrates, each bears many important physical similarities to Earth. Those similarities reflect the basic conditions needed for Human life to flourish.

The Planets

Of course Human beings have proven they can survive in rather hostile places: the moon base, the L-5 free space colonies, the Martian observation stations, and the deep space observatories on Ganymede and Titan have all existed for several centuries. Yet these outposts in our Solar system do not provide satisfactory homes for large numbers of people. The handful of scientists and technicians who occupy these bases must live within the artificial pressure of tightly closed structures, and artificial means are required to sustain their lives. Since the number of inhabitants is small, they must concentrate on specialized forms of space industry, importing virtually all consumer goods.

Human beings cannot exist indefinitely on food, water, and air alone. We require open spaces, sunlight, plants, animals, and large numbers of our fellows with whom we can trade ideas as well as objects. We can't make any artificial environment large enough to satisfy these needs, so even the "permanent" colonies of our Solar system serve only as way stations for our expansion into space. Any large, permanent, self-sustaining Human colony needs a world similar to our Earth in many respects.

ATMOSPHERE

Such a world must have a breathable atmosphere. It must contain no poisonous gases, like chlorine or flourine. Its partial pressure of oxygen must be at least 0.17 bars, the minimum with which Human lungs can function. Pure oxygen doesn't do either, since it causes rapid burning that destroys our carbon-based life. The inert portion of the planet's atmosphere must contain substantial nitrogen, for without this element, proteins, the essential building blocks of our bodies, cannot be formed. Yet if the partial pressure of nitrogen is too high, it reacts with our body chemistry to cause sickness and death. Carbon dioxide is also needed for the plants that comprise our foods, but too much carbon dioxide makes a planet's atmosphere a "greenhouse" that would be unbearably hot.

A habitable planet must have large oceans, since water is the primary component of our bodies and the foods we eat. Studies of many planets have led scientists to believe that unless 40 percent of a world's surface is ocean, no portion of it will have enough rainfall to support our form of life.

Figure 3.1 illustrates two other important similarities of habitable worlds. Each of the Human colony planets orbits about the same distance from its sun as does the Earth from its sun, and their suns are the same size as the Earth's sun.

Though it's not immediately obvious, the reason suns of all habitable planets must be about the same size relates to the formation of a breathable atmosphere. The oxygen-nitrogen atmosphere needed for Human life never occurs on totally lifeless worlds. Such an atmosphere is a by-product of a long evolutionary process. Though the intricacies of this process remain a mystery, scientists know with certainty that it takes billions of years.

The giant stars of our galaxy do not live that long. For reasons explained by complex physics, large stars consume their fuel quickly and burn out, while small stars, like Earth's sun, remain the same size and brightness for billions of years. In fact, the smaller a star is, the longer it will live.

Stars can be too small for habitable planets as well. To be warm enough for Human life, a planet must orbit its sun closely enough to receive the right amount of energy. If Earth's sun had 30 percent less mass than it has, then the Earth would have to orbit it so closely that the sun's gravitational attraction would stop the Earth's rotation with respect to the sun, just as the Earth has stopped the relative rotation of its moon. This would make one side of the planet too hot for life, while the other side would be too cold.

The requirement that stars be neither too large, nor too small, has allowed GAILE scientists to rule out all but 12 percent of the stars in the heavens as possessing habitable planets. In addition to the relationship between star life and size, there is also a relationship between star temperature and size. Though notable exceptions exist, large stars tend to be hotter than small stars. It is therefore quite natural that most of the stars of a suitable size for habitable planets also have about the same temperature and emit the same color of light. Stars are typed or classified by letter according to temperature, and virtually all suns that might have habitable planets fall into three of the seven major types—F, G, and K.

TEMPERATURE

Since nature has dictated that stars with habitable planets must be the same size and temperature, then the planets habitable by Humans must orbit at about the same distance as the Earth orbits the sun. Human beings and the life forms we consume for food can exist only within a very nar-

PLANET PHYSICAL DATA COMPARISON

	EARTH	WYZDOM	POSEIDOUS	BROBDINGNAG	MAMMON	GENESIS	YOM	ROMULUS	ATHENA
Planet Diameter (X Earth's dia.)	1.0	1.14	1.11	1.04	0.78	1.01	0.88	0.91	0.96
Gravity at Surface (9.8 m/s^2)	1.0	1.27	1.18	1.07	0.69	1.01	0.82	0.87	0.94
Atmospheric Pressure (average at mean sea level) (X Earth's atmospheres)	1.0	1.81	1.32	1.12	0.34	0.55	0.70	0.77	1.05
Fraction of Surface Covered by Land (%)	29	26	1.5	19	48	21	35	29	24
Obliquity* (degrees)	23½	18	2	15	5	27	36	19	21
Length of Day (Std. Earth Hrs.)	24	13.3	19.6	14	22.1	12	26.9	44	25.2
Length of Year (Std. Earth Days)	365	341	225	369	349	347	345	360	539
Distance from Sun (astronomical units)	1.0	0.98	0.70	1.09	1.02	0.96	0.92	1.05	1.45
Mass of Sun (X mass Earth's sun)	1.0	1.08	0.91	1.26	1.16	0.98	0.87	1.19	1.40
Distance from Earth (light years)	—	4.3	20.9	32.3	39.3	43.5	50.2	58.2	77.6

*inclination of rotating axis toward the plane of orbit.

Table 3.1

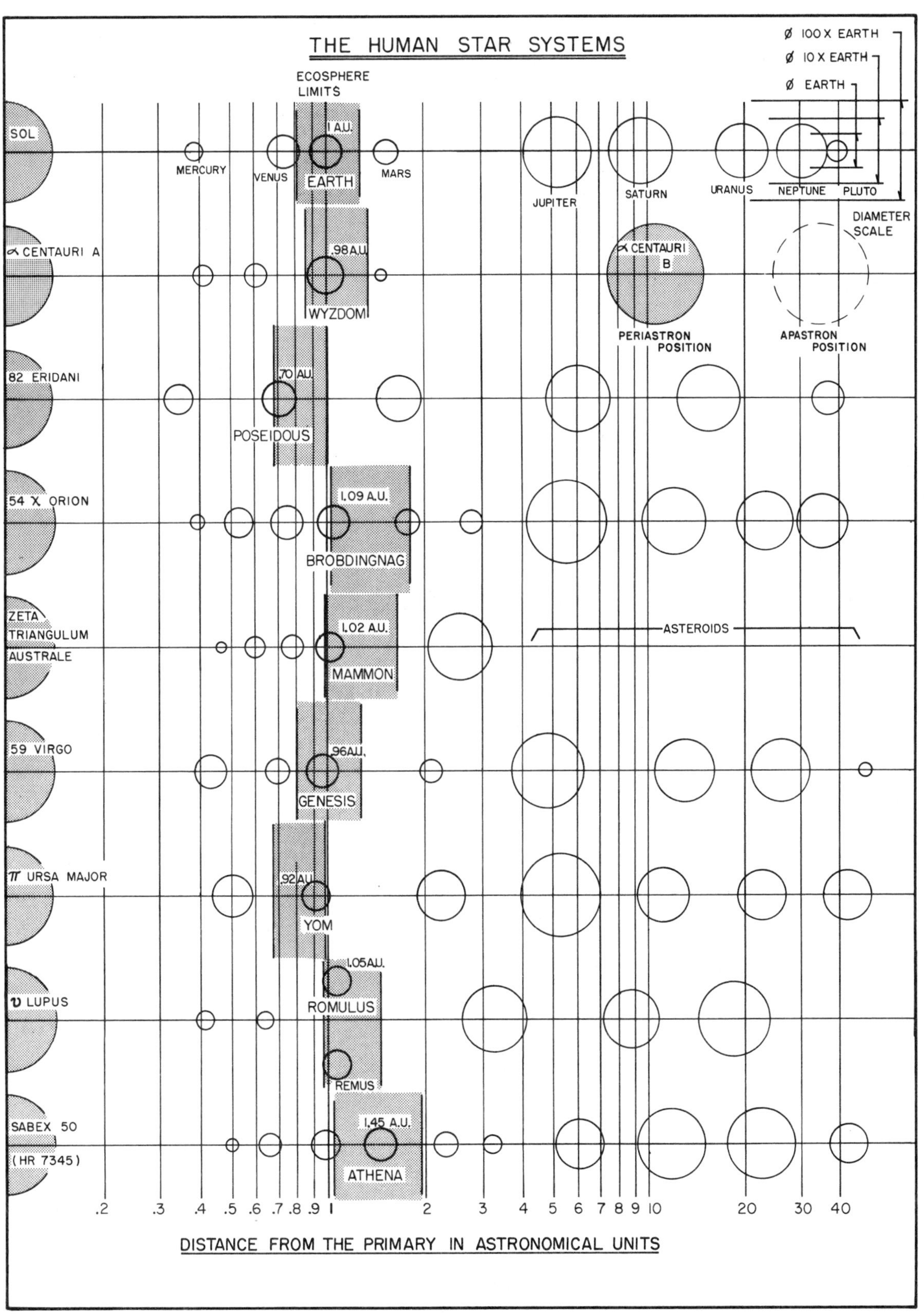
THE HUMAN STAR SYSTEMS
ECOSPHERE LIMITS
Ø 100 X EARTH
Ø 10 X EARTH
Ø EARTH
DIAMETER SCALE
SOL
MERCURY
VENUS
1 A.U.
EARTH
MARS
JUPITER
SATURN
URANUS
NEPTUNE
PLUTO
α CENTAURI A
.98 A.U.
WYZDOM
α CENTAURI B
PERIASTRON POSITION
APASTRON POSITION
82 ERIDANI
.70 A.U.
POSEIDOUS
54 X ORION
1.09 A.U.
BROBDINGNAG
ZETA TRIANGULUM AUSTRALE
1.02 A.U.
MAMMON
ASTEROIDS
59 VIRGO
.96 A.U.
GENESIS
π URSA MAJOR
.92 A.U.
YOM
υ LUPUS
1.05 A.U.
ROMULUS
REMUS
SABEX 50 (HR 7345)
1.45 A.U.
ATHENA
.2 .3 .4 .5 .6 .7 .8 .9 1 2 3 4 5 6 7 8 9 10 20 30 40
DISTANCE FROM THE PRIMARY IN ASTRONOMICAL UNITS

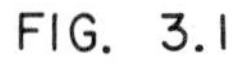
FIG. 3.1

row range of temperatures. Above 65° C. or below -60° C. we must wear environmental suits to avoid perishing within minutes. Our food crops are confined to an even narrower range. For practical purposes, no place can be considered habitable whose temperature falls consistently above 45° C. or below 0° C.

DAYS AND YEARS

Days and years on habitable planets cannot be much shorter or longer than on Earth. The length of a year is dependent on the mass of a planet's sun and its distance from it. The length of a planet's day cannot be so long that the noon or midnight temperatures exceed tolerable limits. Nor can it be too short, lest the planet be torn apart by the force of its spin.

GRAVITY

Planets suitable for Humans may be neither too large nor too small. If they are too small, then, like the planet Mars, their gravitational attraction is too weak to hold a breathable atmosphere. If they are too large, the gravity at the surface is too great for Humans to withstand. The Minutae, who may soon join the Galactic Association, come from a planet whose gravity is more than 2½ times that of Earth. They are a very tiny species and are therefore better able to withstand high gravitational fields. The upper gravitational limit acceptable to Human beings on a continuous basis is between 1.3 and 1.6 times that of Earth (written 1.6 g). The lower limit is about 0.7 g for retention of an atmosphere. The density of most "terrestrial" planets (planets made of rock and heavy minerals) varies slightly with diameter but is nearly universal from star system to star system. Thus, once we have set limits on acceptable gravitation, we have in effect imposed limits on diameter. All habitable planets must have diameters somewhere between 0.78 and 1.3 that of Earth. In fact, those discovered to date range from 0.78 to 1.14 times Earth's diameter.

LIFE FORMS

Humans as organisms can't be isolated from other life forms, since we need a variety of them as food. We require plants to convert sunlight into food energy, to convert our exhaled carbon dioxide into oxygen, and to convert nitrogen into proteins. In a closed environment we require organisms to break down our wastes. Although all of these processes can, on a limited scale, be accomplished synthetically, only the natural processes provided by other living organisms are economically viable on a large scale. The life forms of habitable planets are, in most cases, as varied and plentiful as the life forms on Earth, but since they evolved independently, each species is totally different from any that exists on Earth.

Some general statements can be made about all corporeal life. In all biosystems, be they carbon-nitrogen-water based, sulfur-chorine-hydrofluric based or whatever, there is a basic food chain. The chain begins with life forms that convert sunlight to chemical energy. These forms are eaten by other creatures who may in turn be eaten by still others. Each life form occupies an ecological "niche" or position in the food chain. In general, there are more individual life forms, or at least more masses of life forms, lower on the food chain than higher.

More and less complex forms of life can be found in all life systems. The more complex are sometimes said to be more "evolved," but, in fact, some very simple life is the product of as many stages of evolution as *homo sapiens*. "More complex life forms" is not really a precise, scientific term, but is used here to mean that a particular species has evolved a higher degree of structural specialization. Mammals and flowering plants are among the most complex life forms of the animal and plant kingdoms on Earth.

The basic classification schemes used on Earth have been found to be useful in studying and classifying the life of the colonial planets. The primary classification categories are the kingdoms: plants, animals, protists, and synergists (viruses). Within kingdoms are grouped numerous "phyla." Some phyla such as tracheophytes (vascular plants) and chordates (animals with spine-like nervous systems) appear common to most of the colonial worlds. Each phylum is divided into several classes. Angiosperms (flowering plants) are a class of tracheophytes; mammals are a class of chordate. Similarities exist between classes of the various colonial worlds, but the differences are far too great to consider the classes identical. For example, many of the colonies have warm-blooded flying animals resembling the Earthly class aves (birds). These "birds" may bear their young alive or they may not have feathers, yet having feathers and laying eggs are two important characteristics of the group of animals classified as aves on Earth.

The similarities between life forms on Earth and on the colonies sometimes shock pioneers, but they should come as no surprise. The laws of physics apply throughout the known universe. Chemical elements found on Earth are the same elements that constitute all matter. Biological phenomena can't violate these laws. In addition, since certain forms of biological organization (such as cell structure) are more efficient, they win out over less efficient forms conceivable within our physiochemical framework. Life evolves in similar patterns on each of the different worlds, yet the individual species, their genetic organization, biochemistry, habits, etc., are unique.

MICROORGANISMS

Because the life forms of other planets are quite alien, they are potentially very hazardous. The greatest hazard springs not from large, carnivorous creatures, but from microorganisms such as protists and viruses. Such creatures cause most of the diseases on Earth, but Earth's population has some immunity to most of them. Humans and our domestic animals have no immunity to the microorganisms of another planet. Diseases there have the potential for wiping out an entire colony in a very short time. Fortunately, during the last 500 years, microbiology and immunology have become well understood and fewer devastating diseases have struck in the colonies than were originally anticipated. Scientists speculate that this is because few parasites native to the colonial worlds can adapt to Human hosts. Rather than exposing us to harm, our alien body chemistry may offer the best source of protection from disease. Pioneers have dealt with most serious viral outbreaks which have hit the colonies by using life support systems to maintain the victims until medical scientists isolate the virus and produce antitoxin. In the last 200 years of colonial development, deaths from viral epidemics have been few. Yet danger from these outbreaks still exists, particularly in the newer colonies. The microbiology laboratory remains as vital to new colonists as their computer and power reactor.

SIMILARITIES AND DIFFERENCES BETWEEN THE COLONIES AND EARTH

Many potential pioneers ask if intelligent life has been encountered on colony worlds. The answer, by definition, is no. GAIL's strict non-interference policy precludes this, although the colonies of Yom and Romulus come closest to being exceptions.

The similarities between the colonial worlds and Earth are interesting, but their differences ignite the imagination. Each planet is a world unto itself, diverse in geography, life forms, scenery, and climate. Each has its swamps, deserts, forests, grasslands, mountain ranges, and fertile valleys. Each has enough land to satisfy the needs of today's colonists, future immigrants, and their descendants for dozens of generations.

Yet the greatest contrast between the colonies and the Earth springs from the pioneers' lifestyles. Today, 15 billion souls pack the Earth, and as a poet wrote so long ago, not one of them is an island. Eighty percent of the world's population crowds into megapolises stretching for hundreds, sometimes thousands, of kilometers, and very little of the Earth's prime land is left unoccupied.

In contrast, the colonies are vast empty worlds. One statistic illustrates the difference: The most densely populated colony, Poseidous, has a population density of only 2.1 people per square kilometer, compared to the Earth's population density of more than 101 people per square kilometer! Colony cities are manageable, numbering in most cases less than a million people. As on Earth, the cities serve as centers for commerce, manufacturing, and trade, but it takes only minutes to zip out of them into the open countryside. Cities can afford the luxury of open space within their borders and crime and pollution are almost unknown.

The crowding and the age of Earth's societies have also crippled individual activity. Great bureaucracies and political power structures control most economic and social life. Forms of government vary from state to state, but certain patterns are universal. No building can take place in urban areas, no business enterprise can be started or expanded, and no drastic changes can be made in any activity without securing a myriad of permissions from governments and from one's neighbors. Such regulations are a necessary evil of life in a crowded world, yet these necessary evils have spawned stagnation and corruption. Favoritism, politics, and prejudice too often play the major role in determing who will be permitted to undertake an activity and who will not.

In contrast, colonial land is so plentiful and colonial needs so great that few controls are necessary or possible. Population growth is welcomed not discouraged, for there is much to be done and few hands to do it. In every enterprise from basic industries to the arts, innovation rather than regulation is encouraged. Bureaucrats, managers, and a class of idle rich don't fit in well where almost everyone must be willing to roll up their sleeves and do manual as well as intellectual labor. Some people still primarily organize while other primarily carry out specific tasks, but class distinctions between these groups are not as pronounced in the colonies as on 24th century Earth.

Earth is wealthier than the colonies. It has more factories, facilities, consumer goods, and luxuries even on a per capita basis than any of the colonial worlds. Most Earth residents can retire at some age between 60 and 80 and live the remainder of their 120 to 150 years engaging in leisure pursuits. Yet such wealth means little where no virgin forests survive, no air is unpolluted, and crowds fill every desirable place.

Pioneers on the colony worlds own fewer consumer conveniences, fewer luxuries, and most of them work past the age of 100. Yet a colonial family can have a five-hectare spread of land with trees and fields, perhaps even a lake or beach, for their personal retreat. Families can develop their own 400-hectare farm and raise not only food for sale but grow fresh fruits and vegetables for their table that cannot be equaled on Earth. Colonial craftsmen build genuine wooden furniture, an art lost centuries ago on Earth. Natural materials—wool, silk, and leather—supplement factory synthetics. These items can scarcely be had on Earth at any price.

The Earth is a planet of conveniences. No one must perform the drudgery of manual labor. Machines move earth, harvest crops, lift and carry, assemble machinery, weld and fabricate, and take care of domestic chores like cooking and cleaning. On the colonies, machines are in short supply. Many assembly operations must be performed by hand until machines can be built to do them. People build vehicles, electronic equipment, and structures by hand, aided by simple cranes and lifting tools. Many farmers must clear their land with manual disintegrators and plow with manual power-assist drives. Personal robots are non-existent in many colonies and semiautomated cooking and cleaning are part of the daily chores. Many pioneers find these manual tasks satisfying and challenging. Some even choose to return to a totally natural life, making all of their clothing on simple hand-operated machines and growing their own food using natural methods. Yet no pioneering society has ever delayed making automated equipment to do manual labor, and most pioneers strive to create Earthlike levels of convenience as soon as possible.

FOOD

Colonial food differs little from food on Earth. Few colonies possess native staple grains containing the appropriate food value and sufficiently high yield per hectare for a modern society. On Earth, selective breeding techniques developed over five centuries have produced grains with high yields, resistance to disease, the ability to withstand climatic extremes, and high food value per unit weight. Therefore, most colonies have imported the seeds of basic food crops such as rice, corn, wheat, soy, barley, kelp, and hybrid algae, and all use the latest agricultural and aquacultural techniques. Similarly, colonists raise highly specialized domestic animals. Dairy cattle, pigs, sheep, trout, shrimp, and halibut comprise the most common species.

After several generations, the older colonies have turned to the adaptation of native plants and animals for food. The same selective breeding techniques used on Earth have been applied to produce desirable food qualities in species native to the colonies. The new foods usually lend themselves to new culinary methods, and the older colonies have already created unique cuisines that impress the most discriminating of Earthly gourmets.

MEDICINE

The quality of medical service on the colonies equals that of Earth, but is not so widely distributed geographically. A fully-equipped hospital capable of treating or curing all Earthly ailments forms an essential part of the equipment brought by the first colonists, and GAILE constantly updates medical information in the colonies' central computers. As colonies grow, medical facilities become more widely disbursed. Most colonies, though, find it more expedient to concentrate hospitals that specialize in sophisticated techniques such as organ rejuvenation and cancer cure in the central cities. Rapid transportation is available to carry people in need of such treatments from the planet's remotest outposts. Pioneers who live far from the cities generally learn more first aid than the average Earth person so they can cope with emergencies until help arrives.

The cost of medical care is provided in a variety of ways on various colonies, but only Genesis and Poseidous employ any sort of socialized medicine. Doctors generally provide medical services on a contract basis; for a monthly fee, all medical care is provided. This practice encourages frequent check-ups and early diagnosis. The same automated techniques for daily and weekly monitoring of body functions commonplace on Earth are available on the colonial planets, but individual families do not often have their own personal somascanners.

In general, medicine is of less concern to pioneers than it is to Earth people, although doctors are always welcome and needed. Colonists seem to be much healthier than people on Earth. Medical researchers speculate that three factors may account for this. First, immigrants start out healthy. Few of the chronically ill attempt to make the trip. Second, colonial diets contain fewer highly-refined foods, since refined food production requires both more raw food and processing machinery. Lastly, many doctors believe that many real physiological illnesses are psychosomatic in origin. Pioneers tend to be busier and happier than Earth people, and, as such, have no psychological reason to become sick.

EDUCATION

Education has always been of great importance to pioneers on every colony. Indeed the foundation of the colonial program rests upon Humanity's desire to expand its knowledge, and to educate itself about the universe and the planets on which we live. Though pioneers find that every day they learn a great many things about their world, their jobs, and themselves, formal education is also important to both children and adults. More than 50 percent of the adults who emigrate to the colonies ultimately find themselves pursuing a job, trade, or profession different from the one they pursued on Earth. Conversely, pioneers consider it extremely important to educate their children not only about their new world but about the Human race and its long, difficult history. Pioneers know they must not forget Earth's past, lest they make the same mistakes that were made on Earth.

COMMUNICATIONS

For centuries, any person on Earth has been able to converse at will with any other person anywhere on the globe through various forms of electronic communications equipment. All such equipment employs some form of electromagnetic radiation to transmit the message. All electromagnetic radiation travels through space with the speed of light, yet even at this speed, a sentence would take years to reach even the nearest of the colony planets. This fact of physics makes conversation between pioneers and Earth residents impossible. Instead, colonists must record messages by voice or writing which ships or space probes then carry to Earth. (See Interstellar Transportation and Communication for details.) Although pioneers can continue to communicate with Earth, most of them quickly form new relationships on their new home planets.

INTERPLANETARY TRAVEL

Except for starship crew members, few people ever make a round trip between Earth and the colonies. Few are wealthy enough to afford the great cost of such a trip merely for a vacation. Occasionally, Earth's government will appropriate funds for a prominent scientist or government official to make a round trip, and about one percent of those accepted for the GAILE sponsored emigration program find they cannot tolerate life on the new planet and return home. The GAILE screening process insures high probability that every pioneer will adjust to his new world, but no method of selecting people for anything is perfect. Those who can't adjust are repatriated and GAILE attempts to secure them a job and to return a portion of their Earthly assets. However, potential colonists should recognize that GAILE does not take these actions lightly, and few who decide to abandon their new life ever recover fully the income, assets, and opportunities they gave up on Earth to make the trip.

MONETARY SYSTEMS

Colonial monetary systems differ little in form from the money of Earth. Records of financial credits are maintained in computer accounts and transferred as obligations are incurred. Only two of the colonies, Genesis and Romulus, rely on the planetary government to control the monetary system. The rest of the colonies maintain several privately operated exchange systems.

As there is so little trade between the colonies and Earth, there is little convertibility between Earth and colonial monies. It is very difficult for a rich person to emigrate to one of the colonies with much of his wealth since the securities and property he owns on Earth have little or no value to the colonists. A wealthy person can liquidate his Earthly wealth, buy capital equipment, valuable to the colony, and pay GAILE for his passage and the cost of transporting the equipment. Few have ever opted for this course, but one notable exception, Luigi Albertino, son of a wealthy industrialist, designed the first torroidal reflux plasma fusion reactor and could not get a permit to operate it on Earth. He had the prototype plant shipped to Brobdingnag, where power was greatly needed. There he operated the plant as a highly successful private enterprise.

Most pioneers find themselves quite penniless when they arrive on their new worlds, but they soon rectify this situation. New arrivals live in the temporary housing afforded by the ship's spires (see Interstellar Transportation and Communication) and are given an advance to purchase basic necessities until they find employment. Within a few days of arrival, most immigrants find work, the advance is repaid, and they begin to accumulate property.

LAND CLAIMS

On most colonies, unoccupied land can be claimed by anyone, but colonial governments set limits on the amount any person can claim for a specific purpose. For example, one cannot generally claim 1000 hectares for a personal residence, but one can claim that amount for a farm or mining site. To maintain a claim, the land must be developed for the specified purpose. Claimed land usually becomes available again if undeveloped within a specified period. As long as the land remains under development it may be sold. Most urban land falls in this category.

GOVERNMENT SYSTEMS

Governments play far less important roles in the lives of pioneers than they do in the lives of Earth people. The colonies field no standing armies and police forces are small by Earth standards. The existence of nation-states and their struggles for land have been the traditional causes of war throughout Earth's history. All colonies, except Poseidous, have global governments subordinated by regional and local governments of various kinds. Also, the planets are big enough for their small populations to have all the land they want. Thus, the two principle causes of war do not exist. Furthermore, the high mobility and ease of communication between all pioneers on each planet, makes it unlikely that separate, competing nation-states will ever arise.

CRIME

The same reasons that reduce the chances of war on the colonies also reduce individual violence. By Earth standards, the colonies exhibit low per capita incidence of theft, fraud, murder, and assault. Yet in any Human society, some individuals prefer to violate others' rights rather than work honestly to satisfy their wants. Consequently, most of the colonies maintain investigative bureaus to assist local crime-fighting organizations. These bureaus primarily combat organized crime and large-scale fraud. The central computer bureaus maintain their own force of detectives to combat computer crimes.

Colonies have fewer laws to break than Earth. "Victimless" crimes such as drug abuse, homosexuality, thought crime and sedition infract no colonial statutes, a fact that alleviates much of the burden of law enforcement agencies. Rights to property, freedom of speech and assembly, and the rights to travel, worship, transact business, and choose one's legal guardian are guaranteed by the colonial constitutions.

INTERPLANETARY WAR

Prospective pioneers often ask about the possibility of interplanetary war. No planet under the protection of the Galactic Association has had to worry about invasion from space since GAIL was founded. Even so, the possibility exists that one day GAIL will encounter a technologically advanced, yet hostile race. The chances of that happening are greatest for colonies on the outer reaches of Association explored space.

While the remotest colonies make no specific provision for interstellar war, each has some line of defense. Every colony possesses at least two orbiting space stations. These are used primarily for observing the planet's weather, map making, mineral surveying, relaying communications, and scientific experimentation. The stations also serve as astronomical observatories, scanning and mapping nearby space and studying the planet's sun. While performing these functions they act as early warning systems for the approach of starships and can identify non-Association ships. The stations are equipped with w-field disintegrators (see Interstellar Transportation and Communication) to protect them from meteors. These w-field beams are effective weapons in the event of attack. Should a gang of extortionists try to seize a station and turn the beam on the planet, the stations can be disabled from the planetary surface.

POLLUTION

The problems of pollution which have concerned Earth's governments for centuries pose little problem for colonial governments at present. In the colonies, the most modern of Earth's technology is employed as a starting point, and most industries are therefore pollution-free. In their cities, pioneers leave open spaces for parks, so that future generations will have the open space they need.

LANGUAGE

All but two of the colonies speak Standard International English, or a slight variation of it. SIE is often criticized as cumbersome and difficult to learn, but it is understood, if even rudimentarily, by most of the people on Earth. Consequently, early pioneers drawn from all over the Earth had to use SIE to communicate with each other. Poseiduous and Genesis are the exceptions to this rule. Because a large percentage of southern Asians numbered among Poseidous' early pioneers, Poseidons developed a unique dialect known as Sinoenglish. On Genesis, the planned and idealistic nature of the society led founders to adopt Esperanto, an ancient Earth language specifically designed during the national period to be an international tongue.

Anthropologists and linguists speculate that someday colonies will evoive their own planetary languages. In fact, every colony more than fifty years old has already evolved native words which make ordinary conversation confusing to newcomers. Yet experts doubt the languages of the colonies will ever become totally alien. An interdependence of worlds exists, like the interdependence of continents on Earth. As long as frequent communications remain necessary, languages cannot become too diverse.

The following sections describe each of the planet colonies. Each section contains a map, a data table, a brief description by the GAILE staff of the planet's physical and social characteristics, and an account of life on the planet by a resident chosen for his or her broad knowledge of the planet's life and lifestyles.

MAPS AND TIME CONVENTIONS

A number of conventions to aid the Earthbound reader have been incorporated into every section. All planet maps are drawn so that their rotation proceeds from left to right. The top of the map is by definition the north pole and the bottom the south, regardless of the polarity of the planet's magnetic field. On Genesis and Poseidous, place names have been converted from native spellings to SIE equivalents to facilitate pronunciation and the accounts by residents have been converted to SIE.

As the data tables show, each planet has a different year and day. Consequently, all colonies have developed their own calendar and time scale, different from the 12-month, 24-hour convention used on Earth. Most colonial time is based on "hours, minutes, and seconds" in multiples of eight or ten. To avoid confusing Earth readers, all references to time have been converted to their equivalents in Earth Standard Time and dates have been converted to the Gregorian calendar.

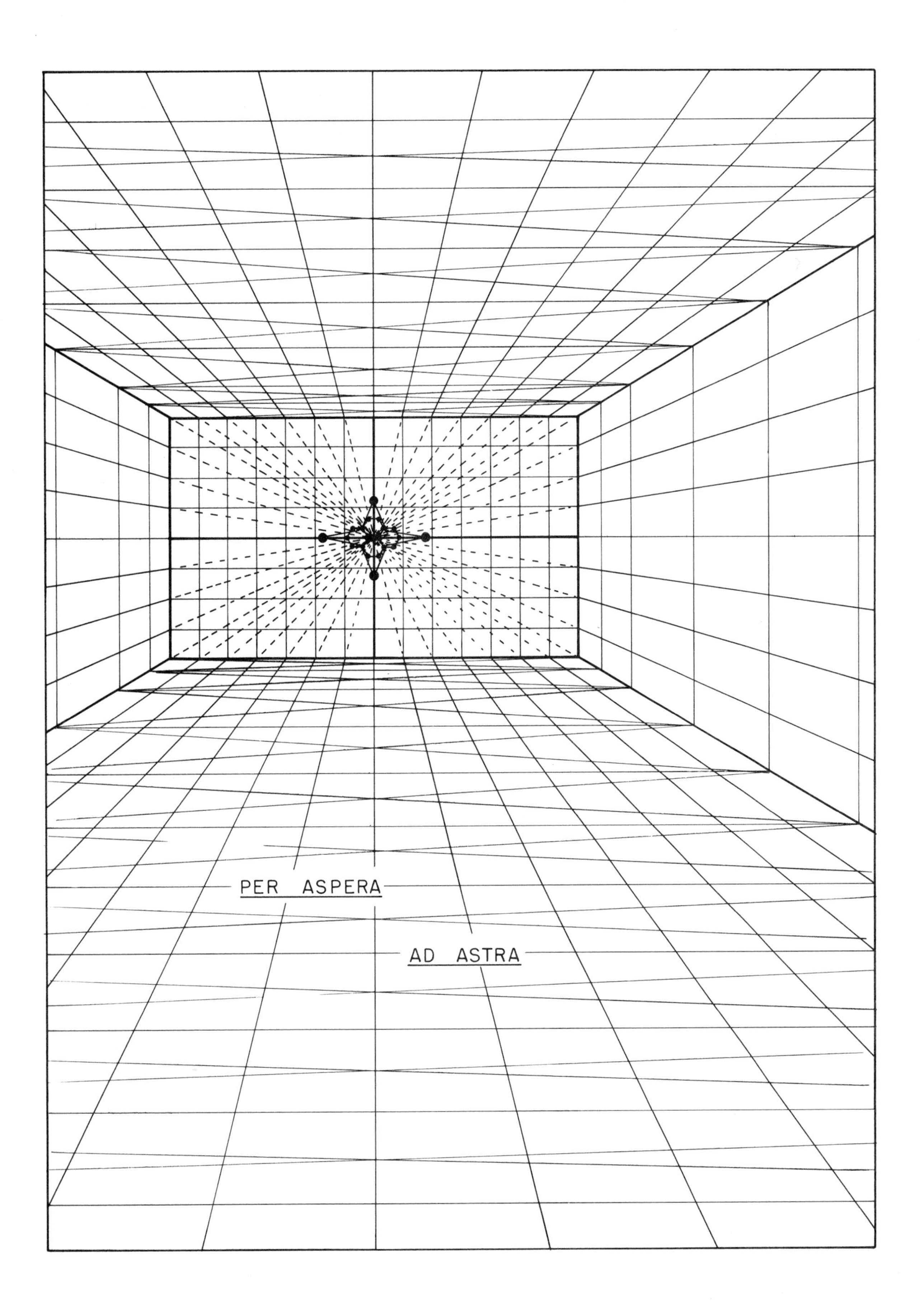
PER ASPERA
AD ASTRA

WYZDOM—First Frontier in the Stars

IMPORTANT STATISTICS

Planet name: Wyzdom

Equatorial diameter: 14,542 kilometers

Mass (Earth = 1): 1.65

Surface gravity: 1.27 g

Escape velocity: 13.5 km/s

Albedo: 0.35

Atmospheric pressure at mean sea level: 1.81 bars

Fraction of surface covered by land: 26%

Maximum elevation above sea level: 12,110 meters

Length of day: 13 hours 18 minutes

Length of year: 341 Earth standard days

Obliquity: 18 degrees

Von Roenstadt habitability factor (Earth = 1): 0.83

Current population: 42.6 million

Number of major population centers: 15

Year settled: 2122 adtc

Number of moons: zero

Star name: Alpha Centauri

Star type: G2V

Distance from Earth: 4.3 light years

Distance from planet: 0.98 AU (Earth-Sol = 1 AU)

Star Mass: 1.08 (Sol = 1)

Travel time from Earth

Ship's time: 59 days
Planet time: 85 days

Atmospheric Composition

Oxygen: 17.2%

Nitrogen: 66.0%

Carbon dioxide and inert gasses: 16.8%

Table 3.2

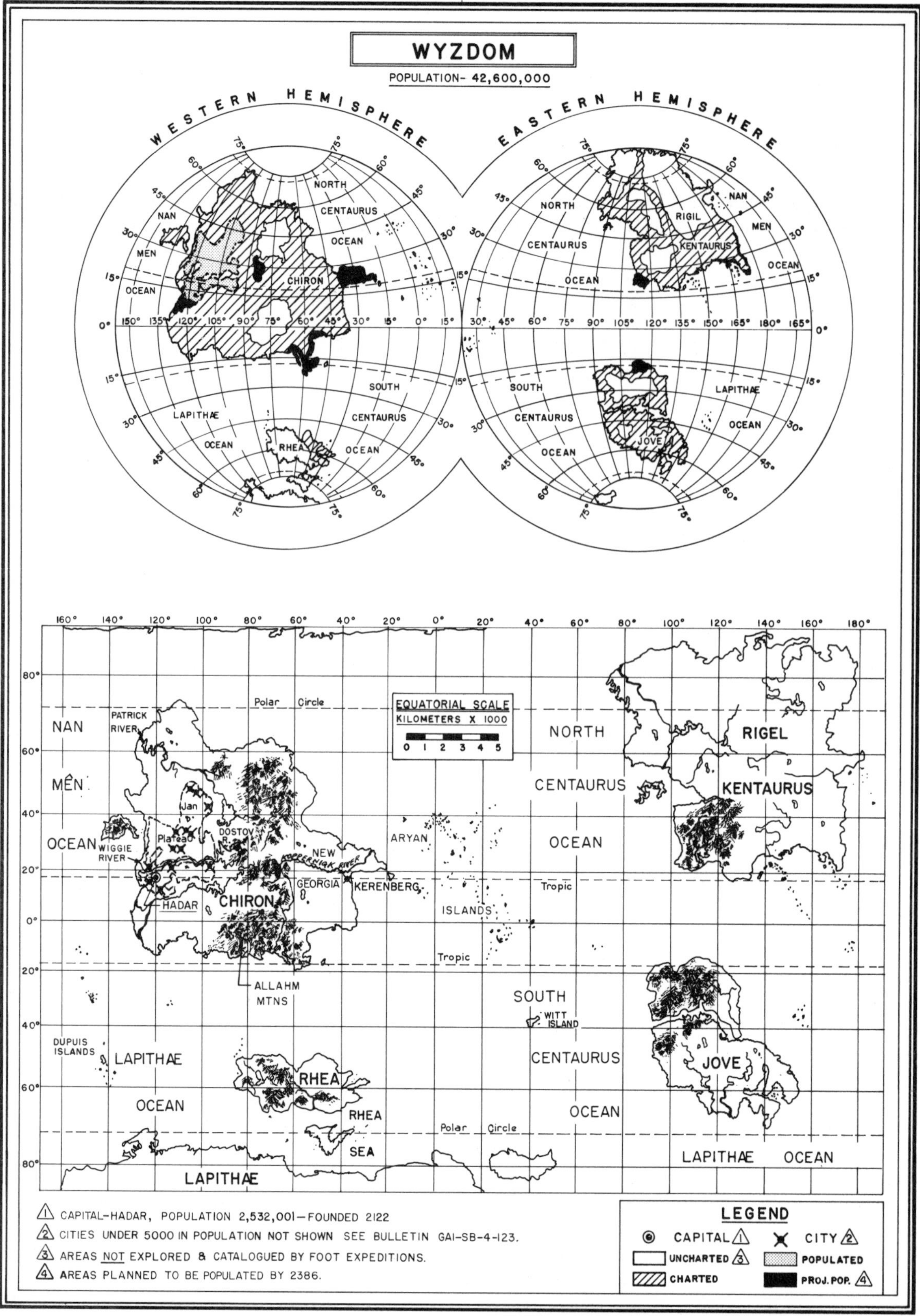
WYZDOM
POPULATION- 42,600,000
WESTERN HEMISPHERE
EASTERN HEMISPHERE
NORTH CENTAURUS OCEAN
CHIRON
NAN
MEN
OCEAN
SOUTH CENTAURUS OCEAN
LAPITHAE OCEAN
RHEA
RIGIL
KENTAURUS
JOVE
Polar Circle
EQUATORIAL SCALE
KILOMETERS X 1000
PATRICK RIVER
WIGGIE RIVER
HADAR
Jan
Plateau
DOSTOV R.
NEW GEORGIA
KERENBERG
ARYAN ISLANDS
ALLAHM MTNS
Tropic
RIGEL
WITT ISLAND
DUPUIS ISLANDS
RHEA SEA
LAPITHAE
1 CAPITAL-HADAR, POPULATION 2,532,001—FOUNDED 2122
2 CITIES UNDER 5000 IN POPULATION NOT SHOWN SEE BULLETIN GAI-SB-4-123.
3 AREAS NOT EXPLORED & CATALOGUED BY FOOT EXPEDITIONS.
4 AREAS PLANNED TO BE POPULATED BY 2386.
LEGEND
CAPITAL 1
CITY 2
UNCHARTED 3
POPULATED
CHARTED
PROJ. POP. 4

FIG. 3.2

WYZDOM—First Frontier in the Stars

PHYSICAL ENVIRONMENT

Wyzdom, the third of four planets orbiting Alpha Centauri A, an identical twin of Earth's own sun and its nearest stellar neighbor, lies a "scant" 4.3 light years from Earth. Wyzdom's night sky appears identical to Earth's except that the constellation Cassiopeia contains an additional bright star, Earth's sun.

Unlike Earth's sun, however, Alpha Centauri A is but one star in a triple star system which contains a smaller K type star, called Alpha Centauri B, and a tiny M type star known as Proxima Centauri. Stars A and B orbit each other once every 80 years in a highly eliptic path, at a distance that varies between 35 and 11 astronomical units. Since star B lies as far from A as the planet Saturn lies from Earth's sun, its radiation comprises less than six percent of the total falling on Wyzdom and therefore has little effect on the planet's surface temperature. The companion star is the second brightest object in Wyzdom's heavens and, visible both day and night, appears as a brilliant dot some 1500 times brighter than Earth's moon. During half of a Wyzdom year, the companion star appears in the day sky, while during the other half, it shines in the night sky. On those nights, it obliterates all but the brightest stars, creating a continuous twilight. Wyzdomites call these months "am" months, and the months when the companion star rises during the day they call "nocht" months. As each 80-year period passes, the am and nocht months shift from winter to summer, and back again, creating a complex pattern of seasonal variations that profoundly influences the native life forms.

The third star of the Alpha Centauri system, Proxima, orbits the other two at a distance of about 11 thousand astronomical units, more than 270 times the distance that the planet Pluto orbits Earth's sun. Though invisible to the naked eye from Earth, Proxima appears in Wyzdom's sky as a moderately bright star of magnitude 3.6.

More than one planet in the Alpha Centauri system possesses life, though the second holds greater interest for scientists than for potential pioneers. Graves' Planet occupies the second orbit of the companion star, Alpha Centauri B. Its life is chlorine/silicone based rather than oxygen/carbon based. It is quite primitive in comparison with the sophistication of life on Wyzdom. Yet the discovery of life on Graves' Planet during the second expedition to Alpha Centauri excited scientists almost as much as the discovery of life on Wyzdom, for it proved that elements other than carbon, oxygen and nitrogen could form the basis of life and greatly increased the probability of encountering other intelligent life in the galaxy.

GENERAL CHARACTERISTICS

The largest of the colonies, Wyzdom dwarfs the Earth. Its diameter measures 14 percent greater than Earth's, while its mass bulks 1.65 times Earth's. Consequently, gravity at the planet's surface pulls with 27 percent more force than it does on Earth, exceeding that of any other colony planet.

While oceans cover the majority of Wyzdom's surface, the land masses spread across an area 17 percent larger than Earth's land area. Most of the land is divided into six large, widely-spaced continents. Chiron, the largest, and the only one inhabited by Humans to date, has an area of approximately 102,000,000 square kilometers, making it nearly twice the size of Earth's Eurasian land mass. Two features stand out on Chiron's map: the great Jan Plateau and the Allahm Mountains. The Jan Plateau, with a mean elevation of 4600 meters above sea level, covers an area half the size of Earth's South American continent. To the east stand the Allahm Mountains, taller than any mountain range on the habitable planets. Their tallest peak, Mt. Alfus, breaks all records; topping out 12,110 meters above sea level.

Of the other major continents, Jove and Rigel Kentaurus display a wide variety of terrain and climates similar to those on Chiron. Rhea, the smallest continent, consists primarily of extremely mountainous or desert regions, except for a small strip of temperate rain forest which lies along the west coast. Lapithae, like Earth's Antarctica, lies entirely within the southern polar circle of Wyzdom and therefore contains no land area that Humans wish to use.

CLIMATE

The first time a nearby crack of Wyzdom's thunder leaves an immigrant's ears ringing or a 50 km/hr breeze knocks him down, the newcomer learns why Wyzdomites take the weather seriously.

Wyzdom's atmospheric pressure, about 80 percent higher than Earth's, causes 80 percent higher wind pressures at comparable velocities. In addition to packing more destructive force on land, Wyzdom's winds generate huge ocean waves that can devastate shoreline structures, discouraging both waterfront building and the use of small pleasure boats on the open seas. Fortunately, Wyzdom's more temperate climate mitigates the power of its weather. The planet's lower obliquity (angle of inclination toward its ecliptic plane) causes smaller seasonal variations in temperature, and its shorter day causes smaller daily temperature extremes. The lower temperature differences cause lower differences in atmospheric density, the principal cause of thermally induced winds. This reduces wind forces below what they would be if Wyzdom possessed Earth's seasonal extremes. Yet the immense size of Wyzdom's continents creates greater extremes of temperature near their centers than are found on Earth. North central Rigel Kentaurus has recorded temperatures as low as -98 degrees, surpassing Earth's Siberian region.

Wyzdom's high-density atmosphere thins out rapidly with increasing altitude. Thus, the atmospheric density on the Jan plateau approximates that on Earth at sea level, while at the top of the Allahm Mountains the air is thinner than at the top of Mt. Everest.

LIFE FORMS

Wyzdom hosts a variety of wholly alien life forms. Only a fraction, some 238,000, of Wyzdom's estimated million species have been classified. Most is known about the life forms on Chiron. The oceans and the other continents remain to be studied in detail. Studies to date indicate that life evolved quite separately on the five major continents outside the arctic regions. This fact is not surprising, for the continents are far more isolated from each other, and have been so for a much longer period, than the continents of Earth. Totally distinct classes fill the different ecological niches on the various continents, but certain basic physiochemical similarities exist between all of Wyzdom's native life, possibly because, as on Earth, the life of Wyzdom originated in the sea.

Vegetation

Plants constitute the majority of individual species on Wyzdom, as they do on Earth, but there are no forms as advanced as the angiosperms, the majority of Earth's plant species. The class of angiosperms includes the flowering, fruit-bearing plants of greatest importance to Humans, since all our food grains, edible fruits, and most vegetables come from them. Angiosperm seeds develop in an ovary at the base of a flower, but no plant native to Wyzdom uses this kind of reproductive mechanism. In some native species, seeds develop in cone-like structures, while in others, seeds develop on the underside of leaves, very like large spores. Many species employ true spores, in lieu of seeds, similar to Earth's ferns, while others have no reproductive organs at all, but multiply by dividing. These plants send out a long root which pushes up new foliage. After the foliage develops, the connecting root dissolves.

The plants of Wyzdom resemble the plants of Earth in superficial ways, so natives call them "grasses," "shrubs," and "trees." Most plants employ a green substance resembling chlorophyl for photosynthesis. Hence, from a distance a forest of Wyzdom looks like its Earthly counterpart. Only on closer inspection do the differences emerge.

Brown bark covers some Wyzdomite trees while others possess great green stalks up to 25 centimeters in diameter. At lower elevations, the trees are shorter than those on Earth in order to withstand the high wind forces. The strength of the wood of many of Wyzdom's trees far exceeds that of any wood on Earth, while other Wyzdomite trees contain soft, flexible cores that allow them to bend without breaking in the high winds. Fewer species of native trees lose their leaves each season and some broad-leafed plants remain green throughout freezing winters.

One class of plants unlike any other on Earth are the chlorofungi. These large plants, with the structure of mushrooms, reproduce using spores. Colored green, they derive half their nourishment by absorbing decayed or living organic material and half by photosynthesis. Some grow quite large and resemble the giant toadstools of *Alice in Wonderland*.

Animals

Animals on Wyzdom also appear less evolved than those of Earth. The only warm-blooded creatures are flying animals that resemble birds. Zoologists divide them into four classes; some sport feathers, others do not. Although some flying species lay eggs, others birth their young alive. Although some possess mouths with teeth, others have horned beaks and no teeth. At lower elevations, large predatory "birds" fill the ecological niches of large predatory mammals, like lions and tigers on Earth. The largest of these captured to date weighed more than 45 kilos. These extremely dangerous birds can fly only in the dense air at low elevations. Lowlanders protect themselves against them by encircling their living compounds with ultrasonic fields. When venturing outside protected areas on foot, all lowlanders, even children, carry ultrasonic beam generators that repel the birds. Although most Wyzdomites frown upon wholesale slaughter of these magnificent predators, on extended exploratory trips, pioneers carry laser guns as well.

Smaller, cold-blooded animals, similar to Earth's reptiles, and amphibians, live on the ground. They fill a wide variety of niches in the food chain, but most are relatively harmless. Few have scales like Earth's reptiles, and few are found at latitudes greater than 50°. Even smaller creatures, with structures resembling Earth's arachnids or insects, can be found on all land masses. They generally lay eggs that can survive freezing winters. The most common class, called "inocts," walk on eight legs and have evolved into a wide variety of species. Inocts aren't harmful to people since they have little interest in Humans as food and do not carry Human diseases.

The nocht months and am months influence the behavior of Wyzdom's land life almost as much as do the seasons. Many plants grow only in the am months and retire to a semi-dormant state in nocht months. A variety of inocts and cold-blooded animals remain active around the clock in am months. Many inocts emerge from their eggs only during am months. The constant light of the am months influences even the warm-blooded birds. They use this period to force-feed their young to rapid maturity.

Marine Life

As on Earth, Wyzdom's oceans teem with life. More than 70 percent of the photosynthesis that produces breathable oxygen takes place in the seas, but little is known about sea life. Most people on Wyzdom don't live near the water and the oceans have little commercial importance. Wyzdom's "fish" resemble their Earthly counterparts, but no warm-blooded sea creatures have been found to date. Several varieties of 12-legged "octopus," nine-pointed "starfish," and creatures resembling giant sea slugs, sea urchins and anemones are present. A highly-mobile abalone-like monovalve is among the most interesting shellfish. Seaweed abounds, but as on Earth, microscopic plants in the sea produce most of the planet's oxygen. Of greatest interest are a class of photosynthetic, swimming "animals." Although they produce part of their energy from sunlight, they derive the rest from filter feeding. These small creatures extend filigree arrays of light-catching tissue several times the size of their bodies while floating motionless in the water. If a potential predator approaches, they can retract it and swim away so quickly that they appear to vanish in the wink of an eye.

Microorganisms

Microorganisms ranging from viruses to highly mobile, one-celled plants and animals flourish in Wyzdom's air and water. Their structure strongly resembles their Earthly counterparts, but when scientists probed their genetic code, they found a fundamentally different organization. Consequently, Wyzdom's microorganisms have not yet posed a health hazard to Humans. Ironically, the low incidence of native diseases may prove harmful in other ways. Wyzdom's natives have begun to lose their immunity to Earthly diseases. Should such diseases then be brought from Earth by immigrants, they could have as devastating an effect as common measles had upon the Polynesians when Europeans first settled the Pacific Islands. Viruses indigenous to Wyzdom also may mutate into forms harmful to people. For these reasons, the Wyzdom Institute for Disease Control constantly researches the microbiology of Wyzdom and monitors disease control activities on Earth. When infectious outbreaks occur, Wyzdomites should be able to contain them without any large scale plagues.

BRIEF HISTORY

Three hundred and two years ago, Captain Jan De Wyze and the crew of the *Freedom 4* returned from the first manned interstellar voyage to report the discovery of the first extraterrestrial planet whose natural environment could support Human life. Since space warp had not been invented, the round trip had taken 15 years in Earth's time frame plus the additional 46 months *Freedom 4* had spent studying the new planet. Forty-one years passed before the first pioneers left for Wyzdom to establish the first permanent colony. During that period, two larger exploratory parties travelled to the Alpha Centauri system to gather additional data needed to plan the colony. The uncertainties of building a colony so far from Earth, the difficulties posed by the high gravity and atmospheric pressure, the violent weather, and the potential for disease appeared overwhelming in those days, so limited was people's knowledge of ecology, disease, and the limits of Human physiology.

Two thousand and seventy pioneers established their home on the Jan Plateau 254 years ago. They chose the Plateau because its atmospheric density approximates Earth's, and they feared the effects of violent winds at lower elevations. Despite the years of planning, the colony did not proceed as smoothly as new colonies do today. The group contained a preponderance of scientists and political undesirables, and neither sort makes particularly successful settlers. Because of the high cost of early interstellar travel, colonists skimped on the supplies and equipment they brought. They planned to rapidly produce the additional food and machinery they needed to survive. Had they failed, resupply would have been impossible, for the absence of large starships with warp capability made travel times unacceptably long.

All did not go according to plan. Although Earth's grains grew poorly in the foreign soil and climate of Wyzdom, mercifully, disease did not prove to be the hazard it could have been. The colony's scientists, writers, and bureaucrats could not be motivated to do the manual labor required on the new world. Machinery broke down frequently, but the colony had few skilled mechanics to repair it. During the second winter, rations were reduced to a mere 1000 calories per day. Forty-three newborn infants and more than 200 older immigrants died of malnutrition and cold caused by frequent power failures at the central plant.

A council elected by the colonists governed the original colony according to socialistic principles, but bickering among council members brought practical decision-making to a standstill. At the end of the second winter, the pioneers rebelled against the elected government and in desperation turned to Medhat Hadar.

Hadar, a young businessman from the ancient city of Beirut, operated the only successful farming/ranching enterprise on Wyzdom. In theory, private enterprise was illegal, but Hadar operated his farm so efficiently that his 150 hectares, a mere fifteen percent of the land under cultivation, produced 40 percent of the colony's food, much of which was sold on the black market. The colonists offered to make Hadar absolute dictator if he would get them out of their predicament. Hadar agreed and immediately abolished the socialistic economy. He parceled out land to individuals willing to work it and loaned farm machinery to those who could best use it. Farmers repaid loans in the form of goods to a central trading company operated by Hadar himself. Medhat abolished all scientific research and put the personnel to work at mundane tasks. He loaned the machine shop to a group of engineers and mechanics who operated it for profit and paid back the community for its use. Within eight months after taking charge, Hadar's government had sold all community-owned capital equipment to the highest bidders who could demonstrate the ability to repay a loan. Within three years the colony prospered and profits from the loans of communal machinery began to flow into foundations supporting scientific research and the first school system on Wyzdom.

After seven years, Hadar voluntarily stepped down as dictator and restored a democratic form of government. Before abdicating he drafted most of the present consitution guaranteeing not only personal rights, but property rights as well. He had become the wealthiest man on Wyzdom and remained so for the rest of his life. Upon his death, he left the bulk of his fortune to found the University of Wyzdom, which remains the largest and most important institution of higher learning on the planet. He also had firmly established Wyzdom's pattern for growth and development, a pattern that has served most of the subsequent Human colonies. In recognition of his contribution to the planet, the pioneers posthumously named the capital city after him.

From Hadar's time on, the Wyzdom colony grew, both from the birth of natives and from the influx of immigrants which accelerated after the building of transports with warp capability. During this time the population's standard of living increased as well. The colony achieved self-sufficiency in all physical goods about 160 years after it began, though even today Wyzdom relies heavily on information from Earth for everything from engineering designs to movies and books.

CURRENT STATE OF DEVELOPMENT

Population

At this writing, Wyzdom's total population stands at 42,600,000, approximately the size of a large Earth city. Seventy percent of the people live on the Jan Plateau, yet the population density there remains slightly less than four persons per square kilometer. All immigrants land on the Plateau, the most Earth-like region of the planet. A majority of these remain there permanently. About five percent of Wyzdom's population consists of immigrants, the balance being native born; this contrasts sharply with the newer colonies. The average income of immigrants is only six percent lower than that of the population as a whole, a remarkable fact when one considers that immigrants arrive with no real assets save their knowledge and skills. Fewer immigrants earn incomes in the lower 20 percent of the planet's population, which suggests that immigrants have more consistent economic success than the native population.

Food

Despite shaky beginnings, Wyzdom has developed an industrial base that offers its residents as high a physical standard of living as on Earth. Food consists primarily of plants and animals brought from Earth and cultivated on the colony. Corn, wheat, and barley comprise the staple crops of the Plateau, although rice is increasingly popular in the warmer lowland regions. The Plateau's climate is varied enough to support most Earthly fruits and vegetables from apples, cherries, and beans in the north to citrus, grapes, kiwi, and squash in the south. Earth's heavy beef cattle have not adapted well to the planet's high gravity, but smaller dairy cattle and sheep are raised for their milk as well as their meat. Goats and rabbits have been deliberately excluded from the Plateau because they become wild so easily. In the absence of natural enemies, these harmless creatures would devastate the countryside's vegetation.

Wyzdom is slowly developing its own unique cuisine that makes heavy use of spices and vegetables native to the planet, as well as of native "bird" eggs. Colonists have domesticated several species of these flying creatures and produce their eggs commercially. Since Wyzdom lacks native grains and fruits, leaves and roots provide the vegetable input to Wyzdomite cooking. The broad, crisp leaves of the Panjam Pad, a kind of water lily, form one of the most popular vegetables. Generally served raw, they may be lightly sauteed or deep fried. Dishes partaking of both native and Earth foods are most common. Kharbog bulgar, for example, consists of a mixture of bulgar wheat and kharbog, a soft and nutty tasting root, cooked together with a variety of native spices.

Housing

In an expanding society, shelter remains one of the most important consumer industries. Most Wyzdomites live in single-family dwellings sprinkled sparsely across the landscape. Virgin land remains available for claim in all major industrial regions, and economical transportation allows Wyzdomites to work relatively far from their homes. Most homes are built from modular sections manufactured by more than 20 companies, although in a few cases, individuals construct their homes themselves from traditional materials such as wood, stone, and glass. About 20 percent of the population maintains homes in major cities, usually in high-rise structures designed for many families, though most city dwellers also own country homes they can escape to for more privacy.

Consumer Goods

Almost every type of manufactured consumer good found on Earth from bioscanners, somafields, autoservers, and androids to furniture, clothing, toys and games can be purchased on Wyzdom. The variety of styles is somewhat more limited than on Earth, however, and Wyzdomites appear less style-conscious than Earth people about all things from clothing to computaplex terminals.

Transport

Because of its low population density, Wyzdomites rely primarily on antigravity cars and trucks for personal as well as commercial transportation. Most vehicles employ advanced designs including fully-automatic or semi-automatic guidance control. Large commercial vehicles travel without any Human operator over preprogrammed routes between cities. In the larger cities, antigravs prove impractical, and high density guideways transport people and freight. For long trips, such as from Jan Plateau to New Georgia, supersonic antigrav transports fly over regularly scheduled routes. At this writing, construction is proceeding on the planet's first bulk transport conduit under the Allahm Mountains to reduce the number of bulk-carrying antigravs making the trip between the Plateau and New Georgia.

Medical Services

Despite the high gravity, Wyzdomites boast excellent health records in comparison with people on Earth. Injury and sickness have not vanished. Falls, even from modest heights, can prove quite serious; so medical services remain a necessity. These services, provided solely by private individuals or companies, span a wide variety of medical theory. No licensing laws exist on Wyzdom; anyone is free to give medical treatment. Because of this, Wyzdomites choose their physicians very circumspectly, and degrees from the famed Hadar School of Medicine are valued highly.

Education

Education for everyone from children to centegenarians is common. Unlike Earth people, Wyzdomites purchase schooling like any other service from a variety of sources, ranging from private tutors to not-for-profit cooperatives. Ready access to computer-aided instruction allows self-education to a very high degree, even for young children. Of course, some parents do not provide well for the education of their offspring, and, without doubt, these children are disadvantaged. Since Wyzdom is a young and growing society no stigma is attached to adults who catch up on education missed in their early lives. Retraining for new jobs or professions also is quite common as economic needs change and new opportunities arise.

Industry and Technology

An industrial infrastructure as comprehensive as Earth's, if not as large, supports the wide range of Wyzdom's consumer goods and services. Energy and materials production, capital equipment, manufacturing, transportation of goods, communications, banking and distributing employ almost 50 percent of Wyzdom's working population. Wyzdomites are beginning to develop technology that has improved the efficiency and quality of Earth's industries. The bipolar computer-link technique; preconceptual, real-time DNA manipulation, field-reinforced space frames, and the formula for Centauri stones number among the most important of the many industrial techniques and processes invented on Wyzdom. During the coming century, the importance of Wyzdom's contributions will no doubt increase dramatically.

Government

Governments play smaller roles in the daily lives of Wyzdomites than of Earth people. An excerpt from Medhat Hadar's address to the colony on the day he stepped down as dictator best summarizes this tradition.

> "For five years I have held absolute power over this planet as few have held it in Human history, and I have learned much about what such power implies. Political power cannot exist without economic power. Those who hold the weapons can be no more than lackeys of those who control the grain silos. Once a government wields power over the livelihood of its people, its power will grow inexorably, until its control is absolute. This is true of all forms of government, whether elected by the people or not. . .I have (therefore), in consultation with the wisest people of the colony, devised a constitution which I hope will forever divide the force of arms from the power of industry."

Since that time, though Wyzdom has grown from a single settlement to a widely dispersed industrial society, Wyzdomites have maintained the maxim that government's sole responsibility is to protect the rights of individuals against harmful actions by others, including not only acts of violence and theft, but breach of contract, fraud, and wanton environmental destruction.

The constitution of the planetary government drafted by Medhat Hadar remains the supreme law in effect today. Hadar knew that as Wyzdom grew, its need for regional and local governments would grow, and he therefore allowed for the formation of these. He entrusted the planetary government with only those powers required to guarantee rights, such as freedom from search and seizure and the right to a writ of habeas corpus. Property may not be confiscated, except for payment of a lawful, contractual debt. No government may subsidize any industry, control prices, restrict output, or set arbitrary standards for any product, service, or commodity. Taxation in all forms may not exceed two percent of any person's income, and no government with the power to tax may enter into any business, save operating courts of justice, police, and planetary defense.

The planetary constitution divides the government into two parts: a policy-making and legislative arm, called the High Council of the Executive, and a judicial branch culminating in the Supreme Court of Justices. The general population elects only the High Council members. Taxpaying citizens are accorded votes in proportion to the taxes they pay. The High Council consists of 11 members who serve for six-year terms expiring on a rotating basis. The Council elects one member to a year-long term as Chief Executive and appoints key members of the executive branch: Chief of Police, Chief Prosecutor, and Director of the Treasury, GAILE liaison officer, Chief of Administration, and Community Liaison Officer. The High Council makes most laws, although any legislation may be introduced by petition of voters and passed by popular vote. A simple majority of the Council or the voters is required to pass regulations. A two-thirds majority is required to sell bonds or levy taxes, and a unanimous Council vote, combined with 90 percent approval of the population, is required to amend the constitution. It has been amended only five times since its writing.

The Supreme Court of Justices, appointed by the High Council for 30-year terms, forms the highest court to which any decision of a public or private court may be appealed. The Court may review the constitutionality of any existing law. Justices whose terms have expired may request to be admitted to vote with the present court on all matters relating to constitutionality. The High Council also appoints judges to lower courts, but the constitution permits all disputes to be submitted to nongovernmental judges acceptable to both parties. This system prevents clogging of the courts by allowing people to pay to have disputes resolved.

Wyzdom's limited government appears to work. Crime is low, perhaps because opportunities exist for anyone willing to work, perhaps because "organized" crime is small. Of course, Human society will always harbor those who prefer to take their livelihood from others, but Wyzdomite governments have managed to cope with them. Justice is "retributive" rather than "punitive," meaning that emphasis is placed on the criminal's repayment of his crime and associated costs. For example, someone guilty of computer embezzelement would be required to pay back all that he stole, plus interest, tracing fees, and the cost of his trial. A person found guilty of assault would have to pay damages to his victim, plus the cost of tracking him down and trying him. Victims have some say in these judicial matters, and may request that the criminally insane be confined to institutions.

Perhaps as Wyzdom grows older and more crowded, the current governmental institutions will prove inadequate. Yet Hadar's constitution has served through many changes in the society, and, if anything, the system of free enterprise has proved its adapability.

ACCOUNT OF LIFE ON WYZDOM

Editor's Note: Alex Kotorovich, 43, Director of the New Georgia Bauxite Recovery Operation and Mayor of New Georgia, gives an account of life on Wyzdom. A native, born of immigrant parents, Alex moved with his family to New Georgia and became acclimated to the higher atmospheric pressures at age 13. He took his Bachelor of Science Degree in Geology from the Wyzdom University College of Science, then went to work for the Kerensky Mineral Development Company as an exploration geologist and spent nine years as part of a mineral survey team. The team explored much of Wyzdom's uninhabited land mass, making detailed surveys of mineral deposits. During his travels, Alex saw much of the new planet that Human eyes had never looked upon. His personal observations, taken in the form of copious notes and pictures, formed the basis of his textbook on Wyzdom's physical geography, which remains the definitive work on this subject today.

After his travels, Alex returned to New Georgia, where he resides today with his family. His experience as an explorer, a civic leader and a professional author makes him eminently qualified to write of life on his home planet.

I have never been to Earth, though I am Human. My ancestry dates back 100,000 years to the first *homo sapiens* that stood erect, yet in another sense I form the first of a new species. With my kind begins the evolution of homo wyzdom, a different being adapted to a different world.

As I emerged from the womb, my first breaths drew thin air much like Earth's, for my birthplace, the Jan Plateau, lies in the rarified atmosphere 4600 meters above the sea. I found learning to walk a painful experience, for Wyzdom's high gravity is unforgiving of the mistakes of children. Falls come fast and hard. Yet I survived and grew, and my muscles acclimated to the stronger pull of our world. My legs grew thick and my back broad and strong. Today, my proportions make even my Ukranian parents seem slender by comparison.

The first major event of my life occurred at the age of 13, when my father announced we would be moving to New Georgia. The Lowlands! In my early years I was not an adventuresome child. Few people lived in the lowlands then. They were thought of as wild and dangerous places, far from the civilization of the Plateau. Rumors abounded of giant storms that leveled whole villages and of giant birds that carried people off whole to devour them in their lairs.

My father is not disposed to argue, especially with his children. The job that awaited him at the new organochem complex offered a monumental leap for his career. For us there would be more money, more land, a large new home, and college educations that he never had. Our howls of protest fell on deaf ears. We packed our belongings and moved.

My mood changed from petulant depression to awe and wonder as we streaked across Chiron on the hypergrav. From my vantage point 30,000 meters up, beneath a purple sky, all of the continent's great natural features passed in review. The green forests of the eastern Plateau ended abruptly in great cliffs dropping away to the maze of ridges of the upper Dostov River. Then the Allahm Mountains rose before us. When I asked my father the difference between a mountain and a hill, he replied, "A mountain makes your palms sweat when you stand at the bottom and look up at it." He had seen the Allahms before. Most large mountains have a timber line, above which vegetation cannot live, and their tops are capped with snow. The Allahms are so tall that a second line appears, above which no snow can fall! Their great rock crowns thrust out of white mantles of perpetual snow; giant guardians of our planet's treasures.

The hypergrav swooped down from the western slopes of the Allahms into the port of Kerenberg, and from there slower antigrav cars carried us to the bustling construction camp at the plant. The first months in the camp dragged interminably. There was little to do except read or organize impromptu field games with the other children. Our parents took note of this and began to organize a school to continue the educations that most of us interrupted when we moved. In those days, educational terminals were nonexistent in the province. A few parents had teaching experience and took it upon themselves to use one of the general access terminals after business hours to read and memorize a lesson to present to the children the next day. This was the rude beginning of a tradition of voluntary education that continues to the present in New Georgia, although now ample teaching facilities exist and most children have their own terminals.

During those early months, Mother poured through the prefab catalogs, picking out a house for us. On weekends when father could get away, he took us over the surrounding wilderness for hundreds of kilometers in every direction looking for the perfect homesite. In those days, personal land claims in the lowlands could total ten hectares. We soon settled on a gently sloping hillside at the edge of a large lake. A forest of short, evergreen trees covered the countryside around our new home. Most of them bore green compound needles. I called them "Christmas Balls" because each needle consisted of a circle of needles, each bent from its root to its tip and connected again to form a tiny, green cage.

Since deliveries on modular homes ran to three months in those days, Father decided we should clear the land for the house ourselves. Each weekend he gathered up the entire family and we flew to the homesite for a weekend of work. Father rented a large laser cutter/disintegrator and a couple of antigrav "skyhooks." One by one we cut down enough trees to make a circular clearing 40 meters in diameter. We used the skyhooks to move the logs to a great pile so the wood products company could pick them up, then burned out the stumps with the disintegrator beam. We worked very hard, even though machines did most of the physical labor. It was very satisfying to see our progress each weekend. At night mother cooked outdoors, and we slept in a temporary dome in the clearing. It all seemed like a great adventure.

Since the first day in the lowlands, my parents admonished us not to wander off alone, and to always carry our ultrasonic generators to repel the great predatory birds. These creatures never appeared near the organochem plant, perhaps because the noise and bustle of the construction frightened them, but on the second weekend at our wilderness homesite, I made the first of many encounters with these magnificent beasts.

Father had just cut a large tree, and we were in the process of stripping branches and attaching antigrav hoists so we could move it to the log pile. Suddenly an enormous shadow flashed across the clearing, and all heads turned upward to see a giant bird circling overhead. Its wings spanned nearly six meters, and an awful head, armed with a jagged, hooked beak protruded between them. As it circled, the sun glinted off its brilliant plumage which varied from forest green on top to pale sky blue beneath. Golden feathers covered its head like a helmet, in contrast to its blood red beak and feet.

Upon seeing the bird, Father ran to get his laser rifle, and the rest of us scrambled for our sonic guns. After circling the clearing a few more times, the bird alighted near us and began to approach on foot, walking awkwardly but deliberately on its stubby legs. It stood about 1.5 meters tall and appeared more curious than hostile. My adrenalin surged, and my heart pounded as I looked from the bird to my father and back. None of us had ever confronted such an animal before, and we didn't know how dangerous it was. The bird and my family faced each other for several long moments, when suddenly, as abruptly as it came, it turned, leaped into the air, and streaked across the clearing to snatch a smaller bird about the size of a chicken from midair. It continued up over the tree tops, clutching its prey, and soon disappeared from view. The instant it vanished, we all burst out talking at once, and the bird consumed our interest as we worked for the rest of the afternoon.

My teens passed quickly in New Georgia. Despite the informality of our school, I learned enough by age 17 to pass the entrance exam for the Wyzdom University at Hadar. Despite my interest in biology, I majored in geology because I thought it would help me to get a better job. By graduation, my emotional and intellectual horizons had broadened, and I yearned to see more of my fabulous world. I accepted an offer to become an exploration geologist and thereby began a ten-year adventure that took me over four continents.

Wyzdom's crust is quite poor in the heavier industrial metals, such as chromium, copper, iron, lead, nickel, and zinc. Even the lighter structural metals, aluminum, titanium, and magnesium, exist only in very low-grade deposits. Because of the scarcity of transportation, early pioneers used the poor deposits of the Plateau, but as the population spread out and the capacity to build transports grew, Wyzdom's mine operators began looking far and wide for potential mining sites. Satellites and shuttle flys give valuable information about the probability of finding useful ore deposits, but only old-fashioned drilling and sampling methods can prove that commercially minable deposits exist.

My first trip out as a fledgling geologist took me to the western foothills of the Allahm Mountains in southern Chiron. There, the same geologic forces that created the mountains pushed up iron and nickel deposits of minable quality. We traveled in a convoy of low flying levitrucks which provided living quarters for the crew as well as carried surveying and drilling equipment. Going out we flew directly south of our base on the Jan Plateau, passing into a broad flat region covered by dense, tropical forest.

Though these were winter months, temperatures ranged into the low 30s and it rained daily. As in most lowlands, the forest trees were short and thick, though some long and slender species resembled giant field grasses. These strange plants grew in patches, driving out the broad-leafed, low trees beneath. They sported bright white or purple structures like pom-poms at their tops. Though they ranged up to 20 meters in height, their trunks varied between 10 and 20 centimeters in diameter, and they could bend up to 90 degrees in strong winds. Over most of the plains and foothills, vegetation lay so thick that we couldn't land without burning a small clearing. Despite the fact that our laser defoliators stripped the ground bare, the vegetation grew to more than a meter tall again within a week. The southern plains may someday serve as incredibly productive farmland if people can learn to cope with their unrelenting heat, violent winds, and prolific natural vegetation.

The next five years took me to the four corners of the Chiron continent; from the southern rainforests to the central deserts ruled only by giant inocts, to the frozen tundra in the north where the wombat moles bore their intricate mazes hundreds of meters beneath the surface. Despite great variations in climate and life forms, gradual transitions separated one biome from the next. When my exploratory missions began to take me to other continents, I saw startling changes from what I had known on Chiron.

In those days Vladimir Kerensky still ruled Kerensky Minerals Company with an iron hand. His business genius lay in his ability to sense an economic need long before most people began thinking about it. He reasoned that people would live in the pleasant climates of the uninhabited continents of Jove and Rigel Kentaurus long before they occuped the tropic regions of Chiron. When these pioneers moved to new continents, they would need minerals and Kerensky would already have located the best deposits and made plans for mining them. To this end he dispatched me to lead an exploratory team to Rigel Kentaurus, and for much of the trip we penetrated virgin country. As part of our duties, Kerensky wanted us to observe and photograph the landscape, to classify interesting and potentially dangerous life forms and to make weather observations in addition to our usual surveying and sampling tasks.

To get to Rigel K, we had to cross the North Centaurus Ocean. We traveled in extra large levitrucks designed for these missions. Even at our cruising speed of 800 kilometers per hour, the ocean crossing took more than 14 hours. In midwinter, great storms sweep the North Centaurs. Winds reaching nearly 300 kilometers per hour drive giant waves to heights of 30 meters. Because of Wyzdom's high gravity, such waves pack more energy than comparable waves on Earth, and therefore travel much farther after they leave the storm area. Even in calm winds, the winter ocean remains restless. During the trip we passed through one great storm, flying high, surrounded by a wall of grey clouds. When the weather cleared, we skimmed the water just above the waves, hoping for glimpses of interesting sea creatures. At such high speeds, we didn't have much luck, but we did catch sight of a "Buffington's swimmer," a large, cold-blooded water-breathing animal, shaped somewhat like the elasmodon of Earth's Loch Ness. All across the sea, long winged birds, resembling enormous versions of the Earthly albatross, crossed our path, occasionally dipping into the sea for food.

The map of Rigel Kentaurus (figure 3.3) shows the path of our journey. We crossed the coastline at 43 degrees north latitude and headed northeast along the shore of the Gulf of Boheme. We soon came upon the most unusual coastal formation I have seen, a beach made of solid black glass. The

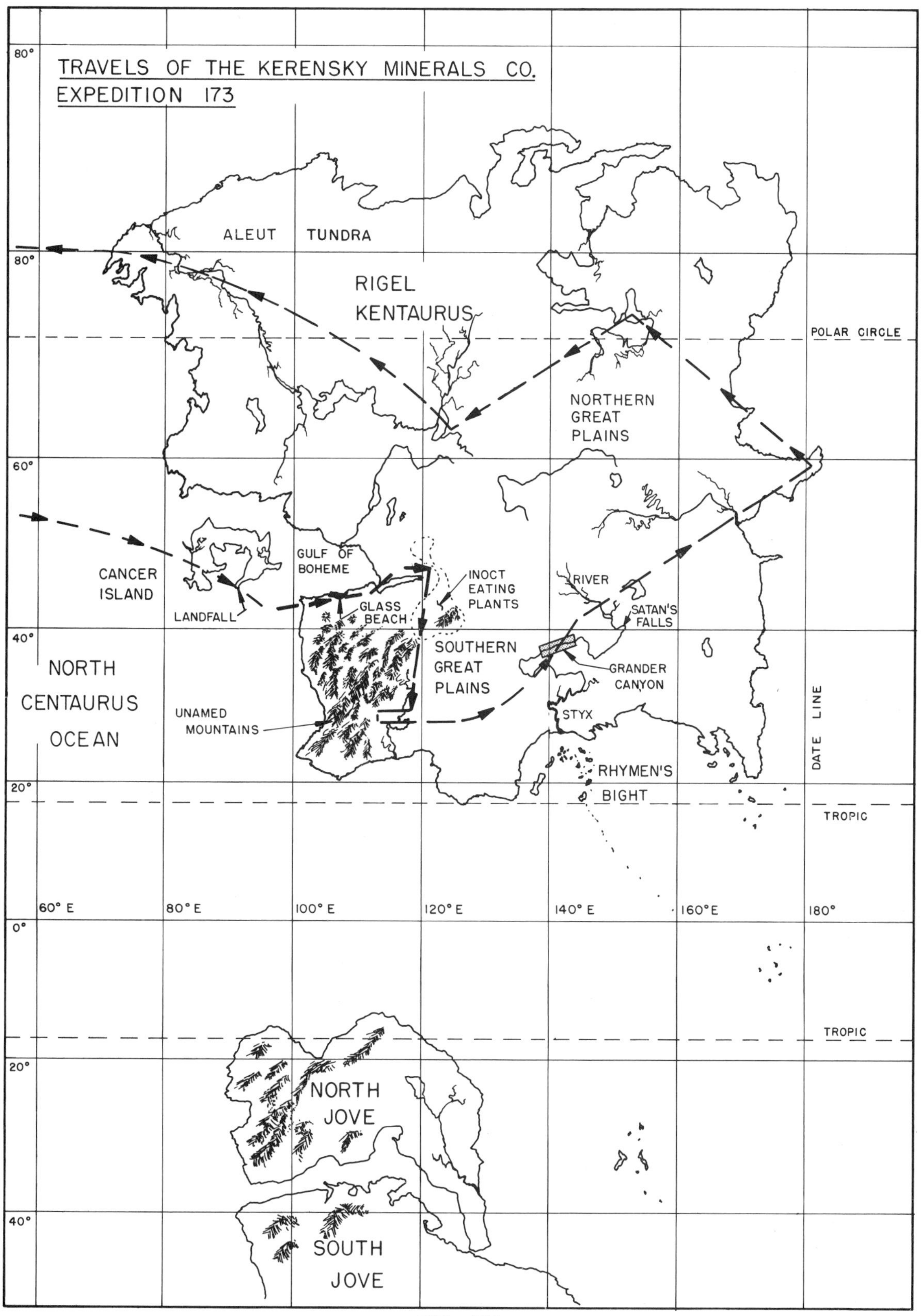
TRAVELS OF THE KERENSKY MINERALS CO.
EXPEDITION 173
80°
ALEUT TUNDRA
RIGEL
KENTAURUS
POLAR CIRCLE
NORTHERN
GREAT
PLAINS
60°
CANCER
ISLAND
LANDFALL
GULF OF
BOHEME
GLASS
BEACH
INOCT
EATING
PLANTS
RIVER
SATAN'S
FALLS
40°
NORTH
CENTAURUS
OCEAN
SOUTHERN
GREAT
PLAINS
GRANDER
CANYON
UNAMED
MOUNTAINS
STYX
RHYMEN'S
BIGHT
DATE LINE
20°
TROPIC
60° E
80° E
100° E
120° E
140° E
160° E
180°
0°
TROPIC
20°
NORTH
JOVE
40°
SOUTH
JOVE

FIG. 3.3

view from the air was so extraordinary that we stopped to have a look. In shape the beach looked like any other sand beach, but instead of sand, it consisted of a solid sheet of jet black glass, slightly rippled, but overall, remarkably flat and smooth. The black tint, apparently caused by a variety of trace elements, prevented us from seeing far beneath its surface, but I could discern many small air bubbles trapped in it, indicating that it was a natural formation. Closer inspection of the landward edge revealed a giant fissure which long ago had flowed hot lava, a sort of linear volcano five kilometers long. Apparently the glass flowed from this crack and cooled, but many unanswered questions remain. Why didn't the glass shatter upon contacting the water? Why hasn't the ocean eroded it into sand? How did it happen to flow so evenly and smoothly?

With the strangeness of the glass beach lingering in our minds, we pressed on to the northeast, rounding the northern edge of a large unnamed mountain range that dominates the southwest quadrant of the continent. In the mountains' foothills, along the edge of a great plain stretching to the opposite shore, we began our search for minerals. Immediately, we began to notice important, but subtle differences in the basic structure of Rigel's plant life. Gone were the compound needles and broad leaves of Chiron, and in their place grew leaf structures more reminiscent of Earthly ferns. Further south lay forests of inoct-eating plants that devour their prey like Earth's Venus's flytrap. Unlike the flytrap, however, these partially carnivorous, partially photosynthetic plants live in large groves. We couldn't imagine how such large numbers of plants could exist without decimating their inoct food supply, but several years later biologists noted that the inocts consumed by the plants were scavengers, thriving primarily on decaying leaves dropped from the trees. Thus, like all things in nature, the inocts and the trees formed a continuous, self-renewing process wherein neither one could exist without the other.

The birds of Rigel Kentaurus fascinated us even more than the plants, for they did not bear feathers but wore instead a light coating of fur. Their wings consisted of skin stretched across complex arrays of bones and their heads bore bird-like beaks containing rows of jagged teeth. They flew quite well in the heavy air, though they did not glide as well as their feathered counterparts on Chiron.

The Styx River traverses the south Rigelian plain, meandering like Earth's Mississippi. The river basin is one of the well-explored regions of the continent. The first Humans set foot there during the second exploratory expedition more than 285 years ago. Although commercially of little interest, the river has formed one of the largest and most beautiful canyons in the Human worlds, dwarfing Earth's Grand Canyon and matching the intricacy of its formations. As geologists, all of us itched to finish our work and then see this great natural wonder.

We crossed the plains, arriving at the canyon's edge during the night, and waited until the next morning for a good view. Since the forest runs right to the canyon's edge we camped in a clearing a kilometer away. The next morning, after an early breakfast, we set out on foot for the canyon rim. Little had prepared me for the view we found. Beyond the edge, the world seemed to drop away. All perspective was lost amidst the great valley spreading below us. Within it, huge columns of rock rose singly, in great arches and in buttresses against the canyon's walls. Unlike Earth's Grand Canyon, this Grander Canyon lies in an area of heavy rainfall. Green foliage clings to the walls and rock outcroppings, though as the sun rose, we saw that many of the sheer rock walls were exposed. The magnitude of this formation can not be appreciated fully from the canyon's edge. Indeed, we didn't have a complete view of it until we took our levicars almost two kilometers above the rim. Later in the day, we hovered next to the sheer walls, examining the different strata that contain a record of much of Wyzdom's history. Someday many geologists and paleontologists will spend their lifetimes studying the important records of the canyon which will help piece together a history of his planet.

In this space I can recount but a few of the many experiences of discovery I had during my years as an explorer; the attack by a pack of Jovian griffins, the largest of the planet's flying predators; the eruption of Mt. Isis, largest volcano on the planet, which dwarf's Earth's Mauna Loa; the exploration of Fairbank's caverns, extending for 20 kilometers beneath the south Jovian plains. Yet after ten years in the field, I felt a desire to settle down. I transferred from exploration to the company's operating division and once again found myself in New Georgia. The province had grown populous enough to justify opening a bauxite mine, an aluminum refinery, and a shaping mill. I entered the project early, performing the detailed preconstruction drilling and analysis necessary to design day-to-day mining operations. As the project unfolded, I learned more and more about the mining business, until over the next decade I worked myself into the position of Director of Bauxite recovery operations at the Koppernick River Complex. I am now responsible for every aspect of mine operations, from taking ore out of the ground through transporting it to the inlet of the adjacent refinery.

Since ending my career in exploration, I have indeed "settled down." My wife, Sonja, and I are raising our children, and we own a commercial orchard. I am an elected public official and manager of a large company. Though settled, life seems far from static to me. New Georgia encompasses an area about half the size of Europe on Earth. When my parents first moved here, it contained less than 300,000 people. When I returned from my world travels, 1.2 million people lived here. Today the population has again more than doubled. The principal city of Kerenberg has grown from an outpost to a major metropolis and the province that once had to import virtually all consumer and industrial goods now boasts many major industries. Life in the region is changing and diversifying so rapidly that, though I'm staying still, I often think I'm traveling around the world, living in a new community each year. Sonja and I actually have more freedom to travel when and where we want to than we did when we were younger, but we have the advantage of feeling that we are a part of a local community too.

My home has a 40 hectare commercial claim next to it where my wife operates a profitable citrus orchard. We've left a strip of virgin forest between our house and the "farm," so Sonja can forget about it when she wants to. Our house (figure 3.4), like most other rural houses, is constructed from modular housing components flown in by levitruck and stuck together like giant branches on an artificial tree. Like modern houses on Earth, ours has full spectrum force field walls that keep out dust, cold, and intruders, yet can be adjusted in intensity to admit light or fresh air as desired. Though the structure is built to withstand strong earthquakes and the high forces of lowland winds, a ring of native trees around it provides an extra safety factor by serving as a windbreak. The low density of dwellings in most parts of the Province makes central water, power, and waste disposal systems impractical. Water comes from our own well, and waste water is treated and used on our garden and orchard. We compost organic wastes and save recyclable trash in a large storage bin that the recycling company collects each month. All else we burn in our smokeless plasma incinerator. The whole process is automated and concealed underground.

Power comes from our laser fusion reactor that uses heavy water as fuel. We consume about a liter per month, delivered quarterly by the service company that maintains our power plant. Our communication link with the outside world is via satellite through our roof-mounted antennae. It provides a channel for videophone, recorded correspondence, news, weather forecasts, entertainment, books, computing service, banking, computer lessons, and catalog shopping. Like many Provincials who grow much of their own food, we shop for the many grocery items that are processed or can't be grown here through a remote order system. Our groceries are then delivered by a robot levitruck that regularly covers our neighborhood.

Because people are spread so thinly on Wyzdom, neighbors don't seem to get on each other's nerves. Physical contact can be avoided unless a person wants it. Perhaps because of this, Wyzdomites are highly tolerant toward each other's life styles. Even on the Jan Plateau, the most densely populated area of the planet, few universal standards of moral, social, sexual, or religious conduct are held. Family organization ranges from homosexual hermits through nuclear families to large communal families encompassing up to a dozen adult men and women and several dozen children.

Physical proximity does not link communities on Wyzdom as much as do common interests, friendships, and lifestyles. Groups of people drawn from hundreds of kilometers in all directions often undertake projects of common interest, such as building an athletic center or a lake, or operating a school of their particular design.

I, myself, am a member of several communities: Legally, I'm a resident of the New Georgian Province and active in its politics; professionally, I am a miner; socially, my family and children form an

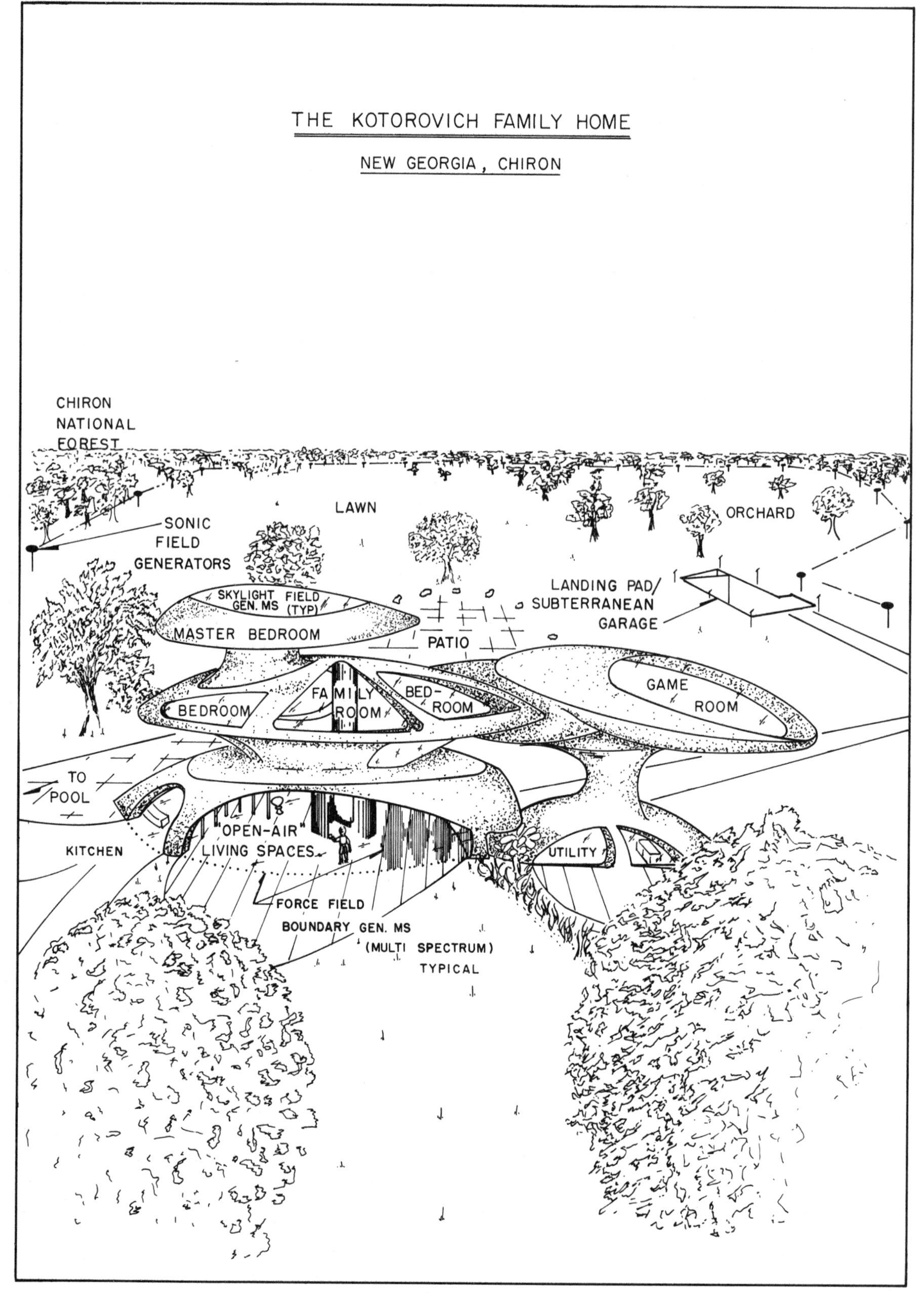
THE KOTOROVICH FAMILY HOME
NEW GEORGIA, CHIRON
CHIRON NATIONAL FOREST
LAWN
ORCHARD
SONIC FIELD GENERATORS
SKYLIGHT FIELD GEN. MS (TYP)
MASTER BEDROOM
PATIO
LANDING PAD/ SUBTERRANEAN GARAGE
BEDROOM
FAMILY ROOM
BED-ROOM
GAME ROOM
TO POOL
"OPEN-AIR" LIVING SPACES
KITCHEN
UTILITY
FORCE FIELD BOUNDARY GEN. MS (MULTI SPECTRUM) TYPICAL

FIG. 3.4

important part of my life, so I am part of a community of parents that operates a cooperative school along traditionalist principles; and finally my family is active in sports, so we are members of the Pan Wyzdom Track, Field, and Swim Association. All of us participate in competitive events and games that could culminate in a planet-wide contest of champions. Many of our sports trace their origins to Earth, but Wyzdom's high gravity and, in some instances, the high atmospheric pressure, change the dynamics of the games and therefore their strategies. Wyzdom's track and field statistics appear unimpressive alongside Earth's, but in swimming contests, where the effect of gravity is reduced, athletes of Earth wouldn't stand a chance.

I currently serve as "mayor" of the Provincial Government of New Georgia. A province is a rather vague governmental body, sandwiched in between highly localized "commonwealths" and the larger planetary government. The Province operates local police patrols and has original jurisdiction over minor crimes, such as burglary, petty theft, and minor environmental pollution. It may not make any laws abridging the rights enumerated in the planetary constitution, but it makes a variety of laws of local interest, and it administrates provincial land claims in accordance with general policies created by the planetary government and GAILE. Some people in well-defined geographic areas have created local commonwealths to provide more service and protection than the Province offers. Industrial groups, such as the West Koppernick Electronics Complex, and commercial groups, such as the City of Kerenberg, as well as groups of individuals have formed legal commonwealths. These perform all the functions within their areas that the provincial government would perform.

As mayor of New Georgia, I chair the seven-member Provincial Council, the elected legislative and executive body of the Province. My job consumes only 10 to 15 hours of my time per week, for I am not actually an administrator. I leave day-to-day government operations to five full-time, salaried officials: Chief of Police, Provincial Prosecutor, Chief of Administration, Provincial Judge, and Liaison to the Planetary Government. I act as a reviewer and final authority, as well as a lawmaker. To discourage "professional" politicians, no person may hold a position or be elected for more than one term, and I will be relieved when mine is over.

Wyzdom's growth during our first century has been slow and sometimes painful. The cost and complexities of space travel in our first century served as the principal barriers to large-scale migration, but now, with the most difficult years behind us, we stand poised to make great strides in learning and industry which will benefit all intelligent life. As an advanced, highly industrialized society, Wyzdom offers the conveniences of life on Earth; yet Wyzdom has lost neither the frontier's sense of adventure nor the friendly intimacy of a small community. We welcome to our young planet all people who are willing to work hard for themselves and to enjoy life with us.

*****

OPPORTUNITIES FOR IMMIGRANTS

Wyzdom offers more opportunity than any other colony for Earth people with skills to employ them in their new home, because Wyzdom's economy is the most diversified and advanced. Wyzdom actively seeks all skilled tradespeople. Unskilled and semi-skilled workers will also be accepted, provided they are of sound ethical character and have not been convicted of violating any law on Earth that would be punishable on Wyzdom. The following skills are in greatest demand at the time of this release: electrokinetics technicians, doctors, robot mechanics classes 1 to 4, teachers of mathematics and sciences, skilled cooks and chefs, and engineers and designers of all types.

Wyzdom's high gravity subjects newcomers to significantly higher physical stresses than they have become accustomed to on Earth. While such gravitational forces have no harmful effects on people in normal health, grossly overweight individuals will not be accepted as immigrants.

Virgin land claims may be filed for unclaimed land anywhere on the planet. Maximum personal residence claims vary depending on the province from three hectares on the Jan Plateau to 20 hectares on the continent of Rigel Kentaurus. Farm claims and industrial claims are unlimited, provided all space is put to work in three years. Special claims are available on petition to the provincial governments. Claimed land may be transferred via sale.

GAILE provides transportation for 15,000 immigrants annually in a starship of the single hex configuration. Travel time requires 8.4 weeks of ship time and 12.1 weeks of planet time. Twelve information probes are exchanged between Earth and Wyzdom each year.

Poseidous — World of Oceans

IMPORTANT STATISTICS

Planet name: Poseidous
Equatorial diameter: 14,204 kilometers
Mass (Earth =1): 1.46
Serface gravity: 1.18 g
Escape velocity: 12.9 km/s
Albedo: 0.40
Atmospheric pressure at mean sea level: 1.32 bars
Fraction of surface covered by land: 1.5%
Maximum elevation above sea level: 5486 meters
Length of day: 19 hours 37 minutes
Length of year: 225 Earth standard days
Obliquity: 2 degrees
Von Roenstadt habitability factor (Earth = 1): 0.88
Current population: 19.4 million
Number of population centers: 90
Year settled: 2177 adtc
Number of moons: 3

Star name: 82 Erida i
Star type: G5V
Distance from Earth: 20.9 light years
Distance from planet: 0.70 AU (Earth-Sol = 1 AU)
Star Mass: 0.91 (Sol = 1)

Travel time from Earth

Ship's time: 60 days
Planet time: 86 days

Atmospheric Composition

Oxygen: 22.7%
Nitrogen: 75.1%
Carbon dioxide and inert gasses: 2.2%

Distance (km)	*Period (Earth days)*	*Mass (Earth's moon =1)*
18,450	0.23	8.33 E-7
226,000	10.2	0.27
42,000	0.82	0.009

Table 3.3

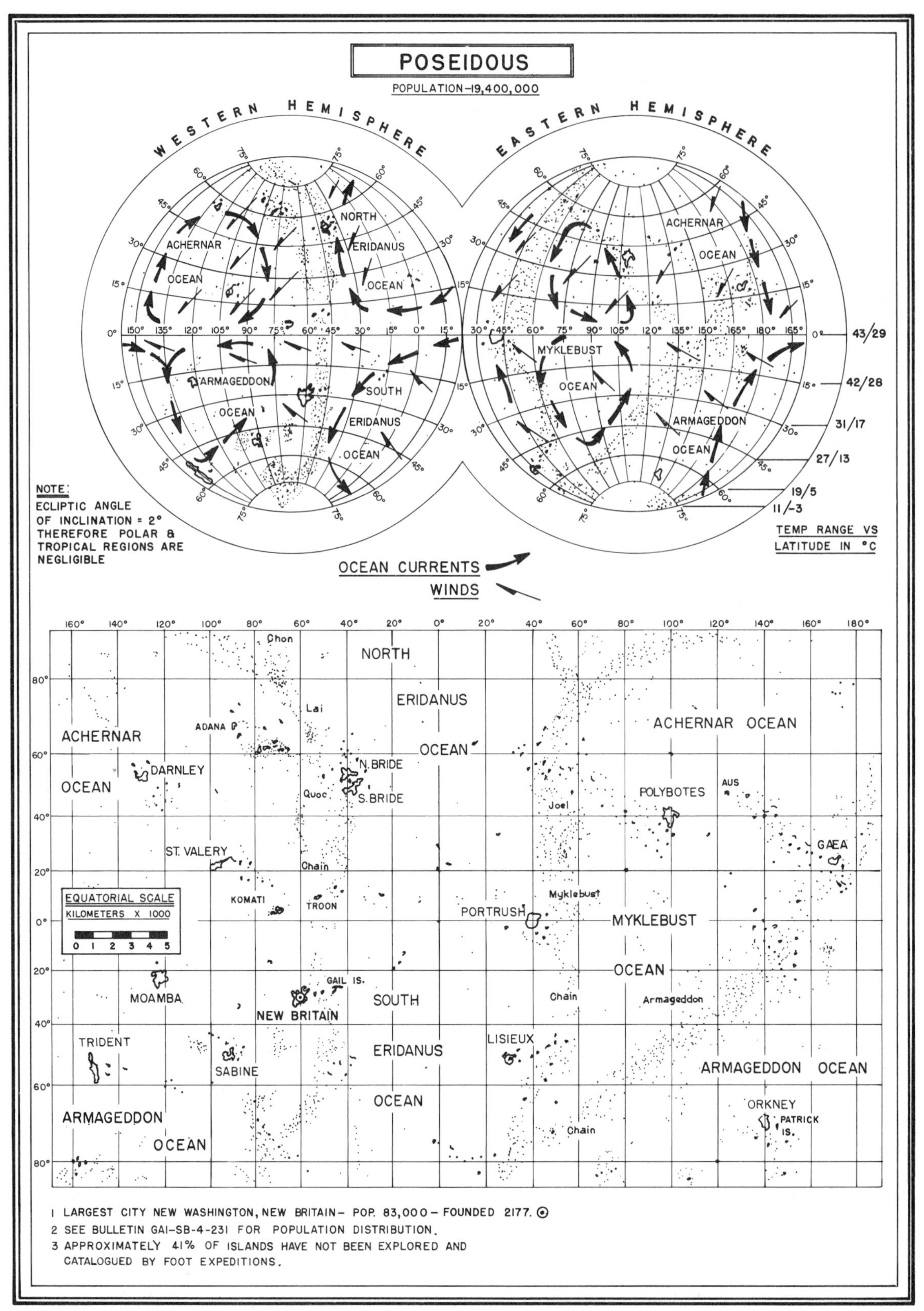

POSEIDOUS
POPULATION–19,400,000
WESTERN HEMISPHERE
EASTERN HEMISPHERE
NORTH ERIDANUS OCEAN
ACHERNAR OCEAN
ARMAGEDDON OCEAN
SOUTH ERIDANUS OCEAN
MYKLEBUST OCEAN
43/29
42/28
31/17
27/13
19/5
11/-3
TEMP RANGE VS LATITUDE IN °C
NOTE:
ECLIPTIC ANGLE OF INCLINATION = 2° THEREFORE POLAR & TROPICAL REGIONS ARE NEGLIGIBLE
OCEAN CURRENTS
WINDS
Chon
Lai
ADANA
DARNLEY
N. BRIDE
S. BRIDE
Quoc
Joel
POLYBOTES
AUS
GAEA
ST. VALERY
Chain
KOMATI
TROON
PORTRUSH
Myklebust
EQUATORIAL SCALE
KILOMETERS X 1000
0 1 2 3 4 5
MOAMBA
GAIL IS.
NEW BRITAIN
Armageddon
TRIDENT
SABINE
LISIEUX
ORKNEY
PATRICK IS.
1 LARGEST CITY NEW WASHINGTON, NEW BRITAIN– POP. 83,000 – FOUNDED 2177. ⊙
2 SEE BULLETIN GAI-SB-4-231 FOR POPULATION DISTRIBUTION.
3 APPROXIMATELY 41% OF ISLANDS HAVE NOT BEEN EXPLORED AND CATALOGUED BY FOOT EXPEDITIONS.

FIG. 3.5

Poseidous — World of Oceans

Editor's Note: George Soonge, 57, Marketing Director for Kunitani Manufacturing Company, describes life on Poseidous. A native, born into one of the oldest families on the planet, George graduated cum laude from Vallois University with a Bachelor of Science degree in meteorology. He spent five years working for the Poseidous Weather Service, a private company. During that time he developed his theory of the effects of small moons on Poseidous's weather. Publication of this theory lead to the development of a computer program for predicting the occurrence and paths of tropical storms.

After five years, George tired of meteorology and took a job selling Kunitani's appliances. He proved so successful at his new career that he rose rapidly through the ranks to a position of top management within 20 years. He has traveled widely on Poseidous, visiting literally every island that contains any sort of retail store or co-op outlet.

George's hobby, Poseidous's history, forms the subject of his very entertaining and popular book, Waterlogged Spacesuits. *He is active in the local politics of his home island, New Britain, and was serving as member of the Council of Governors at this writing. George's extensive knowledge of the many communities of Poseidous, his understanding of its unique weather and his experience as a writer make him highly qualified to tell Poseidous's story.*

Take one look at the map of Poseidous, and you'll notice something's missing. Poseidous has no continents. If we could strip away the water on Poseidous and Earth, the two planets wouldn't look very different. Poseidous too has high areas surrounded by deep basins, but the plateaus don't rise quite high enough to show above the sea level as do the continents of Earth. Hundreds of thousands of islands freckle the face of Poseidous. Most of these, the tips of subsea mountain ranges, lie in long chains stretching thousands of kilometers. In some cases the same geologic forces that pushed up Earth's mountains created those of Poseidous, while others, like Earth's Iceland and the Hawaiian Islands, spring from volcanic origins. Still others have been formed by the "batisatoll," tiny creatures like Earth's coral that build large rock-like structures for their homes.

It's natural that a planet with so much ocean attracted people of seafaring heritage. The ocean shapes the lifestyle of every Poseidon. It influences not only our hobbies, our commerce, and our institutions, but also subtly weaves itself into the subconscious instincts of every person. The ocean draws us magnetically to its shore to build our homes where the pungent smell of sea air, the constant roar of breakers, and the empty expanse of sea and sky permeate all our senses.

The antigravity field obsoleted the great floating ships of Earth's oceans before the founding of the Poseidous colony, yet Poseidous' seas still serve important commercial functions. Mariculture provides a major food source and offshore minerals fulfill all of the planet's raw material needs. The sea creates a giant playground for 95 percent of all Poseidons who engage in at least one form of water sport. Swimming, gill or snorkel diving, sailing, submarining, fishing, and surfing head the list of leisure time pursuits. Our beaches form the focal point of social activities on almost every island.

Newcomers from Earth might find Poseidous's societies less baffling if they understood some history. Two hundred and twenty three years ago Captain Joel Myklebust and the crew of the *Armageddon* discovered Poseidous while surveying ten stars along a roughly circular route through space. The find of a second habitable planet elated Myklebust and his crew and the fact that it contained mostly water didn't diminish their enthusiasm. Epsilon Eridani was the last scheduled stop on their mission and supplies had run dangerously low, yet despite these circumstances the *Armageddon* remained for 21 months on Poseidous, surveying and studying it. The explorers lived off native foods and, though lacking spare parts, jury-rigged their critical equipment to keep it functioning. They returned to Earth more than a year late, having been given up for lost, and their fantastic news made Myklebust's superiors overlook the fact that he had broken every rule in the space pilot's book.

Despite the enthusiasm that greeted the *Armageddon's* return, 24 years passed before the first pioneers set foot on Poseidous to make it their home. Earth's space exploration has historically proceeded in cycles and Myklebust's return coincided with a downturn. In those early days of space travel, the exploratory missions to nearby stars and the large-scale colonization of Wyzdom had cost the people of Earth dearly. They were less than eager to finance a second extraterrestrial outpost while so much land remained on Wyzdom. Yet the proponents of space travel found allies in the states of eastern Asia who felt they weren't getting a fair share of places on Wyzdom-bound starships.

A mixed bag of Human beings from all of Earth's major nations comprised the first party of pioneers. Since Poseidous's largest island lay in the southern temperate zone it provided the logical spot for the first settlements. The majority of the settlers hailed from what were then England and America; so they named their island New Britain and its capital New Washington. Subsequent shiploads of immigrants drew heavily from ancient Japan, Singapore, Indonesia, and China.

The success of Medhat Hadar's rampant capitalism on Wyzdom caused it to become the model for Poseidous's social structure, but the diversity of backgrounds among the colonists prevented them from reaching a consensus on the implementation of Hadar's ideals. After nine years of bickering, the pioneers verged on civil war, despite the manifest economic success of the colony. Then a statesman, named Chou Loon, emerged with a compromise. To be rid of the socialist factions, he proposed dividing the existing capital assets and moving them to any island the socialists chose. Though this plan would cost the capitalists dearly, it seemed a small price to pay to be rid of the collectivists, and the two islands, equally matched in initial resources and population, would provide a basis for comparing systems. Thereafter any Poseidon could choose the one he liked best.

Chou's plan was grudgingly accepted and established a precedent that guided the planet's subsequent development. To this day, Poseidous lacks a planetary government. The islands govern themselves autonomously, similar to the former nations of Earth; yet they manage to coexist in peace. Free trade and personal movement exists between islands, and to date all island governments remain more or less democratic.

Many immigrants from a now unified Earth, recalling the awful and bloody history of the mother planet, express anxiety at the absence of a planetwide government. They fear war will one day break out between the islands, transmuting their idyllic paradise into a fiendish hell. I personally doubt that this will ever happen. Poseidons have much to lose and little to gain from armed conflict. The peaceful resolution of the most serious inter-island dispute in history illustrates the economics of peace.

About 150 years ago, the pirates of Polybotes built powerful levitrucks, armed them with laser cannon and began marauding the transit lanes. They stole both ships and cargoes, then ransomed them back to their rightful owners. Sometimes they made use of the cargo themselves, destroying the vessel or converting it to yet another raiding ship. The majority of the people on Polybotes did not belong to the pirate band, but they closed their eyes to the wrongdoing because the pirates brought a lot of money to the island without costing the locals a centime. In time the piracy grew quite costly to the major trading islands, which organized an effective boycott on all trade with Polybotes. After a few months of the boycott, the non-pirates realized that the loss of trade hurt them far more than the pirates' spending helped them. Quietly they organized themselves and began to arrest each pirate crew as it disembarked from its vessels for a night on the town. When all had been rounded up, the Polybotians invited representatives from every island to watch the destruction of the pirate vessels, thus demonstrating their good intentions and obtaining a swift end to the embargo.

The diversity of Poseidous' societies transcends mere politics. Arts, family structure, mores, eating habits, esthetics, and governments follow patterns unique to each island.

For example, the island of Moamba, settled by the dissident socialists from New Britain, retains a peculiar brand of socialism that is about as inefficient as socialism has ever been. Yet Moambans seem to like it and feel a strong sense of attachment to their community and their island. Being rather puritanical and straight-laced, they wear light-weight, but starkly simple clothing resembling coveralls, usually in drab colors. The unisex look is omnipresent; men and women crop their hair closely, and frown upon makeup or bodily ornamentation.

Moambans are big on community projects and often spend weekends working to improve a park or build a stadium. They must like parades too, for I manage to become entangled in at least one each time I visit. Disposable incomes of the islanders appear quite low, despite the official statistics. Appliance sales on the island are not as brisk as they ought to be, considering the size of its population. The sales my company makes tend to be basic items like cookers, waste annihilators, preservators, and ovens. Very few clean fields, domestic robots, or entertainment centers find their way into Moamban homes.

Moambans minimize family life at the expense of community life. Marriages are rare, and sex for pleasure is discouraged. Yet Moamban women have a "duty" to bear children, and so most find a way to get pregnant without having any fun. After their children are born, community child-care centers raise them, and mothers see their offspring only during spare hours. Though some mothers do bring up infants at home, education in state boarding schools is compulsory, and when the children reach 18, they go out into the world to repeat the dreary process.

Life on Moamba stands in total contrast to Troon, without doubt the most purely hedonistic culture that ever existed. I suspect the easygoing attitudes of the Troons have to do with the weather. The nearest inhabited island to the equator, Troon remains very humid and near 32 degrees C. at all times. Nudism, though not universal, is widely practiced and quite obvious in all public places. I have no difficulty telling the girls from the boys there! Troons go in for weird, brightly-colored things. Body paint and makeup are commonplace.

Most Troons don't work very hard, but then, why should they? The beaches are gleaming white, and the crystal blue water measures in the high twenties all the time. Fruits and vegetables from Earth's tropical regions flourish without much attention. Troon's chief exports consist of all kinds of toys, from beautiful polyglass sailboats to strange little dolls that fit one inside the other. Troon produces Poseidous's only genuine hardwood furniture, beautiful, if impractical, and arts and crafts of all kinds abound on the island. The people of Troon prefer simple lifestyles; so for wholly different reasons than on Moamba, Troon also posts rather punk appliance sales. Most Troons reside in houses without walls and engage in simple diversions. They would rather do without something than work too hard for it. Most live in large, loosely structured family groups. Few Troons have a single mate, but somehow they manage to care for their children and to educate them about the things they consider important.

Notable exceptions to the Troon stereotype exist, of course. An old friend of mine, Alex Popodopolus, lives in the center of the island and is the nearest thing to a hermit I have ever met. Most of his time not spent gardening or fishing is consumed in producing his own strange brand of wire art. He purchases his artist's supplies and other things he can't make himself by catalog from the income of a conservative portfolio accumulated during his years as a successful insurance salesman.

Lifestyles on the rest of the island vary between these two extremes. Most communities allow much personal freedom and support modern, industrialized economies. They offer the safety and conveniences of life on Earth, although the variety of goods is somewhat more limited. I wish I had the time to describe each island community, for each has its endearing features, but I think I can convey the spirit of life on Poseidous best by telling about life on my home island of New Britain.

More than half of New Britain's population traces its ancestry to eastern Asia, a fact that profoundly influences our culture, despite our island's Anglo-Saxon name. The island lies in the temperate latitudes, so the air temperature rarely varies from 26 to 22 degrees C. New Britain supports much of Poseidous's basic industry and over 50 percent of the population makes its living from manufacturing, construction, or agribusiness. Laissez-faire applies to business and personal life. New Britains believe that anyone can chose his lifestyle as long as it doesn't harm others. An island government attempts to provide some sort of guidelines for community development, but were it not for the spirit of cooperation that pervades the populace, I doubt that anything would remotely resemble the government's plan.

Although New Britain makes a small blotch on the map of Poseidous, the enlarged map (figure 3.6) shows what a large and uncrowded place it really is. The most fanatic of water haters can find room in the interior to get lost in, and even on the sea coasts leagues of open space remain for future generations.

About a third of all adults live in nuclear families consisting of one man and one woman; another third roams relatively unattached, and the last third are members of extended ring or line families. I am a member of a line family that dates back 170 years to Li Soonge, an upstart young immigrant who had the audacity to marry the Anglo-Saxon daughter of one of New Britain's oldest families. Today the Soonge line encompasses 98 members spanning five generations. I live with the 31 members of Molly Soonge's branch. "Grandmother," as we call her, married into the family, but her daughter, Camile, by Li Soonge's eldest son, is my mother and a direct descendant of Li.

Grandmother staked the claim of our present home 51 years ago along with her two daughters and two younger men who had married into the family. My cousin, Leslie, had just been born, and Grandmother felt that the family would soon outgrow the original family complex near New Washington. She proved an astute planner, for the branch today includes some 16 adults and 15 children, yet the present home (figure 3.7) still offers room for future growth.

Centuries ago Humans learned that by dividing their labors and specializing, they could raise their material standard of living, but it took them much longer to realize that what was good for business was good for their personal lives as well. In a nuclear family, one man and one woman must cope with earning a living in an industrial economy, raising and caring for children, shopping, and taking care of domestic chores like cooking, cleaning, and fixing the house. Each must assume many roles, though most of us aren't well suited to all of them. In a nuclear family two individuals stand alone against a sometimes hostile, always demanding world. If either of them falters, the other must pick up the entire load. Through all these trials, each man and woman in a nuclear family must be the sole friend and companion of the other, always patient and supportive, ready to cope with and adjust to the many changes in outlook, goals, and personal philosophy that each goes through in a lifetime. No wonder marriages that last 75 to 100 years seem as rare as supernova!

"Many hands make light work," and so in an extended family the burdens of domestic life fall on many shoulders. Each member does what he or she does best. Some perform best while pursuing a career. These members earn the money needed to buy the necessities of life. Others enjoy raising children, entertaining and instructing them, organizing and disciplining them, and watching to make sure no harm comes to them. Still others prefer domestic chores: cooking, repairing the machinery, caring for the garden, decorating, building an extension. In our family, Leslie and I earn the lion's share of the money now, though Quon, Daphne, and Roger all intensely pursue promising careers that may project them into highly paid positions when Leslie and I are ready to retire. Grandma remains titular head of the family. Though she sometimes supervises the children, she spends most of her time reading or gardening. My mother, Camile, is retired, along with Wilbur. Together they manage the family investments. Mother reads a good deal, and Wilbur enjoys tinkering with machines of all sorts. Daphne studies toward a Doctorate in marine biology and hopes to land a position at a research institute. Chen works hard at marketing his secret invention, a household gadget that I'm not supposed to describe. One day I'm sure he will be either the greatest success or the worst failure in the family.

The other adults gravitate toward domestic life. Liam, Nancy, Sharon, and Phuong primarily care for the children. Tsen cooks, with help from the others. Peter cares for most of the estate, with help from Sharon with the inside chores. Sharon's holographs have started to bring in fair sums of money

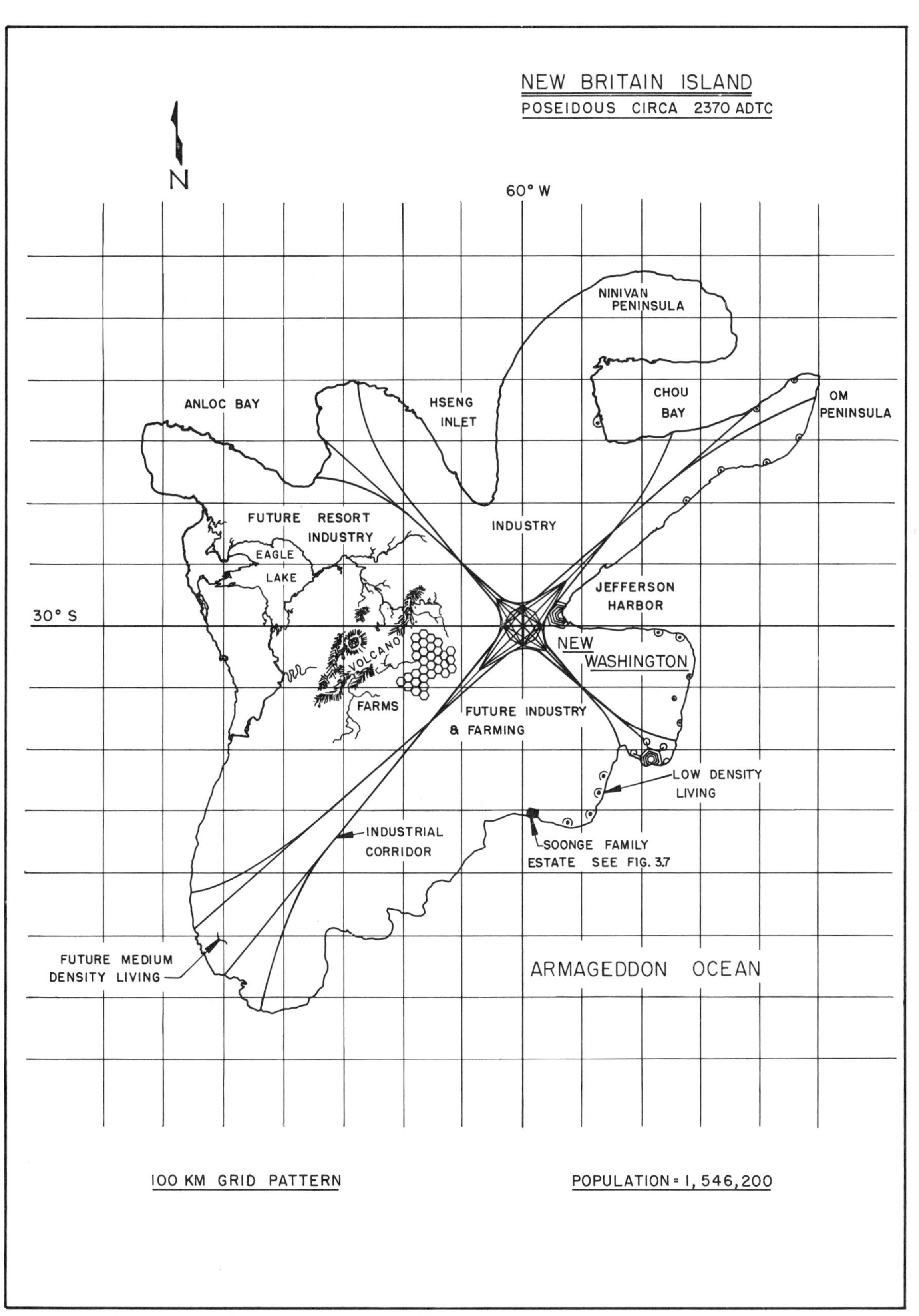
NEW BRITAIN ISLAND
POSEIDOUS CIRCA 2370 ADTC
N
60° W
30° S
NINIVAN PENINSULA
ANLOC BAY
HSENG INLET
CHOU BAY
OM PENINSULA
FUTURE RESORT INDUSTRY
INDUSTRY
EAGLE LAKE
JEFFERSON HARBOR
VOLCANO
NEW WASHINGTON
FARMS
FUTURE INDUSTRY & FARMING
LOW DENSITY LIVING
INDUSTRIAL CORRIDOR
SOONGE FAMILY ESTATE SEE FIG. 3.7
FUTURE MEDIUM DENSITY LIVING
ARMAGEDDON OCEAN
100 KM GRID PATTERN
POPULATION = 1, 546,200

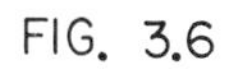
FIG. 3.6

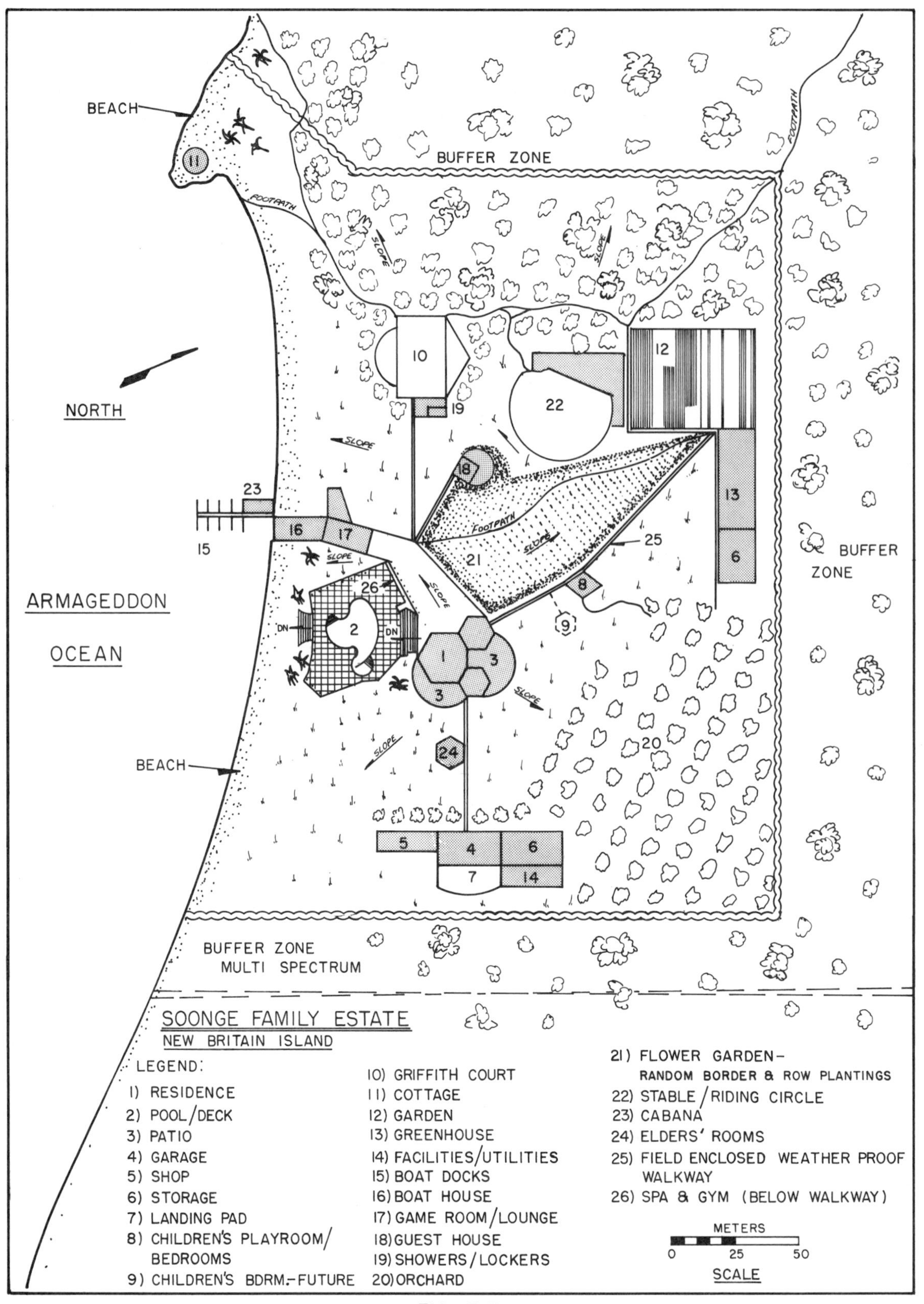
BEACH
BUFFER ZONE
FOOTPATH
SLOPE
NORTH
ARMAGEDDON
OCEAN
BUFFER ZONE
BEACH
BUFFER ZONE
MULTI SPECTRUM
SOONGE FAMILY ESTATE
NEW BRITAIN ISLAND
LEGEND:
1) RESIDENCE
2) POOL/DECK
3) PATIO
4) GARAGE
5) SHOP
6) STORAGE
7) LANDING PAD
8) CHILDREN'S PLAYROOM/ BEDROOMS
9) CHILDREN'S BDRM.-FUTURE
10) GRIFFITH COURT
11) COTTAGE
12) GARDEN
13) GREENHOUSE
14) FACILITIES/UTILITIES
15) BOAT DOCKS
16) BOAT HOUSE
17) GAME ROOM/LOUNGE
18) GUEST HOUSE
19) SHOWERS/LOCKERS
20) ORCHARD
21) FLOWER GARDEN- RANDOM BORDER & ROW PLANTINGS
22) STABLE/RIDING CIRCLE
23) CABANA
24) ELDERS' ROOMS
25) FIELD ENCLOSED WEATHER PROOF WALKWAY
26) SPA & GYM (BELOW WALKWAY)
METERS
0 25 50
SCALE

FIG. 3.7

now that she has begun to earn a reputation, so all of us encourage her to do more art and less straightening and decorating.

Only a few of Li Soonge's genes still float around in the family gene pool, for as each generation grows up, some members born to the family leave it, and others born outside the family marry in. When one "marries in," he or she marries the entire family, not just one member of it. Children, regardless of their genetic parents, are born to the entire family, and the entire family shares legal responsibility for their care and upbringing. By design few of our children are conceived by the same two parents. We also avoid producing too many children from the same genetic lineage, so genetic defects aren't amplified. Leslie and I, who share a common grandmother, have created just one child together. Though IU gene scan can detect genetic disasters early, we loathe abortion and prefer not to increase the risk of it.

Finding new members for an extended family isn't easy. The more people involved in making a decision, the more difficult it becomes. When someone marries in, they pledge to love and support not just one other person, but the entire group. They must find something to love and appreciate about every family member, or the alliance can not be a happy one. Our family encompasses a variety of different personalities, but all share the same basic outlook on life. Though none of us are religious, we all respect nature's works, both grand and humble, and disapprove the wanton destruction of any of them. Though we follow different occupations, all of us believe that what we do we must do well. Like most Poseidons, we enjoy sports and outdoor life. Free and open sexual relations exist between all adult family members, and all of us find something physically attractive in each other. Though we have no hang-ups about nudity, we don't live a nudist lifestyle, and since we aren't into homosexuality, we try to roughly balance the number of men and women. Though some of us spend more time with our children than others, we all enjoy and appreciate their youthful curiosity and energy.

Though I love every member of my family, each is a unique individual and the emotions I feel toward each of them are unique as well. I would risk my life for any one of them. Although I regularly sleep with every one of the women from Grandma to Daphne, I feel closer and more in tune with Leslie than with all the others. One has to meet Leslie only once to be forever branded by the intense fire of her personality. She throws every gram of her drive and creativity into her career as marketing director of Infodyne, the largest information service on Poseidous. Her job takes her traveling as much as mine does, yet on those occasions when we splash down at home together, we usually sit up talking long after the others have gone to sleep. We hash around our latest business ideas, cry on each other's shoulders about problems of corporate politics, and recount tales of business adventures not suitable for the dinner table.

The relationship we share couldn't exist without our extended family. As husband and wife, the demands of our separate careers would constantly clash with the demands of our home and our children and would inevitably lead to unhappiness and divorce. In our extended family we can devote all of our energies to our work, knowing that this is how we can best serve the others, yet within the family framework we can experience the joys of having children and the pleasures of conjugal companionship.

Work has always played a central role in my life. My early love of the sea and sailing drew me to study meteorology in college, and I began a career as a professional weather forecaster. It didn't take long for me to give that up, because weather on Poseidous is downright boring. Water has the highest specific heat of any commonly known substance. It takes more energy to raise the temperature of a kilo of water one degree than it takes for a kilo of almost any other substance. Oceans act as giant thermal "inertias" for nearby land, preventing the temperature of the land from rising or dropping too fast. Water, like all substances, becomes less dense when heated and more dense when cooled. Rising and falling currents of water create a natural circulation that carries massive amounts of heat from a planet's warmer regions to its colder regions. Land can not flow about like this, so differences in temperature are much greater on land than on water. For these reasons the Earth's coldest and hottest regions occur away from the oceans. Poseidous, being all ocean, lacks the temperature extremes of Earth, and the great continents that break up the circulation in Earth's seas cause no irregularities in Poseidous's weather.

Temperatures don't vary much with the seasons either, for Poseidous has a smaller obliquity than any other Human colony. Hence our sun's rays strike every point on the planet with the same angle throughout the year. "Winter," "summer," "spring," and "fall" have no meaning for us at all! The map of Poseidous shows the temperatures at each latitude. Anyone can literally pick the air temperature he likes best, then move to an island at that latitude and rest assured that, 95 percent of the time, the temperature won't vary more than seven degrees either way.

The complete lack of seasons reduces the concept of a "year" to an astronomical measurement. I've suggested to the New Britain Council that we simply make our year coincident with Earth's years to save fooling around with conversions all the time. In principle, the Council agreed with me, but getting every island on Poseidous to agree on anything is practically impossible. The Troon council suggested that the calender be eliminated altogether so that people wouldn't feel pressured by the passing of time! So we continue with a year that equals approximately 225 Earth days.

The map (figure 3.5) also shows that the flow of winds and currents on Poseidous is as regular as the temperature. Most islands have a rainy side where it rains almost daily and a dry side with less rain. Weather forecasters would have had no job at all, were it not for occasional storms that originate in the tropics, but once we solved the problem of predicting these storms, all the challenge went out of the job.

I quit the Poseidous Weather Service, and took a cut in pay to begin selling appliances for Kunitani Manufacturing. My early career took me all over the planet, to tiny islands with one co-op store and to private islands of single families. I found I enjoyed hawking "Neptune" dish handlers, "Trident" environmental conditioners, and "Viking" foodmasters much more than I liked organizing data files to input to the computer. Sales offered a great opportunity to meet all kinds of interesting people and to absorb the variety of fantastic lifestyles adopted everywhere on the planet.

Enjoying a job almost guarantees success at it, and my record sales in the outlying islands soon boosted me up the sales management ladder. As the outlying islands grew in population, sales volumes rose, and it wasn't very hard for me to blow right past the old-timers working the established territories like New Britain and Lisieux. Now I'm responsible for all marketing at Kunitani, and not only sell existing products, but dream up ideas for new products to sell.

My job still calls for a lot of traveling, and I still spend half my nights away from home. Though I have computerized sales reports and a high-priced marketing department full of hotshots to think up new products, I feel the need to get out to the islands and talk with the salespeople. I visit appliance stores too and try to meet as many customers as I can, just to keep in touch with a real world. My own observations of the customer's complaints and desires help me to temper the megabytes of statistical and behavioral analysis that come across my screen with a little common sense. Many times my hunches have not only uncovered new opportunities but prevented pursuing some projects that would have been mistakes.

Meeting new people and seeing new places remains the most enjoyable aspect of my job. In my younger days, I looked upon each potential sale as a challenge, but I've mellowed with age and leave the hard sell to the younger people. I usually travel in my personal levicar. Though at 700 kilometers per hour it is slower than the big interisland transports, I like having my car to make short trips around the island once I've made a transocean hop. I enjoy the solitude of crossing the ocean. Once I set a course, automatic guidance control takes charge and not only steers, but looks out for obstacles and other vehicles as well. From my car's terminal I can read or write reports as conveniently as I can from my New Britain office, yet when I look up, I survey the vast expanse of empty sea and sky. I love to watch the subtle changes in color of the water as sunlight gives way to the clouds of a passing squall. Amidst the peace and tranquility of the open ocean, I find it so natural that Earth's gentlest and most intelligent creatures evolved in the sea. I often wish Earth's great whales and dolphins could become pioneers too and live with us here on Poseidous, for our world is even more suited to them than to us.

One major disadvantage of rising up the executive ladder is that sooner or later you'll be roped into some sort of community service. For some reason people believe that because you know how to organize a business, you can organize the unorganizable . . . government. The fact that someone with an attitude like mine could be elected to the New Britain Council indicates how seriously the average New Britain voter takes his government. The better part of my term of office has been spent hammering out the final details of a land-use plan.

After nearly 200 years, it has become pretty clear what sort of development New Britains want. They wish to live on the seashore and not have a bunch of cruddy-looking industrial complexes mess up the scenery. Industry has, in general, been happy to oblige. Large tracts of land have always been more available in the interior. On Earth, industry has traditionally located near the coast because it needed access to ocean ships and the cooling waters of the sea. The advent of direct fusion energy conversion and antigrav transport eliminated these traditional requirements, but the industrial areas remained near the seas because nobody wanted to be the first to move in next to a steel mill. Now, on Poseidous, industry has located inland, nearer cheap sources of groundwater. The council's plans will do little more than formalize this natural pattern of development.

GAILE finds the absence of an all-planet government on Poseidous quite exasperating. It is much easier for one massive, bureaucratic institution to deal with another giant bureaucracy than to deal with more than 50 distinct island communities and thousands of unaffiliated family units. Nevertheless, GAILE has remained true to its stated policy and has not attempted to coerce the islands of Poseidous into any sort of unification plan.

Poseidons do fullfill their obligations to supply GAILE with scientific knowledge about our planet and with details of technical innovations developed here. Poseidons also have the right to influence GAILE immigration policies so the interests of GAILE *and* the Poseidons are served. To this end, each government sends a representative to the GAILE advisory assembly, which acts as liaison between GAILE and the island governments. The assembly in no way functions as a government and exercises less influence over the New Britain Council than the native Ginko lizards. It meets on GAILE Island, a small island set aside for the exclusive use of the GAILE staff. The island contains scientific laboratories, quarters for the permanent GAILE staff, and the Assembly office buildings. Giant computers on the island organize data received from literally millions of sources around Poseidous, analyze it, and prepare it for transmission to Earth.

Almost every Poseidon participates in the continuing study of our planet. Major industrial companies gladly furnish information about their technical accomplishments and scientific finds in exchange for access to GAILE's own data bases which contain the latest technical information from Earth. In order to insure the continued flow of information, GAILE protects the secrecy of proprietary inventions and formulas from competitors on Poseidous, but distributes the information on Earth. Ordinary citizens of Poseidous may report any observations of new or unusual phenomena directly to GAILE via satellite, through a series of open data channels maintained by the GAILE staff. Though this system accumulates a tremendous quantity of informational garbage (the expected result from the input of millions of untrained observers), GAILE computers manage to sort out the inconsistencies. The GAILE staff feels this is a small price to pay, for last year alone, citizens of Poseidous discovered more than 1100 new species of marine life.

Even more substantial contributions have been made by Poseidous's industries. The bipolar sonic barrier for confining marine animals in mariculture zones has found extensive application both on Earth and on Genesis. A technique for random sequencing multichannel computer inputs using Murphy's statistics, first applied on Poseidous, greatly enhanced the effectiveness of data acquisition computers. Poseidons developed the first nonreflux autoigniter for small-scale laser fusion plants that has made them not only safer, but also less expensive. The list continues through hundreds of minor refinements to industrial techniques and processes that have found application on Earth. Even more importantly, Poseidous's simple weather patterns allowed people to develop the first accurate, deterministic model for predicting weather. The techniques learned on Poseidous provided an essential first step toward the accurate long-range prediction of Earth's weather patterns.

Though Poseidous lacks the potential of the newer colonies, we still have plenty of room for growth. During the coming century, our diversity of thought and life styles will reap an increasingly bountiful harvest of new ideas to help all Humankind.

PHYSICAL ENVIRONMENT

Poseidous, the second of six planets in the star system 82 Eridani, lies 20.9 years from Earth.

Though 82 Eridani is a smaller star than Earth's sun, Poseidous lies so close to it that it receives 14 percent more radiation per square kilometer than Earth. Foutunately the planet's higher albedo, caused by the larger amount of water on its surface, reflects more radiation and prevents the equatorial zones from being unbearably hot.

Not one, but three moons orbit Poseidous. The diameter of the largest measures only 42 percent of the diameter of Earth's, and the other two moons are much smaller. Because it lies much closer than the largest moon, the middle-sized moon appears largest in Poseidous's sky, a full 20 percent larger in diameter than Earth's moon appears from Earth. The largest and smallest moons both look about one half the size of the middle moon, though the smaller moon, orbiting just outside Roche's stability limit, is not round but quite irregular in shape. Because they lack atmospheres and water, none of Poseidous's moons shines more intensely than Earth's moon, and the smallest moon, because of its high iron content, glows with a dull brown light. Orbital periods vary greatly from 10.2 days for the largest moon to 5.5 hours for the smallest. Though none of Poseidous's moons exerts as much tidal force as Earth's moon, the pull of all three moons when aligned with 82 Eridani, exceeds the maximum tidal force on Earth. Yet because no continents exist, no place on Poseidous records any tides approaching those of Earth's Bay of Fundy.

Dry land covers a mere 1.5 percent of the surface on Poseidous. This compares with 29 percent for the Earth. Because Poseidous has approximately 24 percent more surface area than Earth, its total land area measures a respectable 6.4 percent of Earth's. A more impressive comparison results from considering the usable land area of Poseidous. Much of Earth's land falls within arid deserts, frozen tundras, and impassable mountain ranges. Poseidous lacks any of these regions, so virtually all land lying between the 60th parallels may be considered useful.

Though 350,000 islands dot the surface of Poseidous, the ten largest contain more than 70 percent of the land. These ten range in size from slightly larger than Earth's Madagascar to slightly smaller than Great Britain. The map of Poseidous (figure 3.5) shows about 3000 islands each measuring more than 900 square kilometers in area. Of these, some 500 exceed the size of Long Island, which contains one of Earth's largest cities. Literally hundreds of thousands of small islands, not shown on the map, measure a few hectares in area and can support one or two families if equipped with rain catching or desalinating equipment.

LIFE FORMS

The most highly developed indigenous life on Poseidous lives in the oceans rather than on land. Land life, limited primarily to plants, appears quite simple in comparison with Earth's.

Vegetation

Though larger plants have developed vascular systems, similar to Earthly tracheophytes, none have achieved the level of complexity of Earth's angiosperms (flowering plants). Most land species bear superficial resemblance to Earthly ferns. Some of these grow to impressive heights and offer excellent shade. Photosynthetic plants employ green chlorophyl, giving the landscape the same green appearance that it has on Earth. Since Poseidous has no seasons, virtually all species remain green throughout the year.

Plant classes vary considerably from island to island indicating that parallel evolution has taken place. Some species are common to more than one island. These reproduce by microspores so small that they can remain suspended in the air for months. Scientists theorize that storms and tradewinds have carried these microspores between the islands. Poseidous's early pioneers introduced numerous flowering plants on several islands. Most of these consisted of food crops from basic grains like rice and wheat to fruit trees, including oranges, apples, dates, and coconuts. Pioneers planted decorative tropical flowers as well, and in many areas these Earth species have become wild and have started to crowd out native vegetation. Biologists speculate that if measures are not taken to contain imported species, virtually all of Poseidous's native land plants will become extinct within a few thousand years, superseded by the more efficient alien species from Earth.

Animals

Few animal species above the microscopic level live on land. The few that do are cold-blooded creatures resembling Earth's amphibians and small, four-legged creatures of the class "inquadra," similar in structure to Earth's insects. The number of inquad species in no way compares with the vast number of insect species on Earth.

A genus of three-legged amphibians called "ginko lizards" comprises the most unusual of Poseidous's land life forms. Like most advanced animals on the planet, these creatures exhibit a planar symmetry. They possess two forelegs but only one hind leg resembling a tail. They have one head, two eyes, and one nostril centered on top of their heads. Ginkos scurry about quite nimbly and appear to suffer no disadvantages from their unorthodox design. Scientists believe these tiny animals evolved relatively recently from the fish-like animals of Poseidous's seas.

No flying species can be found on land save the inquads. Pioneers have considered importing bird species from Earth, but fear the adverse affects these animals might have on vegetation and the local inquad populations.

Marine Life

The richest diversity of life is found in the seas of Poseidous. Only a fraction of the sea species believed to exist have been identified and classified, and biologists estimate there may be half again as many marine species on Poseidous as on Earth. Even a brief description of each class of sea life would run to hundreds of frames.

Many superficial similarities between the sea life of Poseidous and that of Earth appear to the casual observer. Seaweed grows in many varieties from short grasses to tall, floating structures resembling Earth's kelp forests. Unlike Earthly seaweed, however, many of Poseidous's species reproduce bisexually, employing multi-colored "flowers" to attract small swimming animals that carry genetic information from one plant to another. Unlike Earth's flowers, however, these marine plants do not form the rugged seeds characteristic of Earthly angiosperms.

Among the animals can be found all manner of shellfish and arthropod-like creatures with exterior skeletons resembling Earth's crabs and lobsters. Filter feeders, like limpets and sponges, multi-tentacled animals like squid and octopus, and giant jellyfish all number among Poseidous's sea life. The most advanced forms are swimming animals resembling Earth's fish. These range in size from tiny fingerlings to enormous predators that dwarf Earth's great white sharks. Since they pose considerable hazard to swimmers, Poseidons have ringed their beaches and swimming areas with ultrasonic fields to keep them out.

None of Poseidous's sea animals have achieved the sophistication of Earth's great sea mammals, but some animals are more complex and intelligent than Earth's fishes. The class of "osteicalors" or variable temperature fishes comes in a wide variety of forms. All species swim, but some can fly and others can climb out of the sea onto the rocks of the surf zone in search of food. In shape, most osteicalors have the streamlined form of Earth's fish, but their tails are horizontal rather than vertical. They use gills to take dissolved oxygen directly from water, but their gills function well enough in air to allow them to remain alive out of water for long periods.

The variable temperature fishes derive their name from their ability to elevate their body temperatures above the level of the surrounding water. They are not truly warm-blooded because their temperatures fluctuate within rather wide ranges depending on the water temperature. Temperature variations of ten degrees have been observed without any apparent adverse effect upon the creatures. Osteicalor species range in size from a few hundred grams to several hundred kilos. Most tend to be vegetarians or filter feeders, though many appear omnivorous. Though some wear brilliant colors—red, blue, and yellow—most blend into the dark greens and grays of the sea. Variable temperature fishes comprise the most intelligent of native life forms on the planet; yet their intelligence in no way approaches that of Earth's mammals.

Humans can eat some of the marine life on Poseidous, though it took many years of study to increase the range of edible species to what it is today. More than 500 species of edible fish and 250 varieties of edible sea plants have been identified. Many of these are considered delicacies. A great deal had to be learned about the chemistry of Poseidous's marine life before it could be used as food. As on Earth, many normally harmless sea creatures can ingest microorganisms that turn them from healthful food to deadly poison. Commercial fishers carry real-time analyzers which quickly separate edible species from those containing toxic substances. These devices function so rapidly and precisely that the useless catch can be returned to the sea before it perishes.

The marine life of Poseidous has provided Human beings with more important benefits than food. In the colony's early years, a form of sea grass found in the tidepools of New Britain caused early pioneers to break out in an all-over body rash when they touched it. Concerned about the potential danger of this plant, medical researchers began to study it closely. They soon found that victims of serious cancers who developed the rash began to exhibit complete remissions of their diseases. Prior to that time, treatment of many of these cancers had been only partially successful and very expensive. Scientists later learned that the Human body's reaction to the effects of the alien plant stimulated the production of the very antigens needed to detect and destroy the cancer. The analysis of this mechanism provided Humankind with the ultimate weapon to combat this once deadly form of disease.

CURRENT STATE OF DEVELOPMENT

Population

Most of Poseidous' population of 19.4 million is distributed on just 13 of the 102 major islands (islands larger than 100,000 square kilometers). Cities on Poseidous tend to be small. Over 66 percent of the population are considered to be city dwellers, even though 73 percent of the people depend directly on industries in and around cities for their livelihood. The largest city, New Washington, houses a permanent population of 830,000 yet 1,500,000 hold jobs in it. About 15 percent of the population lives on very small islands not shown on the maps. Most of these lie near enough to larger islands so that one or more of the family members can commute to work there. A few small island dwellers live self-sufficiently, or nearly so. Their needs can be totally satisfied by things they produce themselves from materials of the island and the surrounding sea, though many do enjoy the convenience of power generators and communications service.

Industry and Technology

Poseidous, the second oldest of the Human colonies, possesses all the advantages of an industrialized world. Not only basic necessities, but physical comforts and conveniences as well can be purchased by most people. Poseidous no longer depends on Earth for any physical goods. Even the ship's spires for housing new immigrants are no longer supplied. Poseidons still depend heavily on information from Earth, though many of Poseidous's basic industries use techniques which have been modified slightly to suit the needs of the planet.

Food

Rice forms the staple crop for most of the islands, with wheat a distant second. Climates ideal for growing most types of fruits or vegetables from Earth can be found on the various island communities, and a vigorous interisland produce trade insures the widest possible variety to all. For historical, rather than technological reasons, Poseidons consume more fish in their diet than meat. Currently two-thirds of the fish consumed is produced in aquaculture preserves from species imported from Earth. The remaining portion comes from native species caught in the open ocean. Consumption of native fish has risen sharply in recent years and may surpass maricultural consumption by the end of the century. Poseidon cuisine partakes heavily of its Asian origins, but the dishes based around native seafood taste unlike anything cooked on Earth.

Communications

Poseidons have coped effectively with the most fragmented geography of any colony planet. A ring of communications satellites link the remotest islands to a comprehensive communications network. Competing services offer news, entertainment, shopping, libraries, interpersonal communication, and money exchange with the same high efficiency found on Earth. When electronic communications can't fill their needs, Poseidons flit between islands on high-speed levicars. Many of these are personally owned, though high-speed busses offer the fastest and cheapest links between the large islands. Poseidons lead all colonies in levicar technology. Because they must cover large distances, most cars can travel very fast. Fully automatic guidance, coupled with a continuous link to guidance satellites, insures that collisions won't occur at sea, and that malfunctioning vehicles will be quickly identified and aided.

Housing

Poseidon housing falls into two categories: single-story, modular, prefab houses from the overwhelming majority of personal shelters, although large, Earth-type complexes can still be found in cities. Because of the mild climate on most islands, house construction tends to be very light. Manufacturers make extensive use of glass and field walls which can be completely opened to the outside. Most Poseidons locate their homes on hillside sites with views of the sea. Because housing is cheap to build, more than 85 percent of Poseidon families own homes.

Education and Health

Provisions for children's educations, medical care and other essential services vary radically from island to island. On some islands, governments control all of these services, while on others, all such services must be purchased in an open market. The methods preferred by individual colonists are major factors in determining which island they chose to live on.

* * * *

OPPORTUNITIES FOR IMMIGRANTS

Opportunities exist on Poseidous for all skilled tradespeople and professionals. Android technicians, programmers, telecommunications specialists, doctors, cooks, and engineers and designers of all types appear to be in greatest demand at present. Because Poseidous's immigration program handles fewer people than any other colony, opportunities for unskilled workers are scarce. Immigrants comprise approximately five percent of Poseidous's population and their median income measures 10 percent higher than that of the average Poseidon. All immigrants are advised to learn Sinoenglish before departing for Poseidous. GAILE offers tuition-free classes to all candidates accepted for the program.

Unclaimed land remains plentiful on Poseidous. Claims may be filed with the island government on which land is claimed. On uninhabited islands, no formal claim is required. The amount of land that may be claimed for personal dwellings varies from island to island, and depends in part upon family size and proximity to the sea coast. In general, inland claims may be larger. Many islands require that all industrial claims be made inland. On most islands claims may be transferred via sale.

GAILE provides transportation for 6,000 immigrants each Earth year in a starship of the single hex configuration. Travel time requires 57 days of ship time and 86 days of planet time. Twelve information probes are exchanged annually between Earth and Poseidous.

Brobdingnag—Land of the Giants

IMPORTANT STATISTICS

Planet name: Brobdingnag
Equatorial diameter: 13,255 kilometers
Mass (Earth = 1): 1.16
Surface gravity: 1.07 g
Escape velocity: 11.8 km/s
Albedo: 0.34
Atmospheric pressure at mean sea level: 1.12 bars
Fraction of surface covered by land: 19%
Maximum elevation above sea level: 8726 meters
Length of day: 14 hours 2 minutes
Length of year: 359 Earth standard days
Obliquity: 15 degrees
Von Roenstadt habitability factor (Earth = 1): 0.96
Current population: 9.7
Number of major population centers: 7
Year settled: 2214 adtc
Number of Moons: zero

Star Name: 54 Chi Orion
Star type: GOV
Distance from Earth: 32.3 light years
Distance from planet: 1.09 AU (Earth-Sol = 1 AU)
Star Mass: 1.26 (Sol = 1)

Travel time from Earth

Ship's time: 61 days
Planet time: 89 days

Atmospheric Composition

Oxygen: 18%
Nitrogen: 80%
Carbon dioxide and inert gasses: 2%

Table 3.4

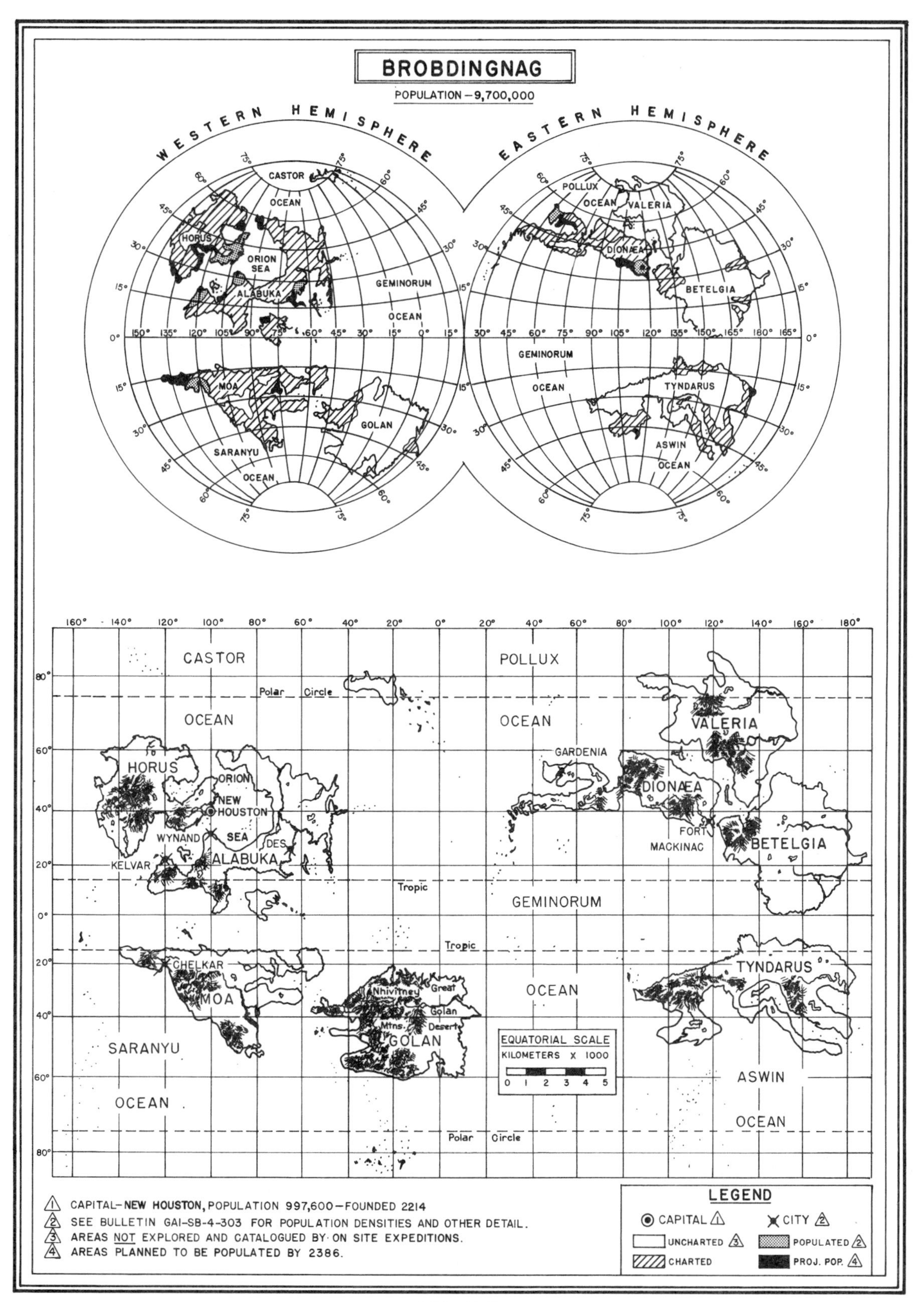
BROBDINGNAG
POPULATION – 9,700,000
WESTERN HEMISPHERE
EASTERN HEMISPHERE
CASTOR
OCEAN
HORUS
ORION
SEA
ALABUKA
GEMINORUM
OCEAN
POLLUX
OCEAN
VALERIA
DIONAEA
BETELGIA
MOA
GOLAN
SARANYU
OCEAN
TYNDARUS
ASWIN
OCEAN
Polar Circle
GARDENIA
NEW HOUSTON
WYNAND
KELVAR
DES
FORT MACKINAC
Tropic
CHELKAR
Nhivitney
Mtns.
Great
Golan
Desert
EQUATORIAL SCALE
KILOMETERS X 1000
CAPITAL–NEW HOUSTON, POPULATION 997,600–FOUNDED 2214
SEE BULLETIN GAI-SB-4-303 FOR POPULATION DENSITIES AND OTHER DETAIL.
AREAS NOT EXPLORED AND CATALOGUED BY ON SITE EXPEDITIONS.
AREAS PLANNED TO BE POPULATED BY 2386.
LEGEND
CAPITAL
CITY
UNCHARTED
POPULATED
CHARTED
PROJ. POP.

FIG. 3.8

Brobdingnag—Land of the Giants

PHYSICAL ENVIRONMENT

Brobdingnag, the fourth planet of the 54 Chi Orion star system, lies approximately 32.3 light-years from Earth. The planet's diameter measures four percent larger than Earth's, and at its surface, gravity pulls with 1.07 times the force of Earth's. No moons orbit the planet, and no nearby stellar objects of unusual prominence distinguish its sky. Its sun, the star 54 Chi Orion, is just slightly larger, slightly hotter, and slightly younger than Earth's own.

Land covers just 19 percent of Brobdingnag's surface, so the planet actually contains 29 percent less dry land than Earth, despite its larger size. Five oceans divide the land into eight major continents and numerous islands. All the continents cover less area than Earth's Asia or Africa, and unlike Earth's continents, all lie almost entirely within the planet's temperate zones. Seven of the continents are gathered into three distinct groups, each group being widely separated from the others by water. The eighth continent, Tyndarus, sits in relative isolation. This geographic fact has caused distinct evolutionary patterns within the four regions.

Brobdingnag's axial inclination to its ecliptic plane of 15° makes its seasonal temperature variations less extreme than Earth's. The planet rotates on its axis once every 14 hours, but its year stretches four Earth days longer than an Earth year. The short day reduces the extremes of daily temperature and keeps the noonday heat over most of the planet from being intolerable.

Brobdingnag's climate is considerably warmer than Earth's. Mean temperatures consistently measure eleven degrees higher than Earth's at comparable latitudes and elevations, and no permanent ice packs cap the planet's poles. Fortunately, little of Brobdingnag's land lies within its tropic zone, since these regions are too hot for permanent Human habitation. The planet's atmospheric pressure at sea level lifts the barometer about 15 percent higher than on Earth, causing higher wind pressure levels at the same velocity. High humidity and daily rains characterize the planet's weather over most of its surface, except for sections of northern Tyndarus and the great Golan desert.

LIFE FORMS

The enormous scale of Brobdingnag's life forms distinguishes the planet from any other.

Brobdingnag contains no warm-blooded native animals. The animal kingdom has no classes resembling birds or mammals, and no flowers decorate any plants. Other species on Brobdingnag occupy the ecological niches filled by mammals and angiosperms on Earth, just as simpler species filled these roles in Earth's prehistoric times.

Animals

Giant, cold-blooded animals that may reach as much as five meters in height and more than 30 meters in length roam every major land mass. These creatures come in an incredible array of forms, but all have two eyes, four legs, and a tail arranged about a single plane of symmetry. Superficially, they resemble Earth's prehistoric dinosaurs. Yet Brobdingnag's "dragons" have several characteristics distinguishing them from reptiles or any other Earthly species that ever lived. Most significantly, they lack reptilian scales, and their skins are smooth and somewhat moist, like a compromise between a lizard and a salamander. Their hearts contain a "3 1/2"-chamber arrangement that differs markedly from the three-chambered heart of Earth's reptiles.

Like Earth's reptiles, the dragons of Brobdingnag have very small brains and exhibit very little intelligence. They cannot, therefore, be trained, but Humans can, by analyzing their instinctive behavior, control and manipulate them. Most dragons are herbivores, but the few dozen carnivorous species comprise some of the more dangerous non-intelligent life forms that Humankind has encountered.

Insects

Brobdingnag's simpler animals also grow to enormous size. Beetle-like creatures weighing up to 500 grams scuttle about on the ground and flying insects spread wing spans up to 45 centimeters. Brobdingnag's insects possess external skeletons and six legs, similar to those of Earth, but there the resemblance ends. None of them have the three distinct body sections—head, abdomen and thorax—that distinguish Earth's insects from other arthropod classes. Important biochemical differences prevent the insects of Brobdingnag from carrying Human diseases, although a few can inject poisons that may be deadly to Humans.

Vegetation

Plants on Brobdingnag match the animals' giant size. Tree-like plants rise up to 80 meters in height. Giant ferns with huge compound leaves spread along the ground. All land vegetation grows very rapidly due to the heavy rainfall and warm temperature.

Marine Life

Cold-blooded swimming creatures roughly resembling Earth's fish abound in the oceans. Many of these are large, some as large as whales on Earth. Seaweeds and simpler animals also exist, though most assume Earthly proportions. Some very large species resembling urchins and jellyfish have been found, and little is known about the creatures that lurk in the great depths of the sea. As on Earth, microscopic plants in the sea perform most photosynthesis which begins the food chain for all marine animals.

Microorganisms

Very small life is as omnipresent as the very large on Brobdingnag. Thousands of species of microorganisms live in the soil, in the water and in the bodies of native plants and animals. These microorganisms produce a variety of diseases in native species, but to date they have produced no serious illnesses in Humans or their imported domestic life forms. Scientists attribute this fact in part to fundamental differences in the physiochemistry of the two worlds. Earth's reptiles transmit few diseases to Humans. Brobdingnag's dragons are far more distant relations than reptiles.

CURRENT STATE OF DEVELOPMENT

Population

Brobdingnag's population of approximately 9.7 million inhabits four of the eight major continents. The majority lives on Horus, the first continent settled. The balance is divided among Alabuka, Dionaea, and Moa, with Alabuka by far the second most populous. The planet boasts six major metropolitan areas with populations of more than 500,000, plus numerous smaller cities and towns not shown on the global map. The largest city, New Houston, serves as the center of government, industry, and finance for the planet and has 1.5 million people living in and around it. Nearly 80 percent of the colony's population depends on industries located in and around cities, but statisticians classify only 60 percent of the population as urban dwellers. The balance commute from outlying rural regions. In addition, most of the urban population maintains vacation homes in the country.

Industry and Technology

Although Brobdingnag's industry lacks the diversity of Earth's, all its people enjoy a high standard of living. Because the planet's population measures less then 0.1 percent of Earth's, a more limited industrial base is unavoidable; yet much of Earth's industry exists to solve problems created by its enormous population. In the absence of such a population and with abundant natural resources, all Borbdingnites have a healthy diet, comfortable housing, medical care, and a variety of personal services and conveniences. Though approximately one million of Brobdingnag's inhabitants are immigrants, little difference can be detected between their incomes and living standards and those of native-born residents. More than 90 percent of the adult population owns land for residential, recreational or commercial purposes. It is hard to weigh the advantages of this wealth against the wealth of Earth people, for though the latter have more diversions and consumer goods, only 5 percent own real property.

Food

Unlike most colonies, Brobdingnag has been blessed with a surplus of food from its beginnings, although farmers cultivate a narrower assortment of crops than Earth's. Most food is produced by natural rather than synthetic methods, because, although these methods require much land (which is cheap), they require less machinery (the most expensive item for a new economy to produce). Because the cost of machinery and chemical preservatives is high, foods tend to be fresher and less refined. Whole grains, fresh fruits, and vegetables can be found at most tables on the planet. Brobdingnites still produce some dairy products from the stock of cattle imported long ago from Earth. These tend to be expensive because most cattle do not thrive in the planet's hot climate and require special care. Most of the milk output goes into the manufacture of gourmet cheese.

Dragons provide the major source of meat protein in the pioneers' diet. These great beasts graze healthily on native vegetation which grows naturally and quickly. They require little care or supervision, and each animal produces an enormous amount of food. Brobdingnites tan dragon hide into leather that looks attractive and competes in price with the cheapest synthetics. Even the blood of dragons, sold fresh in stores, finds its way into a variety of native delicacies that even first-day immigrants find enjoyable. Unfortunately, the dragons bear their young alive and therefore provide no eggs. Imported chickens do lay eggs for Brobdingnag's tables, but their cost is high, relative to meat.

Housing

As on all colonies, housing manufacture comprises one of the oldest and best-developed industries. Most homes are preassembled in modular sections which are carried whole to the field, but over one-third of Brobdingnag's existing houses were sold in kit form and erected by their owners. In the major cities, large, high-rise living units predominate. These structures erected from standardized modules assume a wide variety of forms. All Brobdingnag's living spaces must be protected from the dragons by ultrasonic fields. Native engineers have designed a number of rugged, low-cost, field generators that repel the beasts effectively. On remote living sites, pioneers mount these units to form a protected compound. On foot, people pass under the field unhampered, and when in their cars, the field deactivates automatically for the instant it takes to cross it.

Transportation

Transportation of people and goods depends primarily on antigravity-supported, air-jet-propelled levicars and trucks. These vehicles come in a variety of sizes and types. Small low, and intermediate speed vehicles lie within reach of every family, while high speed, fully-automated trucks and transports whisk people and cargo between continents. Brobdingnites have deliberately designed these vehicles out of their cities, recognizing that they serve as effective transportation only in areas of low population density. All vehicles unload their cargo at the "gates" of the major cities. From there, high density guideways and freight tubes shuttle people and goods to their destinations with greater safety and convenience than could be achieved by allowing thousands of individually operated vehicles to jam the city's skies and to consume its space with landing pads. Although individually programmed automatic or semiautomatic guidance systems control vehicles in the open countryside, all vehicles fall under the control of a master traffic computer as they approach within 100 kilometers of the city gates. The computer directs the vehicle to an offloading point, and from there to a parking space where it awaits its operator's return.

Urban Life

Cities on Brobdingnag serve one major function, social centers. In an age of rapid transportation, no one need live near his job. Factories can locate in remote areas, with plenty of land to spread over and where they don't create eyesores. No one needs to leave their home to shop or transact business. Modern televiewers can bring any two people on the planet face to face in seconds. Any piece of merchandise can be examined on a holographic display and, if desired, ordered on the spot for home delivery. Many workers whose tasks are primarily mental can do their jobs from home. A designer, for example, can make a drawing on his home screen, then with the push of a button, transmit it to the screens of any other designers or engineers he wishes to see it. Yet Human beings desire physical contact with each other. They wish to talk face to face without the filtering medium of the viewer. They wish to touch each other, to smell each other's scents, to gather in familiar groups and to meet new people. Often they wish to examine the things they want to buy personally, trying on new clothes or checking out the latest features of a new appliance.

To do these things, people still go to cities. The cities of Brobdingnag have become nuclei for large industrial regions and great numbers of homes, though like the nucleus of an atom, the city itself occupies but a tiny region within these giant systems shown as dots on the planet's map (figure 3.8). For hundreds of kilometers in every direction farms, factories and settlements sprinkle the landscape, interspersed with open coutryside. Beyond these areas lie thousands of kilometers of virgin wilderness where only the dragons roam.

Family Life

Families tend to be larger on Brobdingnag than on Earth. Every two adults produces an average of 3.4 children. Nuclear family organization, consisting of one man and woman, predominates, although unmarried men and women raising one or more children comprise 26 percent of the adult population. Only 8 percent of all families can be classified as extended or communal.

Education

Communities of pioneers serve important social functions, including the organizing and funding of local schools and the care of young children. Education for children and adults remains a purchased service, and no government-operated schools exist. A compulsory education law requiring all parents to purchase education for their children through age 19 has been enacted on Horus. Today 40 percent of the population continues their education beyond this level, with 15 percent entering universities and the rest attending trade schools.

The Economy

Brobdingnag, like the two older colonies, has achieved material self-sufficiency. The planet imports no physical goods, although it still places heavy dependence on a steady stream of scientific and technical information received from Earth. A system of free enterprise forms the basis of the planetary economy. Few rules govern trade or business activity, and the absence of a labor class has precluded the development of an economic system in which workers and investors vie in open conflict. On Brobdingnag, more than 75 percent of all workers own an interest in the businesses where they work. The government does not sanction corporate forms of investment, but investors create pools of capital to finance industrial projects by a variety of ingenious, voluntary contracts.

Research and Development

The level of basic scientific research sustained by Earth can not be equalled by a young economy such as Brobdingnag's. Nevertheless, the colony's practically oriented research and development efforts have led to discoveries of considerable value to Earth. The most notable of these include: self-exciting guideway bearings; a process for cryogenic preservation of foods that freezes fruits and vegetables in the fields by direct energy transfer; an integrated information coding circuit that greatly reduces the cost of data terminals; low wattage, optical-frequency power distribution for residential structures; the optimal transarch, and crystalline-filled structural resin. GAILE derives substantial revenues from all of these patents, under license to the people of Earth.

As Brobdingnag measures nearly the same size as Earth, study of its geological differences has given Earth's scientists a better understanding of the basic forces that shaped the two planets. Brobdingnag is a younger planet than Earth, perhaps by as much as half a billion years. The study of its geophysical characteristics and life gives a look into Earth's past that has deciphered many of the enigmas of Earth's early development.

A BRIEF HISTORY

Captain Valarie Nhivitney and the crew of the *Golan* discovered Brobdingnag in 2198 adtc. The enormous life forms of the planet so impressed the Captain that she named the planet after the land of the giants in Jonathan Swift's mythical *Gulliver's Travels*. Brobdingnag numbered third among the extraterrestrial worlds found suitable for Human life, though its star system was the 72nd visited by Earth's starships since the first one left the Solar system.

Humans founded the first colony on Brobdingnag in 2214, just three years before making contact with the Ardotians, the first advanced alien life encountered by Humanity. That event, which ultimately led to a greatly expanded program of colonization and discovery, caused all exploration to grind to a halt.

Until then, the Human race had coped confidently with the physical and biological hazards of space travel, but the specter of a technologically advanced, possibly hostile species, frightened people everywhere. For the first time, Earth's nation states subordinated their armies under one command and girded themselves for what might have been the last war. Fortunately, both sides showed restraint and attempted to communicate before hurling laser bolts at each other. The deciphering of each other's language took 20 years of intensive research and might have taken longer if Earth's scientists had not had the experience of learning to "talk" to the whales and dolphins in Earth's oceans a century before. After the dialogue between Humans and Ardotians opened, another 20 years passed before both sides learned enough about each other to develop mutually acceptable programs for continued expansion into space.

This 40-year delay created the 70-year gap between the founding of Brobdingnag and the establishment of any more colonies. The gap separates what are called the "developed" colonies from the younger "embryonic" colonies. If catastrophe destroyed the Earth tomorrow, the developed colonies could continue to grow and maintain space age technologies without the assistance of Earth. The embryonic colonies might not survive being cut off from the rest of Humanity, or they might begin to regress toward a stone age culture, unable to maintain in isolation the complex system of sophisticated equipment that supports modern life.

The Brobdingnag colony made history by being the first to actually follow the plan for its development. The first pioneers carefully analyzed the experiences of Wyzdom and Poseidous, taking note of how previous colonies' plans had gone astray. As a result, Brobdingnag's early history was less turbulent than that of the earlier worlds, although meager resources and limited immigration during the early years made survival far less certain than it is for new colonies today.

The first pioneers came primarily from what were then America, Australia, and Canada. They settled on a broad, low-lying plain on the continent of Horus near the Orion sea. Because the area's climate and landscape resembled the American state of Texas, they named their development after that state's largest city, Houston. The pioneers planned New Houston from the start, and it has grown along the lines of that plan to this day.

GOVERNMENT

Because early colonists had little desire for, or trust in government, they strictly limited its power and authority. The original Constitution, which they wrote before setting foot on the planet, remains in effect today. It allows individuals to develop the colony in any manner they wish, consistent with the overall plan. The result has been orderly, yet rapid growth which assures the planet's continued prosperity.

The Constitution outlines the framework for a planetary government to guarantee basic Human rights, and to deal with GAILE and with other planet governments. To these ends it is empowered to establish criminal courts, a planet-wide police force, and diplomatic missions. The Constitution specifically forbids the government from other activities ranging from urban planning to operating a monetary system. These powers are left to local governments.

Local governments consist of city, regional, and continental governments. These serve primarily as planning and land claims agencies, but some operate service businesses, such as waste disposal, fresh water supply and local transportation. Local governments generally do not maintain their own police forces or courts, but rely instead on those of the planetary government. The jurisdictions of local governments do not overlap. Anyone living within a city or region pays no taxes to the continental government, and it has no direct control over him. New governments can be formed whenever two-thirds of the people in a geographic area vote to approve one.

Account of Life on Brobdingnag

Editor's note: Earle Horne, 62, dragon rancher, titanium mine operator, and chairman of the Dionaea Policy Council, gives an account of life on Brobdingnag. Born a native of immigrant parents, Earle earned a degree in Minerals Engineering from the Ward Institute of Technology and worked for five years as an exploration geologist with the Crawford Minerals Company. He was instrumental in developing the doppler g-wave wide-range scan technique for defining mineral deposits, and shares one of the patents on that process. He became interested in raising domesauri. With the royalties from his patents to support him, he staked a grazing claim and built himself a ranch. He has written two books on the care and feeding of these large beasts which have become "required" reading for anyone who wants to enter this field.

Brob is a big place! Not that the planet's so big—it actually has less dry land than Earth—but the enormous life forms can scarcely be believed! Standing on the plains near a herd of grazing domesauri, I've often reflected on how feebly Humans compare to those great beasts, and yet we have shaped all this great life to our wills. We have prospered and multiplied, and before my great-grandchildren die, a billion Human beings will live on this world. If aliens land here then, they'll observe the planet's Human population, but they will barely notice the great beasts which have roamed here for millions of years.

I'll never forget the first time I saw a dragon. I was almost six. My family lived in a 20 hectare compound about 200 kilometeres from Dawson's Gap, a small settlement in central Horus. As a child, I never ventured outside the compound except when my parents drove me to town or to the compound of a neighboring family. One day my dad said, "Son, it's time you see whose planet this really is." He went to a locked cabinet in the back of the house and got out a strange-looking object. It resembled a misshapen field ball bat with a piece of pipe and an insulator on it, and a strap to hang it over his shoulder. Later I learned it was a laser rifle, which he hadn't used since I was born. We got into the car and headed out of the compound for a while. Soon Dad set the car down in a small clearing, and we climbed out.

Not a sound could be heard. No talk, no activity, no machinery, not even insects buzzing, broke the silence. My dad whispered not to make any noise, and we headed across the clearing toward the trees. We pushed through a wall of ferns at the clearing's edge and started walking in the dense forest. The soft, spongy ground muffled our footsteps, and the cathedral-like canopy of leaves overhead filtered all but green light. After a while we came to another wall of foliage, and Dad told me to be very quiet because we were about to see what we had come for. He pushed the plants aside and guided me on ahead. Nothing had prepared me for that sight.

I looked out of the forest into a larger clearing with a small lake in the center of it. In and around the lake stood 20 of the largest animals I had ever seen. As they grazed, their great necks moved tiny heads slowly over the long grass, cutting a swath as they went. They had great round bodies, long winding tails and legs as big around as tree trunks! Their smooth, leathery skin varied between mottled pink and light brown. The rushing sound of their breathing resembled the roar of distant surf, and the moisture in their great breaths condensed like smoke when they exhaled. So these were the dragons! I had looked at pictures of them on the Infax, and I had gossiped in low whispers about them with my friends at school, but the awe of seeing them in the flesh sent tingles through my body. I stared transfixed for what must have been a long time.

Suddenly a cracking and crunching of vegetation at the far end of the clearing interrupted my trance and another equally awesome and wondrous creature came into view. Its enormous head was dominated by a mouth inset with double rows of pointed teeth, each the size of a man's thumb. It loped along on all fours, though its hind legs were a good deal larger than its front legs. Every few steps it rose up on its rear legs and let out a great hiss that stopped my heart cold. The great grazing beasts raised their tiny heads all together and looked toward the attacker in dumb terror. My father let the rifle slip from his shoulder and held it in front of him. The great beasts began to run from the attacker. Though they moved swiftly, their size made them appear in slow motion. Father pulled me back into the brush, and we walked quickly back the way we came. I heard no screams and did not appreciate fully the fate of one of those big, pinkish creatures near the pond. Twenty years later I would avenge its death in face-to-face combat with a relative to the toothed monster. Though I didn't know it then, the sight in the clearing had marked my soul for life. My hopes and ambitions and the path I would follow to satisfy them had been permanently altered.

Though central Horus had not been settled long in those days, my childhood in the country passed uneventfully. Before my father came to Brob, early settlers had learned to protect themselves from dragons by surrounding their compounds with ultrasonic fields. These fields don't affect Humans, but the dragons find them so repulsive that they won't even approach them. Our house sat in the center of a circular clearing about 250 meters across. At its edge, a green wall of vegetation rose 70 meters in the air. The cleared area around our house contained an assortment of tiny farms and orchards, and a small grassy field for us children to play in.

My parents, quiet people who kept to themselves, had left the crowded bustle of Earth because Brob offered them the rural life of their dreams. I was born the eldest of three children, and despite my parents' natural reserve, we grew to be a very closely knit family. My contacts with other people occurred no more than once a week until I turned five, and Mother sent me to the regional day school at Dawson's Gap. I enjoyed the school and had several friends there. I dabbled in sports but became preoccupied with science at an early age. All fields interested me: astronomy, chemistry, geology, and biology. I had collections of insects, a telescope, a small chemistry lab, and a rock collection. My father worked as a foreman at a nearby chromium mine, and he encouraged my interest in geology. At sixteen, he got me a part-time job as a worker-helper in the mine. I loaded trash, oiled equipment, fetched and carried, and helped out with a lot of assorted dirty work. After a while I even learned to make basic repairs on the autominers. This experience forever solidified my distaste for physical labor. From the first, I showed promise as a good student, and Father began putting aside money early for my education. At 17, I was accepted into the Ward Institute in New Houston.

My parents drove me to Dawson's Gap and put me on the bus to New Houston. Father said he didn't care if he ever saw the "big town" again, and wouldn't go there until my graduation. I had never left the region before. Though I had seen New Houston on the news, it seemed more remote than the planet Earth, which often appeared as the setting of movies and shows we watched at home. Once again I was emotionally unprepared to deal with an unfamiliar aspect of my world.

My bus arrived at Gate Seven of New Houston. The passengers filed out of the bus and down a long tube that led to the main terminal. The terminal floor formed the heart of the gate. From there moving ramps filled with people and cargo rose for sixteen stories and headed in what seemed to be every conceivable direction. Numbers in round signs of varying colors marked the routes, and everyone seemed to know where they were going. Even those that didn't consulted small screens around the main floor of the terminal entering codes on a keyboard and getting directions from an unseen computer buried in the center of the city. I stood staring at all this for half an hour, wondering what I was going to do. At last, I noticed a stylishly dressed fellow carrying a hand-painted sign saying "Ward Tech" pass by with about half a dozen younger people in tow, who looked about as confused as I. He was an upperclass student who had volunteered to meet incoming freshpeople at Gate Seven that day.

After he finished gathering up about a dozen of us, another student came to relieve him, and we headed into town. He took us to a large rack containing rows of barrel-shaped containers about 1.5 meters in diameter, opened one and put our luggage inside. He punched a long number into the key board on the side followed by the words "Ward Tech." The barrel tipped forward into a large chute and scuttled away. We then turned to a nearby self-propelled ramp and started our journey into the city. The older student explained the workings of the guideway system as we went, but I was too preoccupied with the view to listen.

We moved between successively faster ramps until we reached the fastest one travelling at about 35 kilometers per hour. As I looked out I could see dozens of other ramps, some moving parallel with us and others moving in the opposite direction. All moved at different speeds, and now and then one or two would branch off or merge in from an angle. Thousands of interesting people filled the ramps. They didn't look like country folk. They dressed in shiny Virlon or Acrovel outfits in an array of bright colors. Many wore their hair long. Some wore it teased and piled up on their heads in a variety of weird styles. Beyond them loomed the great multi-story structures of the city. The geometric shapes of the buildings seemed incredible. Most country houses are made up of low, flat rectangles, but these buildings contained cylinders, pyramids, and irregularly-piled blocks of rectangular shapes. Graceful arches supported and connected some of the larger structures, and periodically, great silver spires thrust up, crowned with giant arrays of communications antennas. Between the structures stretched long narrow parks containing grassy knolls and tree-lined paths of shiny black material. City people strolled through these parks and sat on benches beside the walkways or on the neatly trimmed grass. As we passed through commercial areas, I could see shops lining ground level arcades, whose display windows filled with brightly colored goods tempted passersby. In other places, bars, theaters, and clubs opened onto the arcades, their signs and displays offering all manner of instruction or diversion from religious lectures to comic opera, from the formally proper to positively sordid.

We arrived at the Ward Institute and descended from the ramps. The campus consists of a rather dense complex of medium-sized buildings on a trapezoidal-shaped piece of ground about 500 meters on one end, 300 meters on the other and 700 meters long. I was shown to a room in a large dormitory building that looked out upon the city from about 50 meters in the air. The dorm buzzed with students newly arrived for the coming quarter. During the next few days I received hundreds of instructions about school procedures: choosing my classes, eating, joining social groups, grades, scholarships, and so forth. In another environment I might have paid more attention to all this, but my mind had been captivated by the wonder of the great city spread out before me, a city that never stopped, never even slowed its endless stream of traffic and activity.

My first quarter at Ward became the all-time low point of my academic career. I spent more time in the city than I did in the classroom or studying. The stores, the shows, and the endless crowds fascinated me. The city's atmosphere seemed electric. I stood for hours in squares or busy intersections just watching all the people pass by. Live entertainment also fascinated me. There is something about watching live people singing, dancing, or acting on stage that just can't be conveyed on the holoscreen at home. I took a job in a bar to earn money. I served alcohol and pills in a variety of forms, and, of course, took to using far too many mind-affecting chemicals myself. The money I earned was supposed to pay educational expenses, but I spent too much of it on entertainment and fast women, my ultimate undoing.

I had dated a few local girls during my high school years, but nothing had prepared me for city women. In the country, men and women dress in similar fashion, wearing loose-fitting jump suits or shirts and pants of rugged material. No one wears makeup and hairstyles tend to be short. City women wear high-styled, tightly fitting outfits, and use a variety of body paints and artificial scents to enhance what nature gave them. I had developed no immunity to these lures, and spent many spare hours in singles bars hustling girls. I guess I was sort of attractive in a country-bumpkinish way, for I managed to spend almost half my nights in the apartment of some unattached female whom I had met only hours before.

Exams came and the results were as expected. My advisor called me in and told me that I had better shape up or I'd find myself permanently out of school and working in the mines again. Fascination with the ways of the city was a common malady among country freshmen, he said, and those who fell so ill that they flunked out regretted it within a year. He suggested that, despite my poor showing in the first quarter of geology, I enroll in his laboratory course, which included a four-week trip to Alabuka. He offered to tutor me in geology during the tween quarters vacation so that I would be prepared for the trip, and he got me a job on campus so I didn't have to work in the bar.

The field trip provided still more revelations. We traveled by levitruck to Alabuka and worked out of a base camp operated by Crawford Minerals. We explored the foothills of the Chebak mountains looking for rare earth deposits and cataloging the planet's geological forms. We also visited the mines and processing plants. There I got to know a couple of the engineers quite well. On weekends we travelled to Wynand to savor the city life of what was then an emerging boom town. The engineers belonged to expensive clubs that offered superb food and outstanding entertainment in far more lavish settings than I had ever seen. I met women too, women so exotic, so sensuous, so worldly wise that I forgot completely about the shop clerks and secretaries I had been chasing back in New Houston. I returned to school with a renewed desire to succeed and I managed to graduate four years later with high honors, despite my early academic probation.

After graduation, I accepted an offer with Crawford Minerals as an exploration geologist. I wanted to see the world and explore territories that few people had seen before. Dionaea and Moa were opening up, and I spent five years on exploratory parties probing unknown sections of those continents in our search for minerals. During those trips I had the chance to renew my interest in the great dragons that roam the planet. I read several books about them and soon learned that people knew very little of these magnificent animals. Though several hundred of the more than two thousand species had been classified, most of those listed lived on Horus. Evolution had proceeded quite independently on the different continents of Brobdingnag, and the animals on each continent adapted uniquely to each environment. People knew very little about their diets, their habits, their natural instincts, or their intelligence. I spent many of my spare hours observing the dragons. I kept careful notes and took photographs of the creatures which helped me to discover more than 20 previously uncatalogued species.

As my studies of dragons progressed, I grew bolder. The majority of species are herbivores that graze on shrubs and trees. Sedentary animals, they don't more around more than they have to. Humans, or anything else smaller than they are, don't seem to frighten them, so I began to walk among them as they grazed. This enabled me to observe their indiscriminate eating habits, and their limited social patterns. In general, females outnumber males by a fairly constant three to one ratio. Their sexual relations occur randomly; they don't take permanent mates, nor does a dominant male maintain his own group of females. Their tendency to herd seems based on instinct and a common desire for food. I suppose nature programmed them to know that the eating is good where the others are eating. A typical group consisting of one to three males and three to fifteen females remains together unless scattered by a predator. Many species are quite adept at sensing natural predators. I've seen large carnivores scatter herds long before I saw or heard the intruder's approach. The first time this happened nearly cost me my life.

I was observing a group of titanisaurs at close range when suddenly, without warning, the group broke into a run. They headed off in a widely scattered pattern, but in the same general direction. I had to step lively to avoid being crushed, but fortunately the creatures move slowly, especially when they're just starting out. I soon found myself alone in a clearing facing a charging icondodon, a large, ferociously toothed creature that stands about six meters tall. I ran at right angles to it, hoping it would not pursue me, but it turned and headed straight for me. I guess it figured I'd be a warm hors d'oeuvre before its main meal. I always carried a laser rifle when walking alone and took careful aim at the monster's head. It looked about 100 meters away when I started firing. The first burn to its forehead didn't appear to slow it down. I fired again at its chest and could see its flesh burning. Yet that didn't appear to hurt it either and its pace barely slackened. It moved about 40 kilometers per hour, so flight was out of the question. Desperately I crouched and took careful aim at one eye, which burst like a balloon of water with the laser's heat, but he had another and simply turned his head slightly to bring me into full view. As he towered above me, I got off a final shot to his other eye, then turned and ran. Blinded, the creature began whirling in circles, and I threw myself to the ground just in time to avoid being swatted by its great tail.

I picked myself up and, rifle clutched tightly, ran as fast as I had ever run in my life. When I stopped, I stood 100 meters from the beast. He had fallen to the ground and was writhing furiously. I don't know if it actually felt pain, or if his nervous system had become totally confused by its lack of sight. I didn't wish to leave it suffering, but I didn't know if my laser rifle had enough energy left to kill it. I returned to my levitruck, a drilling unit that I had borrowed for the day. It contained a 10,000-kilowatt rotary polarized laser drill. I flew back and hovered over the beast. His movements seemed slower now, perhaps from its impressive loss of blood. I positioned the laser carefully and fired a brief burst from overhead. The levitruck was slowly drifting, so the beam cut through the creature's middle in a second, slicing it in two. Blood gushed forth, and although it appeared actually dead for the first time, the halves of its body continued to twitch for half an hour.

Minutes after it stopped moving, hundreds of tiny reptiles emerged from the grass to feast on the large carcass. I returned the next day to find the bones stripped clean of flesh, and tiny bone-eating insects slowly working on the remains of its skeleton. Although I had never been more frightened, I felt sad. Had I been carrying a sonic disperser, the animal's death wouldn't have been necessary, nor my own so likely. Before it happened, I had imagined this battle with myself as Sir Galahad. Since then the image seems foolish. The struggle proved little, except that these huge dumb life forms are no match for Human weapons. It was a needless destruction of one of nature's creatures, a disrespect for the value of life.

I undertook my early years of dragon study as a pure hobby and had been doing it for about three years before it occurred to me that they might be useful. As was typical for early colonies, the first settlers of Brobdingnag brought the seeds of their own foods with them from Earth. Meat on Brob used to cost a bunch because native vegetation wasn't suitable feedstock for Earth-type animals. Ranchers had to fatten cattle, pigs and sheep on grains. The planet's heat and humidity also bothered the animals, so their production had to be confined to northern regions of Horus. People also ate synthetic meats produced from vegetable proteins, but neither the taste nor the texture of these facsimilies equalled the original. The dragons, because of their alien body chemistry, hadn't been considered sources of food for the first century of the colony's history. About the time I was born, dragon meat began appearing in the stores. Most of it came from hunting creatures in the wild, and its quality varied highly from batch to batch. My parents never served the stuff at home.

I got my first taste of dragon while at the Lake Manassis base camp on Dionaea. A newly arrived cook had been working at the camp about two weeks when a young scanner technician lost control of his levicar while landing and accidentally killed a dragon weighing more than ten tons. When the cook heard this, he asked if he could go out and butcher some of its meat for our dinner that night. The camp boss grimaced and complained, but the cook persuaded him. I went along simply because I'd never dissected a dragon myself.

We brought back 150 kilos of meat, plus about 30 kilos of entrails from the carcass of an as yet unclassified obesidon. The cook fileted it and broiled it, and it was the most delicious meat I had ever tasted. After dinner I asked him about it, and he replied that he learned to serve it on his previous assignment. He always cooked and ate some first, because the flavor could vary quite a bit from one animal to the next. As the beast we ate was unclassified, I submitted a classification report to GAILE and christened the species "domesauris." After that we ate domesauris several times a week at the camp. Every month the cook and I went hunting. I identified the proper species and set a trap for it, usually a percussion device that killed them quickly and painlessly. The flavor of wild dragons, even from the same species and the same area, varied considerably and rarely measured up to my first taste. About one creature in four actually made us gag when we tried it, which seemed to me a terrible waste.

From these experiences, a germ of a business idea grew, and two years later I left Crawford Minerals to give it a try. I had assisted another engineer in developing a new mineral survey technique, and, under Brob's law, earned a royalty on its patent. Also, I had saved some money while working as a field engineer. These funds I invested in a second-hand levitruck, some cryopreservation equipment, and a prefab building. In Brob's remote regions, the planetary government handles land claims. I petitioned for a traditional cattle ranch homestead of 8,000 hectares and for a special grazing claim of 286,000 hectares on the continent of Dionaea. I had to persuade the land claims board of the validity of the grazing claim. (Shown in figure 3.9.) Basically, it allows me to have exclusive rights to graze and capture domesauris within a specified area. However, should anyone else wish to use the land for some "higher" purpose, such as mining or crop farming, my claim can be superseded.

As almost no marketing system for dragon's meat existed, I approached traditional meat wholesalers with a proposal to supply them with domesauris of consistently high quality, at prices far below that of traditional Earthly meats. This would enable them to expand their business with a less expensive product. It took quite a lot of salesmanship, including many free samples, to get distributors lined up. Many timid and conservative meat dealers turned me down for a variety of imaginative reasons, but at last I did get enough commitments from buyers to let me start raising dragons.

My early years on the ranch in Dionaea kept me busy. Most of the time I spent alone, keeping watch over the herds and processing meat in my tiny factory. I had to figure out how to keep the beasts from wandering off, how to protect them from predators, how to protect *me* from predators, and most importantly, how to assure a consistent quality of meat. My earliest quality control method was to simply slaughter the animals, sample them, and discard the unacceptable ones. I turned the rejects into fertilizer which I used in my orchard and vegetable garden or sold to a few organic farmers in the area. It seemed a great waste, so I set about finding out why the creatures varied so much in taste. After two years of work, I found the cause lay in their diet. Domesauri will gobble up just about any plant that happens to be in front of their noses. Many of these native varieties impart a foul flavor to the meat if the creature eats them. It takes anywhere from two to three weeks for the taste to be purged out of their systems after they are taken away from the offending plants and fed good food. I could never have identified all the harmful species of plants and figured out what to do about them without the help of my wife.

Although my work fascinated me, the loneliness of life on the range began to get to me after six or eight months. My only contacts with Human beings were with the teenagers from Fort Mackinac (then a tiny outpost) who helped me out at butchering time, and with the people I met on weekly trips to town or on occasional business trips to New Houston. On one of those, I met the college-age daughter of the buyer for a major foodstore. She was a biology major, just about to graduate from prestigious Wheeler College. She had a very quiet, almost hermitic personality, and she said little until one evening after dinner she found out what I did. That evening we talked all night about life in the wilderness and my experiments with the dragons. The week after her graduation, she ran off to Dionaea to live with me. Two months later, her father found out where she was and swore I'd not sell a gram of meat on Horus again. We silenced his objections by getting married and have remained so ever since.

Suzanne has been in every sense a partner in my life. Her interest, inspiration and creativity helped me to develop many of the animal husbandry techniques for which I am known. She is not as disciplined as I, preferring to be off in the forests looking for new plants and animals rather than keeping the records or organizing the feeding schedules. Yet she becomes absorbed in the knotty scientific and technical problems of our work, and I confess her ingenuity often exceeds mine.

It takes 300 hectares of Dionaea's central plains to support one domesauris. Clearing all of this area of unfit food is totally impractical; so we cultivate smaller fields of finishing food and place the animals in them three weeks before packing time. Development of this technique, together with the breeding, guarding, and herding techniques we use today took the next 25 years of our lives. During that time, we watched Dionaea grow from a wilderness to a self-sustaining community. An awful lot of untouched land still remains out where we live. We can fly at top speed for hours without seeing a trace of Humanity, and Suzanne still prefers country life to the city where we also have a home.

A look at the map of my ranch (figure 3.10) gives some idea of how complicated dragon rais-

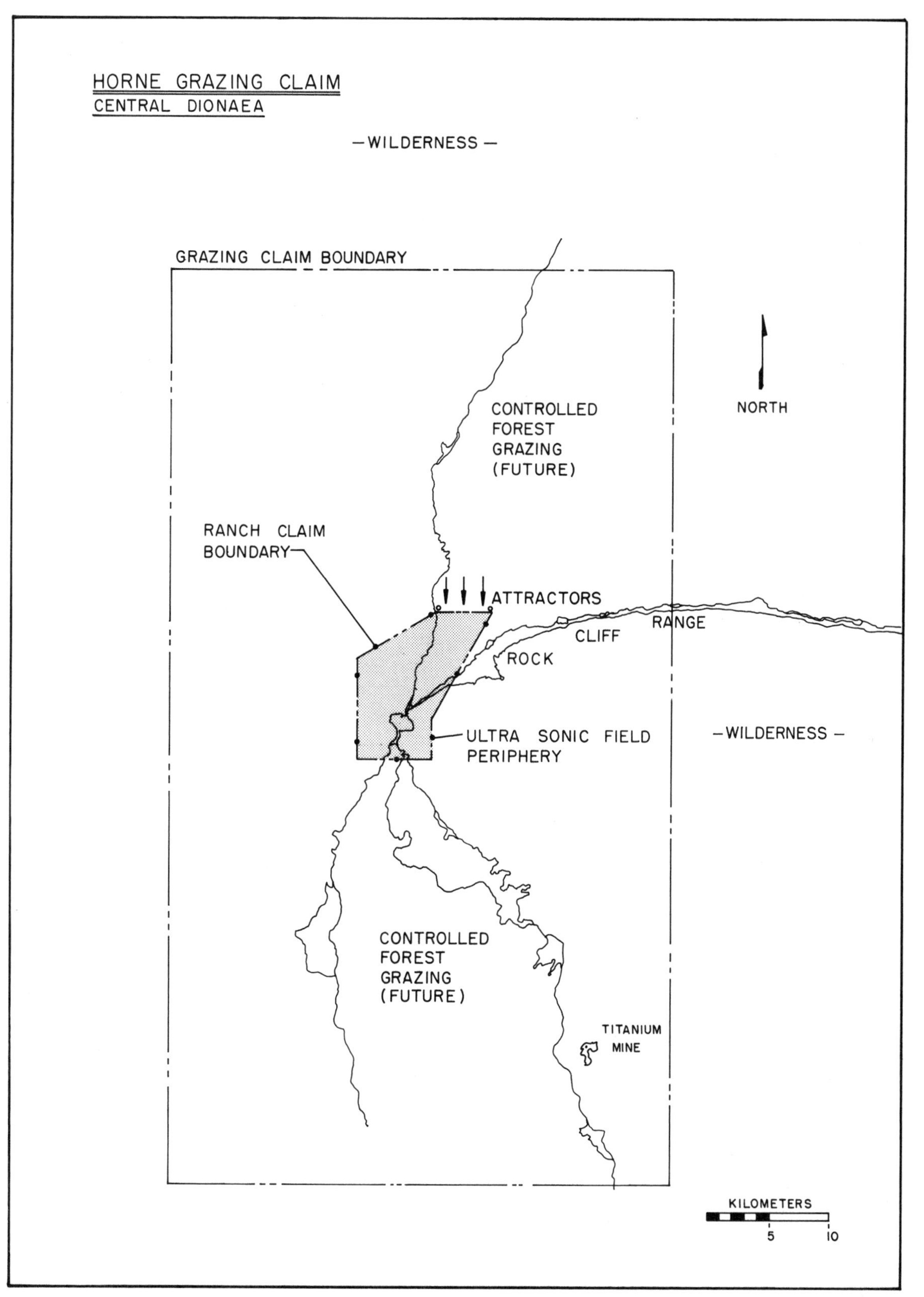
HORNE GRAZING CLAIM
CENTRAL DIONAEA
—WILDERNESS—
GRAZING CLAIM BOUNDARY
NORTH
CONTROLLED
FOREST
GRAZING
(FUTURE)
RANCH CLAIM
BOUNDARY
ATTRACTORS
CLIFF
RANGE
ROCK
ULTRA SONIC FIELD
PERIPHERY
—WILDERNESS—
CONTROLLED
FOREST
GRAZING
(FUTURE)
TITANIUM
MINE
KILOMETERS
5
10

FIG. 3.9

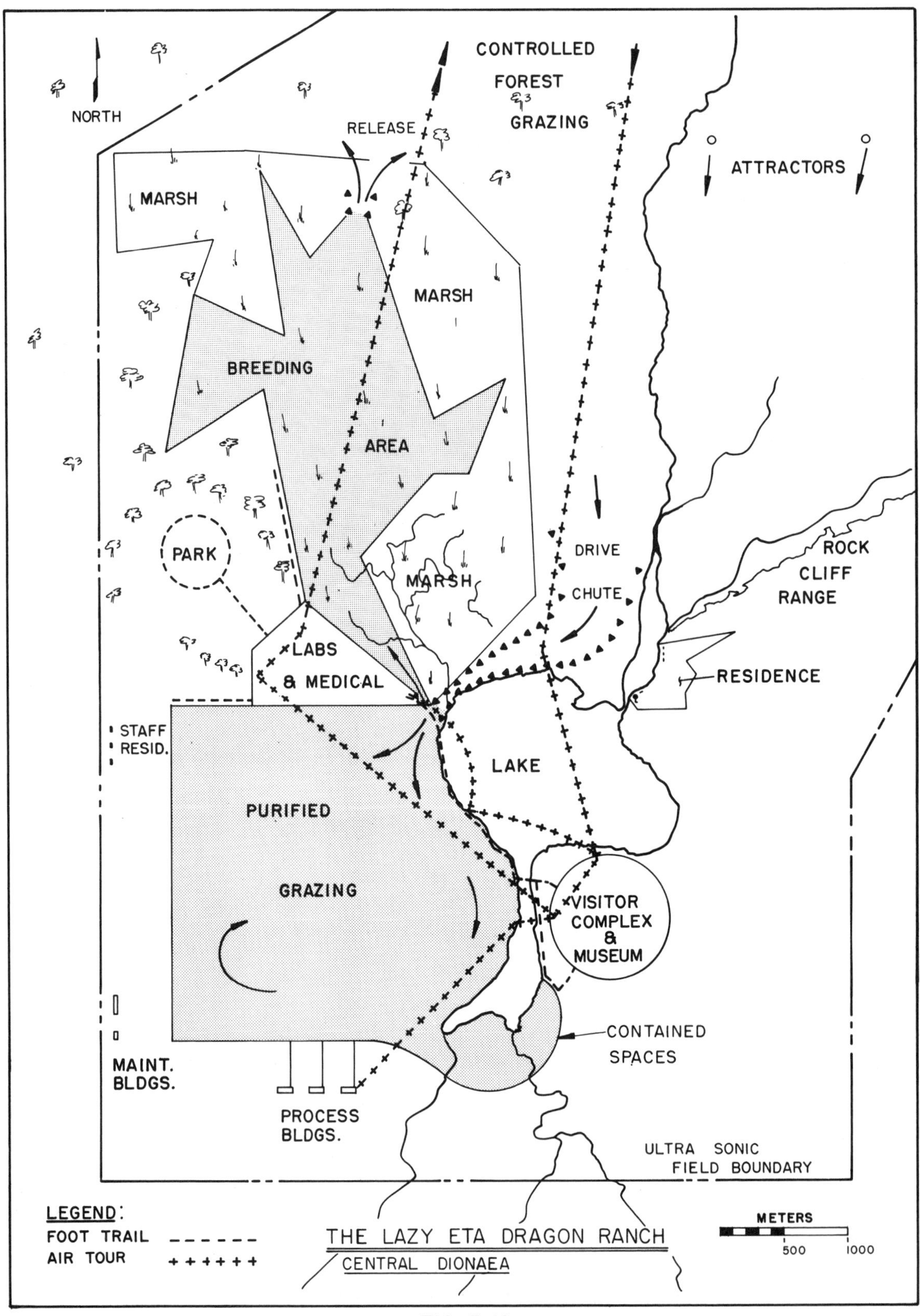
NORTH
CONTROLLED
FOREST
GRAZING
RELEASE
ATTRACTORS
MARSH
MARSH
BREEDING
AREA
PARK
MARSH
DRIVE
CHUTE
ROCK
CLIFF
RANGE
LABS
& MEDICAL
RESIDENCE
STAFF
RESID.
LAKE
PURIFIED
GRAZING
VISITOR
COMPLEX
&
MUSEUM
CONTAINED
SPACES
MAINT.
BLDGS.
PROCESS
BLDGS.
ULTRA SONIC
FIELD BOUNDARY
LEGEND:
FOOT TRAIL - - - - - -
AIR TOUR + + + + + +
THE LAZY ETA DRAGON RANCH
CENTRAL DIONAEA
METERS
500
1000

FIG. 3.10

ing has become. In the old days, I used to have to go out and look for the beasts then herd them in one at a time by levicar. Now, brain wave attractors, based on the principle of the somafield, lure the creatures into the drive chute from all over the grazing area. Once inside, pulsed low-wattage laser fields make it very unpleasant for them to turn around, so they just mosey on down to the purified grazing area to fatten up for a while. Keeping all the bad-tasting weeds out of an area as large as the purified grazing area was no mean trick. Much of that huge space marked "Labs and Medical" is taken up by experimental plots of specially bred feed grass, though the pens for sick dragons take up a good chunk of it too! The scale of the map makes the process buildings look small, but each one contains more than six thousand square meters of space.

After business really got going, all kinds of people from as far away as Horus kept stopping by to see the dragon ranch. I got so tired of explaining how it worked all the time that I installed an automated air car tour that takes visitors all over the ranch and gives a canned spiel by me that doesn't forget something important each time I give it. Though the tour doesn't make me rich, it has paid back my investment, and given me some peace and quiet that I'd have paid dearly to get. Now things run so well, that I have moved my home over to the cliffs above the lake. This not only keeps me away from the tourists, but keeps the ranch hands from sticking their heads in the door every time they have a dumb question.

My success at ranching has attracted many imitators, but I've profited from this too, for my books on dragons bring in a steady stream of royalties with very little work on my part. The meat business has grown so competitive that profit margins have been driven steadily downward. Meat is now so inexpensive that I worry about people eating more of it than is really good for them.

In the 36 years since I filed my claim, Dionaea has grown enormously. It now has a continental government that handles many matters formerly handled in New Houston, and I serve as an elected member of its Policy Committee which formulates laws to protect Human rights. Gardenia and Fort Mackinac have blossomed from small outposts to full-fledged cities offering a wide range of consumer goods and services. Farms have spread on to the central plain, though none have reached as far as my grazing claim. Someday, perhaps five or six centuries from now, there won't be enough land left to graze domesauri. I don't have to worry about that, because I have hedged my bets and gotten back into mining. That little titanium mine in the southeast corner of my grazing claim makes about as much money as the whole ranch.

I've had a good life, and I've enjoyed watching my children grow up to become responsible adults with a world of opportunity before them. My wealth now allows me time for civic matters and philanthropic activities. I think my knowledge of business and the biology of our planet has enabled me to contribute useful ideas as well as money to worthwhile causes. I'm looking forward to helping the newly founded musical society raise money for a performing arts center in Fort Mackinac, so we Dionaeans can enjoy live music and plays just like folks do in New Houston. I've expanded ranching operations on to Valeria continent, not because I need room, but because it will give me a chance to work with the local dragon species and to develop them for commercial use. I'm optimistic about the future of Brob, and I'm confident that pioneers will find the next 162 years twice as exciting as the first ones.

* * * *

OPPORTUNITIES FOR IMMIGRANTS

Opportunities exist for professionals and tradespeople of all types. Especially needed are robot mechanics, electrokinetic technicians, chefs, computer programmers, medical doctors, research physicists, and engineers and designers of all types. Virgin land claims vary depending on the continent. Personal residence claims vary from two hectares in the New Houston region to 25 hectares on Valeria continent. Claimed land may be transferred via sale.

GAILE provides transportation for 15,000 immigrants annually in a starship of the single hex configuration. Travel time requires 61 days of ship time and 89 days of planet time. Twelve information probes are exchanged annually with Earth.

Genesis—The Beginning of Life

IMPORTANT STATISTICS

Planet name: Genesis
Equatorial diameter: 12,850 kilometers
Mass (Earth = 1): 1.03
Surface gravity: 1.01 g
Escape velocity: 11.3 km/s
Albedo: 0.43
Atmospheric pressure at mean sea level: 0.55 bars
Fraction of surface covered by land: 21%
Maximum elevation above sea level: 9127 meters
Length of day: 12 hours 2 minutes
Length of year: 347 Earth standard days
Obliquity: 27 degrees
Von Roenstadt habitability factor (Earth = 1): 0.81 current improvable to 1.02
Current population: 1.4 million
Number of population centers: 7
Year settled: 2284 adtc
Number of moons: 253

Star name: 59 Virgo
Star type: F8V
Distance from Earth: 43.5 light years
Distance from planet: 0.96 AU (Earth-Sol = 1 AU)
Star Mass: 0.98 (Sol = 1)

Travel time from Earth
Ship's time: 64 days
Planet time: 93 days

Atmospheric Composition
Oxygen: 31.4%
Nitrogen: 68.4%
Carbon dioxide and inert gasses: 0.2%

Range of Distances (km)	*Range of Periods (Earth hours)*	*Range of Masses (Earth's moon = 1)*
10,000 - 20,000	2.7 - 7.7	1.06 E - 8 - 1.25 E - 10

Table 3.5

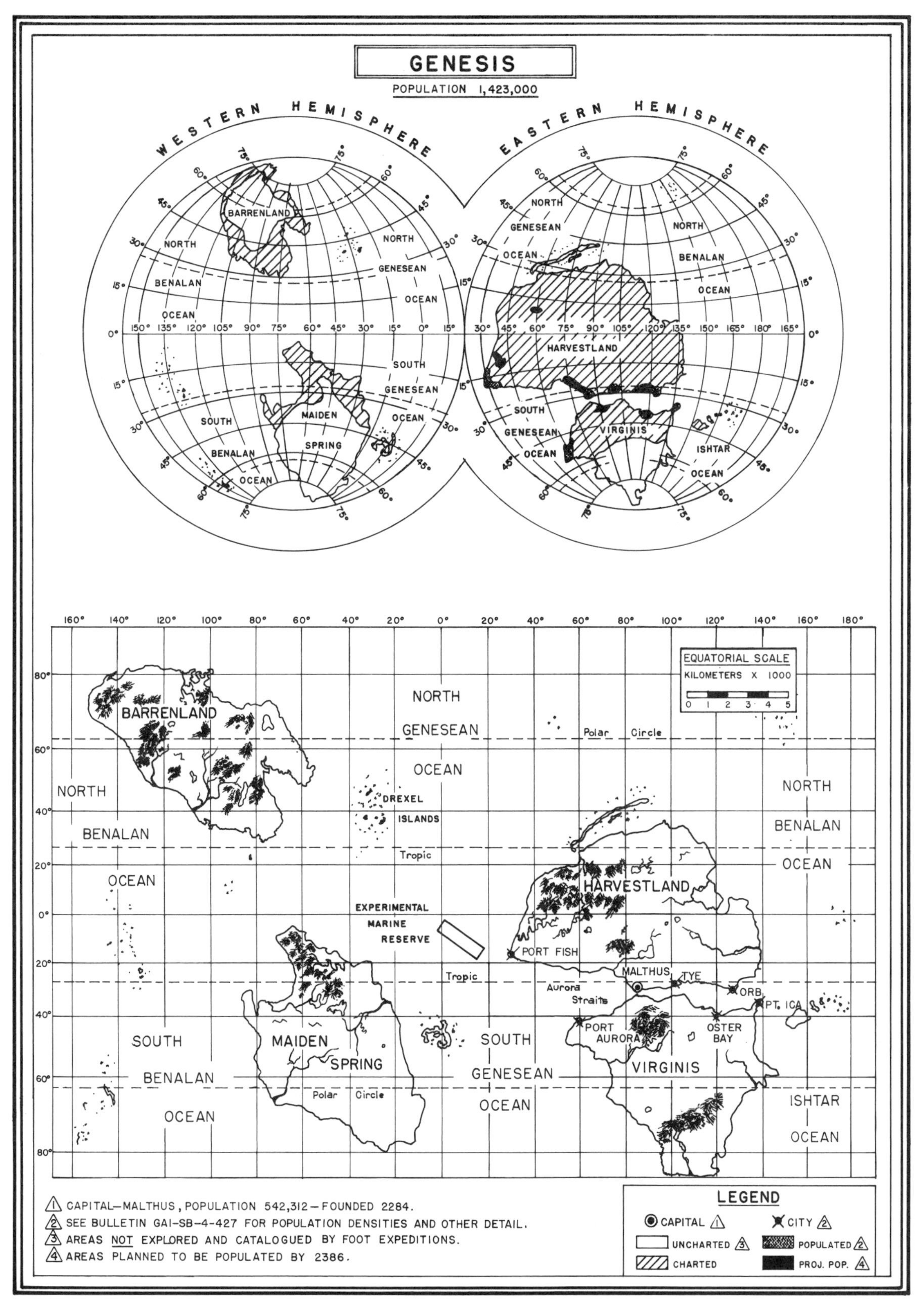
GENESIS
POPULATION 1,423,000
WESTERN HEMISPHERE
EASTERN HEMISPHERE
BARRENLAND
NORTH
BENALAN
OCEAN
NORTH
GENESEAN
OCEAN
SOUTH
GENESEAN
OCEAN
MAIDEN
SPRING
SOUTH
BENALAN
OCEAN
NORTH
GENESEAN
OCEAN
NORTH
BENALAN
OCEAN
HARVESTLAND
SOUTH
GENESEAN
OCEAN
VIRGINIS
ISHTAR
OCEAN
EQUATORIAL SCALE
KILOMETERS X 1000
BARRENLAND
NORTH
GENESEAN
OCEAN
Polar Circle
NORTH
BENALAN
OCEAN
DREXEL
ISLANDS
Tropic
NORTH
BENALAN
OCEAN
HARVESTLAND
EXPERIMENTAL
MARINE
RESERVE
PORT FISH
Tropic
Aurora
Straits
MALTHUS
TYE
ORB
PT. ICA
PORT
AURORA
OSTER
BAY
SOUTH
BENALAN
OCEAN
MAIDEN
SPRING
Polar Circle
SOUTH
GENESEAN
OCEAN
VIRGINIS
ISHTAR
OCEAN
1 CAPITAL—MALTHUS, POPULATION 542,312—FOUNDED 2284.
2 SEE BULLETIN GAI-SB-4-427 FOR POPULATION DENSITIES AND OTHER DETAIL.
3 AREAS NOT EXPLORED AND CATALOGUED BY FOOT EXPEDITIONS.
4 AREAS PLANNED TO BE POPULATED BY 2386.
LEGEND
CAPITAL 1
CITY 2
UNCHARTED 3
POPULATED 2
CHARTED
PROJ. POP. 4

FIG. 3.11

Genesis—The Beginning of Life

PHYSICAL ENVIRONMENT

Genesis, third planet of the star system 59 Virgo, can be found 43.5 light years from Earth. 59 Virgo, a smallish type F8V star, is hotter and younger than Earth's sun. Although it's expected to have a shorter life-span than Earth's sun, it still has several billion years to go.

No large moon like Earth's orbits Genesis, but it possesses an asteroid-like belt of more than 250 moonlets orbiting between ten and twenty thousand kilometers from its surface. Most of these satellites have an irregular shape and tumble slowly in their orbits. From the planet's surface, the naked eye can plainly see more than 100 of them forming a beautiful night time display. The largest measures about twenty-eight kilometers long, but it lies so far out that it appears only slightly larger than a star when viewed from the ground. Composition of the moonlets varies from nearly pure iron to siliceous rock. No one knows if the moons ever formed part of one larger moon, or if they are the captured remnants of a prehistoric meteor shower. The moon belt exerts negligible tidal forces on the oceans, since the moons are distributed evenly around the planet. Principal tidal action is due to the pull of the sun, 59 Virgo.

Genesis is barely larger than Earth, with a diameter less than one percent greater and a gravitational attraction at the surface only one percent higher. Land covers 21 percent of its surface. The bulk of it is divided into four continents: Harvestland, Virginis, Barrenland, and Maiden Spring. Three continents have less area than Earth's average continent, but one, Harvestland, has more surface than Earth's Eurasian land mass. The continents take a more closed form than continents on Earth and are separated by larger expanses of open ocean.

Genesis orbits 59 Virgo once every 347 days in nearly circular orbit. Its day clocks just over 12 hours, the shortest among the present colonies. Genesis inclines 27° degree toward its ecliptic plane, making its arctic regions larger than Earth's and its seasonal temperature variations more extreme. Its short day lessens extreme daily temperature variations, however. The planet's atmospheric pressure at mean sea level gauges just 55 percent of Earth's, but since the atmosphere contains 31 percent oxygen, Humans can breath it. The lower atmospheric density reduces the heat transferred from the poles to the equator by atmospheric convection, causing somewhat greater temperature variations with latitude than Earth's. Lower density also reduces the effective wind pressure, so even though wind velocities average somewhat higher than Earth's, their destructive and wave-generating forces are lower.

Genesis is believed to be much younger than the Earth, perhaps by as much as 1.5 billion years. The planet appears to contain more residual heat than Earth and exhibits much greater volcanic activity. Flying over the surface, one rarely loses sight of active volcanoes.

LIFE FORMS

Except for its blue sky and white clouds, the natural surface of Genesis looks more like Earth's moon or Mars than a habitable planet. Not one tree or one blade of grass breaks the monotonous expanse of cold grey rock, stony rubble and sand. No native animals, not even the smallest insect-like creatures, scurry across the empty waste. There is no food, no soil, no single living thing!

If ancient space travellers had landed on Earth half a billion years ago, they would have viewed a similar scene. Life on all habitable planets began in the sea. Genesis is such a world in its earliest stage of development; so its only life exists in the oceans. Because life evolves from the simple to the complex, most life on this young planet seems elementary when compared with the other colony worlds.

Marine Life

Sea life consists mainly of microorganisms. These tiny monocellular species fill all the functions in the life cycle from tiny one-celled photosynthetic plants, resembling Earth's diatoms, to tiny bacteria-like creatures that consume the dead remains of plants and animals. Larger plants take the forms of simple seaweed while still other small plants containing only a few thousand cells float freely in the water. Animals range in structure from tiny multicellular free-floaters and larger free-floaters resembling jellyfish to numerous species of tiny animals with external skeletons. The latter creatures look like the trilobites which dominated Earth's seas in the Cambrian period, 500 million years ago. Earth's phylum of arthropods, which today includes lobsters, insects, and spiders, traces its ancestry to the trilobites. All animals on Genesis exhibit extremely simple behavior patterns; they have few instincts apart from the desire to eat and avoid being eaten. Complex adaptations like the spider's web, the hermit crab's borrowed shell, and the symbiosis between ants and aphids have not even begun to emerge.

A BRIEF HISTORY

The discovery of Genesis by Captain Ben Alan and the crew of the *Aurora* in 2240 adtc, did not bring universal jubilation on Earth. No statement illustrates the anguish this planet caused the pioneering movement than the following passage from Ben Alan's personal log.

> "Our excitement reached frenzied levels as we made final preparations to land on the planet's surface! The beautiful, blue sphere below us possessed breathable atmosphere, warm temperatures, no radiological hazards, and no evidence of intelligent life. As the landing craft began its descent, we stared intently at the main screen which amplified the view below. We broke beneath a layer of clouds and glimpsed our first clear view of the grey landscape.
>
> "Just our luck! We came down in a desert! Thinking that surely it couldn't go on forever, we pressed forward, travelling at 1200 kilometers per hour, 7000 meters above the surface. Four hours later upon reaching the seacoast, hydrocarbon scanners had not revealed the slightest chemical traces of life. Even Earth's most barren wastes would not have produced such readings. We continued parallel to the shore for fourteen hours more, circumnavigating the entire continent without sensing a living thing on the land. Yet carbon readings in the sea revealed some life there and proved our sensors were functioning. In desperation we touched down and still clad in our biosuits, stepped from the shuttle into an awful landscape littered with ugly black rocks and totally devoid of any life. We took microspopic samples from many places, but even stagnant pools of water in the rocks didn't reveal a single living cell.
>
> "Fatigued and feeling uneasy, we returned to the ship. The next day and for fourteen days after, I dispatched landing parties to the surface. At last the awful reality dawned on us. This planet is a desert. True, some elementary life exists in the sea which probably accounts for the oxygen atmosphere, but how could Humankind survive on those dreadful rock plains below?"

Aurora remained on Genesis for four months, studying what native life there was. When it returned to Earth, Alan's report classified Genesis uninhabitable, but he appended the following comment to his recommendation.

> "Despite the planet's inhospitable environment, I believe that someday Humans will live on it. The planet contains the *fundamental* conditions necessary to support our form of life. When our technology advances enough to allow us to transport much larger payloads across the interstellar space, then we will be able to bring enough equipment and enough supporting life forms from our mother planet to permit life as we know it to thrive there."

He went on to exercise his preogative as captain and named the planet, an unprecedented custom for a world considered uninhabitable.

For 20 years no planetologist challenged Alan's conclusion. The problems of colonizing his barren planet seemed insurmountable. The elementary shellfish of Genesis' seas could have provided some minimal sustenance to a Human population, but nothing approaching a normal Human diet could be cultivated on its barren continents. Humans need more than food too. Few would voluntarily agree to spend their lives in a desolate wasteland. Grass, trees, and other living animals may not seem like necessities of life, but early in the history of space travel, scientists learned how important they could be. After more than two years on the first permanent Martian base, the total bleakness of that planet's landscape began to have serious psychological effects upon the trained and experienced travellers that staffed them. No large and inexperienced group of pioneers could have coped with Genesis indefinitely.

Yet even before the discovery of Genesis, events were under way that eventually made its colonization possible. Contact with the Ardotians in 2217 adtc created a tremendous increase in the level of both Human and Ardot knowledge. Within a few years, the formulation of the Comprehensive Unified Field Theory led to the development of highly efficient matter-antimatter reactors. These reactors allowed people to transport far greater cargoes across interstellar space at a fraction of the cost.

Both Humans and Ardotians soon became interested in Ben Alan's barren planet again. They reasoned that Genesis provided a rare opportunity for an advanced civilization to create a biologically perfect world, a world without disease, pests, vermin, even weeds! It would provide the ultimate test of intelligent life's ability to shape and control its environment. Because of their scientific interest in Genesis, the Ardotians offered to supply engines for the largest starship ever built, if the people of Earth would undertake the planet's development and provide the life-support modules, equipment, and supplies for the vessel. All the Ardotians asked in return for their contribution were detailed reports about the progress of the experiment.

Earth's International Council for Space Exploration began planning for the first colony on Genesis soon after receiving the Ardotian offer; yet another 20 years passed before the launching of the first "Noah's Ark." The magnitude of the project seemed overwhelming. In the space of a few years, Humankind would attempt to leap half a billion years of evolution. Techniques for creating living soil from barren rock had to be developed. The proper mix of desirable Earth species had to be selected, and safeguards to insure that Genesis would not become contaminated by pests and diseases from Earth had to be refined. Finally, a large-scale evacuation plan had to be drawn up, should the entire project fail catastrophically. No detail escaped scrutiny. As a final check, Ardotian computers analyzed the entire plan, independently assessed its probable outcome and made several important recommendations.

Planners chose the southernmost tip of Harvestland for the first settlement, which pioneers named Malthus. Situated at the edge of the southern tropic zone, the climate is warm and the ocean is protected from violent storms. The colony organization followed the lines of a socialist democracy, similar in concept to the highly successful kibbutz used in the 20th century redevelopment of Israel.

The first 3000 pioneers brought food for five years, although the starship made the round trip to Earth annually, bringing still more food, equipment and new colonists. Pioneers lived in temporary housing constructed from parts of the ship that brought them, early precursors to the residential spires of today's pioneering vessels. In the early years, most efforts focussed on cultivation of Earth's native life forms. It took two years to prepare the soil for planting the first crops. At the same time, the pioneers began to develop aquaculture of both native and imported species to hedge against possible failure of the primary food supply. The first pioneers had no resources for manufacturing or processing industrial goods. Most of them were biologists or farming technicians, with a smattering of the mechanics, programmers, and comtechs needed to keep their equipment functioning.

Life's foothold on the planet was assured as food production became self-sustaining at the end of the fourth year. After that the slow process of building basic industries began, first with the importation of mineral recovery equipment, followed by critical manufacturing processes. Development of Genesis has proceeded steadily, if more slowly than on worlds more bountifully endowed by nature. Today, 92 years later, it boasts a modern, industrial society with many of the luxuries and conveniences of Earth.

CURRENT STATE OF DEVELOPMENT

Population

Genesis has not yet achieved material self-sufficiency. Highly complex types of equipment, such as fusion power reactors, computer mainframes, and chemical transformers must still be brought by starship from Earth. Approximately 1,423,000 people now call Genesis home. Of these 500,000 are immigrants. The population is distributed among five "developed areas" along the Aurora Straights. Malthus, the largest of these, numbers 500,000 people. More than 50 percent of the work force engages in land cultivation or creation.

Planning has shaped Genesis' development to a greater extent than any other colony. During the precolonial period, planners developed not only a highly detailed plan for the first 20 years of its existence, but a Master Plan that will govern development over the next 700 years! By the end of this period, Earth's lifeforms will have spread over the entire globe. The master plans calls for Genesis to achieve material self sufficiency in another 88 years, approximately twice the time it has taken the older colonies, better endowed with natural resources, to achieve it.

The living standard on Genesis cannot be compared meaningfully with Earth's or any other colony's. Genesis lacks the wide variety of consumer goods and services available on Earth. Though Genesis hasn't the abundance of desirable land found on other new worlds, the land Genesis offers pioneers is the prettiest and most pleasant environment of any known world, totally lacking in pests and dangers, totally conditioned for Human comfort. More than any physical inducements, however, Genesis offers pioneers the chance to *participate* in the greatest experiment in Human history, the creation of the best of all possible worlds!

The Master Plan calls for maintaining a ratio of 310 persons per square kilometer of developed land. Virtually all of the population lives within developed areas of the planet. Undeveloped land can be claimed by all citizens, but few claims have been filed and most of these have been for weekend retreats rather than permanent homesites. Some people attempt the backbreaking task of developing small plots of land. To do this one must carefully prepare and chemically balance the soil, then plant and cultivate the appropriate mix of plants, animals, and microorganisms to make it a living environment. This process generally doesn't work on plots smaller than three hectares, for the land will not be large enough to sustain self-perpetuating populations.

Housing

Housing is abundant and of high quality. Immigrants use spire structures of modern starships for the first years after their arrival. Despite the shortage of desirable land most people on Genesis live in small, single-family units on individual plots of ground. Large, multifamily dwellings set on large plots can be leased by pioneers who wish to avoid the expense and work of maintaining their own land.

Food

Food crops imported from Earth create an ample and varied diet for everyone. The Genesean diet contains little meat because Earth's food animals require large amounts of hectarage relative to the amount of usable food they yield. Sheep and goats provide some meat and dairy products, and imported fish species are raised in the sea. Since these products cost substantially more than on Earth, the bulk of protein consumed by people comes from grains, beans, and nuts of various kinds.

Communication

Transportation and communication create few problems on Genesis. Most of the population lives within small, well-defined geographic areas, and there are few places to go outside of them. Within developed areas, most pioneers get about in small, low-cost, hydrogen-powered, antigrav cars. These vehicles, with top speeds of about 80 kilometers per hour, are supplemented by large, high-speed antigrav transports that make the large jumps between developments. A large space station in synchronous orbit over the Aurora Straights provides communications channels between all developed areas. Information services and entertainment on Genesis compare in quality with those of Earth.

Energy

Energy comes from large power plants in each development and is channeled to individual users through an optical frequency distribution system. The tiny, individually-owned power plants of the other colonies need not be used because of the relative compactness of Genesis' population.

Consumer Goods

Consumer goods remain limited by Earth standards, but basic items such as clothing, vehicles, preservators, food-preparing appliances, space conditioners, sleep fields, and dataceivers can be purchased by all. Though the centrally planned economy dictates how resources are apportioned between economic sectors, detailed production of many consumer items is left to privately owned and operated enterprises and is sold through private retail outlets.

Health

Medical care, available at no charge through government supported health centers, keeps the population fit. The level of sickness is low, despite the fact that the planet's rarefied atmosphere poses a potential health hazard. The thin blanket of air makes a less effective shield against the short-wave radiation from Genesis's sun. Consequently people who work outside suffer higher incidences of sunburn and skin cancer. Fortunately the latter disease, once a major health threat, is as easily treatable as a nicked finger, and the medical community has developed several innovative techniques for reducing the cost of this treatment still further.

Government

The government of Genesis consists of a Central Committee and of local steering committees, elected by majority vote of the adult population. Agencies of these committees make all major planning decisions for the planet, except for revisions of the Master Plan, which must be ratified by two-thirds majority vote and approved by GAIL.

Research and Development

Scientific research, financed by the government through the Genesis Scientific Research Organization, is essential to the colony's survival. Biology constitutes the principal field of effort, but other natural sciences, such as geology and meteorology, receive high priority as well. While Genesis cannot claim to have originated many significant industrial processes or inventions, it has contributed to a better understanding of the development of species, the origin of planets and the biological interactions between forms of life. For the first time, people have been able to evaluate the consequences of encouraging or eliminating certain life forms in a controlled environment. Preserving the proper balance of nature has proven as important to the survival of Earth as it is to Genesis.

AN ACCOUNT OF LIFE ON GENESIS

Editor's Note: Megan Kearin, 52, director of the Darwin Institute for Evolutionary Studies, gives an account of life on Genesis. Born of immigrant parents, Megan spent her early childhood in Malthus, then moved with her family to Orb in that development's first year. She returned to Malthus to attend the University of Genesis, earning BS, MS and Scd degrees in biology. Her early career as a field biologist for the Genesis Planning Commission took her over much of the planet, cataloging native life forms. She later assisted in the planning of the first hardwood forest at Port Aurora. During her field years, she formulated theories and collected data to substantiate her most important work.

At 32, she was assigned to the Advanced Studies Group, and during the next seven years prepared her now famous report, "A Deterministic Model of Darwinian Theory," which developed the first theory for predicting the origin of species. The importance of this work became evident throughout the Galactic Association, and two years after its publication, the Genesis Central Committee established the Darwin Institute to study evolution and to predict the interaction of life forms brought from Earth. Megan, first named an Associate of the Institute, became its Director five years later. She has played an important role in determining which species will be introduced on the planet, decisions which have profoundly shaped the future of Genesis.

When you grow up on Genesis, you grow up with an awareness that your world is different—special. In my childhood, most of the people I knew engaged in the toil of our planet, creating life from non-life. Even though I was born amidst neatly manicured grass and trees, farms and pastures, I knew my home occupied little more than a green island in a rocky sea. Yet I'll never forget my wonder at first seeing that "sea" myself.

It happened on a lazy Sunday in my seventh year when nobody in the family seemed to have much to do—a rare occurence at our home! I recall talking to my father about his work when he asked me if I'd like to see something of it. Naturally I jumped at the opportunity.

We took the family aircar and headed away from the town center. We rode no more than 15 minutes before I noticed that the houses and occasional trees had stopped and that we were cruising over a flat, seemingly endless plain of grass. Within a few minutes I saw the grass end at a great grey expanse of rock bordered by a "beach" of light grey sand. Father slowed the car as we passed over a cluster of low, mobile shelters and headed parallel to the beach. We soon came upon a cluster of huge, weird-shaped machines working, even though it was Sunday. Most of them had great wide mouths pressed low to the ground. Some carried huge hoppers filled with millions of tiny varicolored pellets. As they moved over the beach they devoured the sand, did something with it inside of them, and dropped it out behind.

Father turned left, and we headed away from the first machines toward a still stranger device that looked like a giant centipede. The machine crawled slowly over a plain of rock strewn with small boulders and coarse gravel. Each of the machine's segments worked on the rock until they had ground it finer than beach sand. Beyond the "centipede" not a single blade of grass or other living thing interrupted the monotonous expanse that vanished in a distant wall of dark, lifeless mountains.

Finally Father climbed slowly upward until we hovered thousands of meters in the air, higher than I had ever been before. As I gazed down upon the stone planet curving beneath me, I didn't speak with my father but shared his pride and confidence. I felt proud of his work, excited about what I could someday contribute and happy to be a part of this greatest scientific experiment.

Yet the experiment of Genesis affected my childhood very little. The first decade of my life remained calm and uneventful. I participated actively in community sports, mostly water soccer, optel, and geomets, and I had quite a few friends. My parents emigrated to Genesis when just barely out of their teens. Neither of them possessed much formal education, but they believed that Humankind could better itself through science and technology, and that if Humanity's greatest scientists and technologists set about to create a better world than Earth then they would succeed. My parents desired, with an almost religious fervor, to help in some small way with the creation of their brave new world, and they wanted their children to grow up with its benefits.

The development at Orb commenced in my 11th year, and the Bureau of Human Resources requested my parents to work on it. The thought of moving alarmed my four brothers and sister, but I looked forward to it with excitement! I imagined all sorts of new adventures for us and was sure the change would be good for the family.

I was right. Orb bustled like a gold rush town. A cluster of buildings roughly a kilometer across sprawled over a flat plain of rock near the seashore. In every direction, work crews toiled at turning the barren rock into soil. Everyone in the community, even children, played a part in the building. Each day during school students got two hours off at lunchtime to make sandwiches for the work crews and take them to the fields. Instead of playing at sports and games, we spent our weekends planting shrubs and trees, or caring for the animals, like birds and worms and bees, that would soon be introduced into the environment.

The work was fun, earned us some money and provided an opportunity to make friends. The greatest satisfaction, though, came from feeling like a part of the new community, and taking personal pride in the progress that we could see advancing each passing day!

In two years I watched Orb grow from a temporary camp to a city. Working at speeds previously unattained, the soil-preparation crews prepared nearly 300 square kilometers of soil for planting. A permanent city center grew from the original town site where the solid rock made an excellent foundation for large structures. Extensive mariculture operations got underway in the nearby ocean, and an electronics plant and an opticable manufacturing facility commenced operations.

Though the early days at Orb appeared confused and disorganized, the area development plan guided every action from the moment the first rock crusher bit into the ground. By the time I moved away, seven years later, the map of Orb (figure 3.12) clearly illustrated how precisely the plan had been followed. My parents still live in the development, and I go to visit them now and then. The area has expanded to several times its original size, but it has maintained the geometrical pattern of its creators.

When we arrived at Orb, Father bought us a portable modular dwelling, which we moved six times during the first two years. He promised that one day we would have a large and permanent home. After two years of waiting we finally got our dream house and two hectares on gently rolling hills. Only prep grass grew on the land when we set the permanent foundation, but we lost no time before plowing that under and setting to work planting trees, shrubs, and a vegetable garden.

At 17 I was ready for college and enrolled in the University of Genesis at Malthus. All students on Genesis whose grades and aptitude tests place them in the upper 30 percent are eligible for admission to the University, although only about 20 percent are graduated from it. The tuition is paid by the government, but except in unusual cases, students or their families must provide for all other expenses.

The University sits on its own 180 hectare park and functions in many ways as a world in itself. My family had experienced little contact with it during my childhood. Now, as a student, I had little contact with the life of "normal" Malthus. Like everything on Genesis, this isolation exists by design. As the principal institution for producing the young scientists and technologists who will control the future development of Genesis, the University plays a key role in the continued survival and prosperity of the planet. The founders desired to establish a self-contained community of scholars, whose presence would stimulate each other to greater levels of creativity than each would achieve in isolation.

My major in biology sprang naturally from my love of plants and animals cultivated by childhood chores at Orb. My interests ran not so much toward physiology or biochemistry as toward macrobiology and ecology. I studied botany, zoology, and the schemes of classification that trace their roots to Aristotle. The interrelationships between living organisms fascinated me most, for these relationships governed the fundamental plan for Genesis's development. As I pursued my studies of these traditional fields, I realized that I needed an understanding of genetics and biochemistry to understand the mechanism that created the multitude of life forms. These disciplines dictated that I learn higher mathematics and computer science.

My years at the University developed ideas and convictions that I hold to this day. Every class I took, every discussion with a professor or fellow student played a part in my intellectual development. Yet two men stand out in my mind as having an overwhelming influence upon me.

I met the first of these, Dr. Herman Zweig, in my fresh year. As a professor of philosophy, Dr. Zweig has analyzed and clearly explained the philosophical underpinnings of our way of life on Genesis. He has shown why a planned society, in a planned and controlled environment, can maximize Human happiness and eliminate suffering, affliction, and inequality. On Earth, planned societies historically have failed. Zweig showed me, and scholars everywhere, how, in the absence of control over nature, all planned societies must fail. But if Humans possess total control over their environment, then planning can create a society whose citizens will live happier and more satisfying lives than is possible on worlds with natural environments.

The first time I heard Zweig speak I felt as if someone had taken a blindfold off my eyes! Suddenly I knew *why* I always had been so proud of what people were doing on Genesis, and I saw

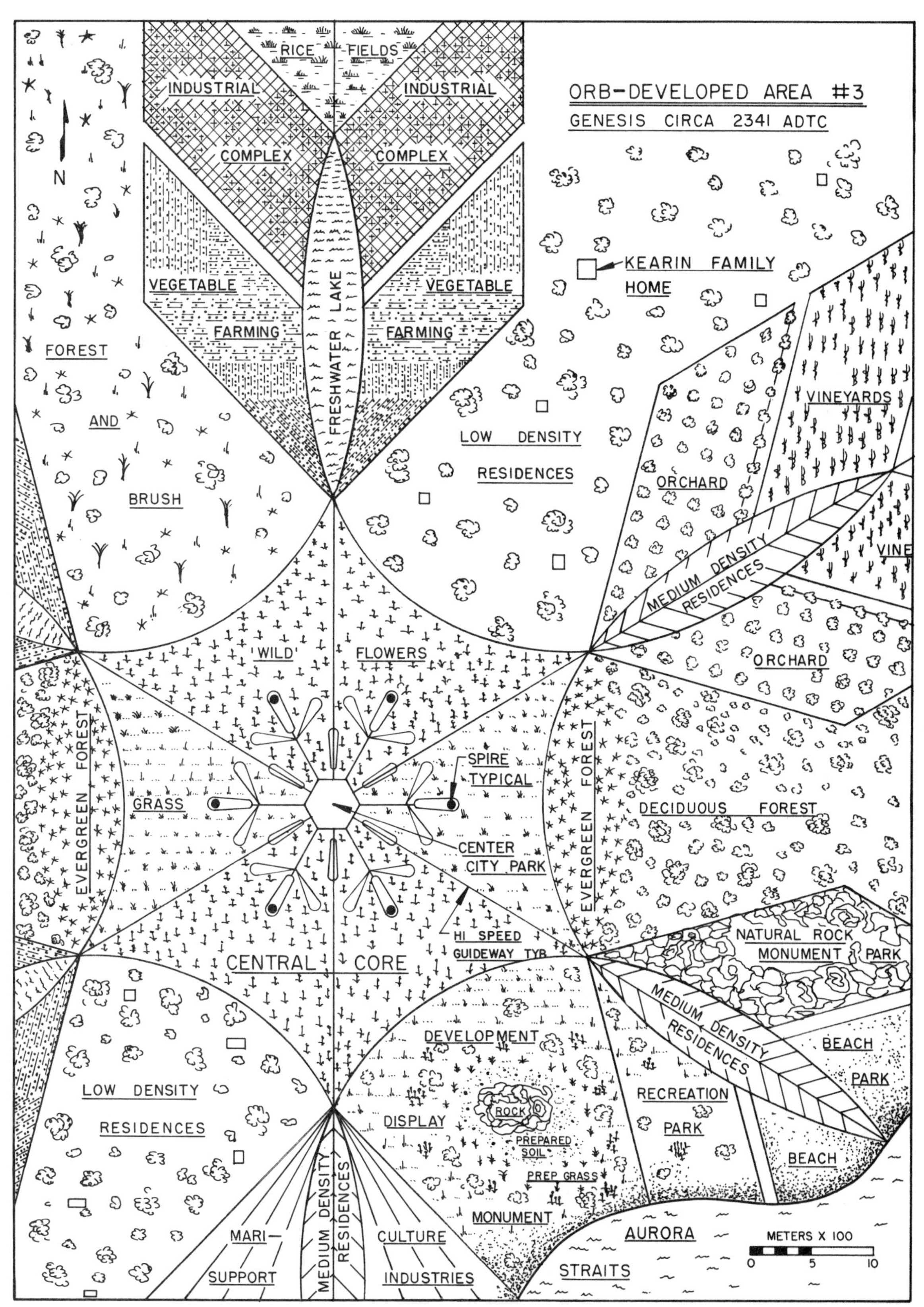
ORB-DEVELOPED AREA #3
GENESIS CIRCA 2341 ADTC
N
RICE FIELDS
INDUSTRIAL COMPLEX
INDUSTRIAL COMPLEX
VEGETABLE FARMING
VEGETABLE FARMING
FRESHWATER LAKE
KEARIN FAMILY HOME
FOREST AND BRUSH
LOW DENSITY RESIDENCES
ORCHARD
VINEYARDS
VINE
MEDIUM DENSITY RESIDENCES
ORCHARD
WILD FLOWERS
GRASS
SPIRE TYPICAL
CENTER CITY PARK
EVERGREEN FOREST
EVERGREEN FOREST
DECIDUOUS FOREST
HI SPEED GUIDEWAY TYP.
CENTRAL CORE
NATURAL ROCK MONUMENT PARK
MEDIUM DENSITY RESIDENCES
BEACH PARK
DEVELOPMENT DISPLAY
ROCK
PREPARED SOIL
PREP GRASS
MONUMENT
RECREATION PARK
BEACH
LOW DENSITY RESIDENCES
MARI-SUPPORT
MEDIUM DENSITY RESIDENCES
CULTURE INDUSTRIES
AURORA STRAITS
METERS X 100
0
5
10

FIG. 3.12

that the importance of our task transcended the limits of one planet. The success of the Genesis experiment will lead Humankind throughout the galaxy to a new milennium of advancement and will change the criteria by which Humans choose habitable planets.

The other outstanding teacher of my University years was Dr. Li Cheng. This widely esteemed professor of biology was the most influential advisor to the Genesis Planning Commission, the agency responsible for implementing the Master Plan. We first met when I took his famous course, "Introduction to Ecosystem Modeling," and later I was honored to have him supervise my doctoral thesis. Dr. Cheng gave me a deep appreciation for the difficulty of the task we faced on Genesis. The great complexity of biological interactions makes it impossible to predict with certainty the effect any one environmental action will have, yet one serious mistake could so disrupt the planet's fragile ecology that Genesis might have to be abandoned. Dr. Cheng, more than any other person, solidified my resolve to contribute to Genesis by increasing our understanding of the interactions between species.

Because he constantly solved real problems for the Planning Commission, Cheng often assigned these problems to his students as exercises. It thrilled us to be working on actual studies that might conceivably affect the future form of our world. In my first course, we modeled the possible effects of introducing a second species of pine tree into the Pt. Ica development. Such work became so heady that I stayed at the University eight years, earning my doctoral degree for a thesis which examined the possible effects of seeding the Genesean ocean with a particular subclass of crustaceans. This and the work of three other students later formed the basis of a large-scale marine experiment in the Gulf of Virginis.

After eight years of relatively easy, though intellectually stimulating, living I craved some variety and excitement. I found both in my first job as field biologist with the Genesis Planning Commission. The Commission needs detailed knowledge of Genesis' biology and geography in order to assess potential development sites and to control the introduction of new life on the planet. Raw data collected by the Commission's field scientists forms part of the data base of the planet's central computer. This is used by all governmental, scientific, and industrial organizations and is finally transmitted to GAIL.

I was assigned to the RV *Genesis Explorer*, one of the Planning Commission's three research vessels. An odd sort of hybrid, it normally travelled in the air supported by antigravs, but it could also float upon the water propelled by hydraulic drives. The vessel's ship-shaped hull with tapered bow and stern measured 130 meters long. Its maximum air speed of 110 kilometers per hour, a crawl by modern standards, sufficed for its mission, and allowed us to outrun storms with ease. Scientific equipment on board included full-spectrum sensor arrays for gathering meteorological, oceanographic, and geophysical data; core drills; biological samplers and analyzers; two submarines; free-diving equipment for 20 divers; a powerful onboard computer for real-time data analysis, and two smaller, high speed air-trucks equipped with data gathering apparatus.

I was one of six staff biologists in a crew of 70. Our work involved solely marine life, since that's all there is in the undeveloped regions, but we were also charged with gathering sufficient information about land areas to evaluate their suitability for Earth species. Typical cruises lasted six months, followed by a return to Malthus for a month of vacation. Life aboard ship was relaxed and informal with superb food, and opportunities to learn something occurred every day.

Though tragedy marred it, I remember our first expedition to eastern Barrenland most vividly. The first two months of the trip we cruised slowly on the sea, studying mid-ocean life. Periodically we stopped for a day to collect live specimens. During this cruise, I learned to dive with a gill, and for the first time encountered face to face the myriad sea creatures I had studied for eight years.

The sea life of Genesis, so small and simple by Earth's standards, amazed me with its beauty and diversity. During the first voyage I personally catalogued more than 150 species of previously unclassified shellfish. The greatest diversity occurs among the microorganisms, which are nearly as numerous as on Earth. The biochemistry of marine life on Earth and Genesis is similar enough so that Earthly species of fish and marine mammals could be introduced into the sea. This has been done only in limited cases with the fish contained by double sonic barriers, for scientists worry that these alien species might so drastically upset the food chain of the oceans that the planet's oxygen supply could be threatened.

Upon arriving on the coast of Barrenland, the work pattern changed. The life forms at the water's edge hold greatest interest for biologists, for these are the species from which future land life on Genesis might have emerged. Their possible similarity to our own distant ancestors who crawled from Earth's primordial seas makes them the subject of much study, even though such investigations have little practical consequence.

We made our landfall on the eastern coast of Barrenland at about 43 north latitude. The dawn's first rays illuminated stark, rocky cliffs plunging hundreds of meters to the sea. After stopping to take life samples from the base of the cliffs, we followed the geologists to the tops to survey the land. We found a broad stone plain that had appeared on the satellite maps of the area, and spent the better part of the day there looking vainly for signs of native life.

As we proceeded northward along the coast, we repeated our alternate examinations of sea and land. Our destination was a place that satellites had identified as a possible development site. The data we gathered there and on subsequent trips to different areas is used to evaluate sites that won't be developed in any living person's lifetime, so long-range is the Planning Commissions effort. Development sites must have several important characteristics: adequate rainfall, mild temperatures year around, a coastal plain bordering on a bay or ocean with gently sloped beaches, an absence of active volcanoes, and an absence of flash flood zones. Though such areas are not too common, the Commission insists all these characteristics be found before the large investment in development is made.

As we approached our destination, the high cliffs bordering the ocean dropped slowly until we cruised past a broad plain ringed by a ridge of mountains barely visible in the distance. The spot where we landed became our base for 60 days while we studied the area in detail. The small levicars made expeditions hundreds of kilometers inland looking for evidence of flash floods. Because Genesis has no land vegetation and therefore no soil, much of its surface is not porous. Most of the rain that falls turns immediately to runoff which heads towards the seas in destructive torrents that would obliterate any Human structure in their path. It rained several times during the observation period, and the entire crew turned to mapping the runoff paths, both during the storm and for a whole day afterwards. We observed the lay of the land in great detail and drew up several provisional development plans.

After completing our study of the development site, the *Genesis Explorer* turned eastward to cross the continent of Barrenland. The biologists had little to do on this part of the trip, except to monitor the life sensors and to learn as much about geology as the geologists would teach us. The color of the landscape changed little as we glided over it. We passed over mountains, high plateaus, and valleys, but no snow and few rivers could be seen at these latitudes. As we approached the western coast we glimpsed a large river coursing between the rock walls of a narrow canyon. This as yet unnamed main artery of the continent is one of the few rivers that maintains some flow year around.

Our fateful return trip to Malthus began uneventfully. We had skirted the southwest coast of Barrenland and crossed most of the North Genesean Ocean when an urgent distress call from Port Fish interrupted our cruise. West of Port Fish lies the largest of several experimental marine reserves, areas of ocean where scientists have introduced varieties of commercially useful fish, along with a mix of other life forms on which the fish feed. The Port Fish experiment had not worked well to date. An imbalance between higher and lower order consumers had greatly reduced the photoplankton activity within the reserve. A double sonic barrier, impervious to all marine life, surrounds the area and prevents this imbalance from spreading to the rest of the oceans, but the unthinkable, happened. A power interruption, compounded by incorrect settings on the back-up power control, had deenergized the barrier for about half an hour. During that period, several schools of fish escaped from the reserve and had to be recaptured or destroyed. The escape of so much as one fertilized female fish into the open ocean might ultimately spell disaster for the entire planet.

The research vessel assigned to the reserve had more trouble than it could handle so the *Genesis Explorer* streaked toward the barrier to help. For a short time pandemonium reigned aboard our ship, but the captain wisely placed Chief Biologist Sandra Lee in charge of the operation. Sandra ordered the technicians to get the submersibles ready and to reprogram the life scanners to detect Earth-like species exclusively. Then she called us to the briefing room.

"You all know what it might mean to Genesis if these Earth-forms escape to the open sea," she began gravely. "When we get to the barrier, those fish will have two hours head start on us, and rounding them up won't be possible anymore. The best we can do is destroy them, and it'll be no easy trick to do that."

She went on to explain about the details of the escape and outlined our plan of attack. We would use the sonic disrupter fields on the ship and the submersibles to kill the schools of fish. These devices, normally operated at low power to stun specimens before taking them, would be readjusted to lethal intensity. She dispatched herself, along with pilot Jack Gavin, in submersible number one and ordered Alfred Abegg, senior staff biologist, to command the number two sub. She assigned the other more experienced biologists to plot probable trajectories of the schools on the ship's computer and relay this data to the ship's helm and the subs. I was to watch the scanner in the undersea operations room and to aim the ship's disrupter fields. Her last grim remarks at the briefing weighed heavily upon us.

"This is going to be a very long night, and we're all going to have to push ourselves to the limit. Remember, one way or another, our lives are on the line. We cannot let up until we've gotten them all!"

When we arrived at the scene of the accident, Sandra was already aboard the sub, and she shot out of the launch tube as soon as we spotted the first school of fish. In the dim light of the undersea room, I watched the large, green display which showed the positions of the subs and the fish in bright yellow outlines. At first things went pretty well. Most of the schools hadn't strayed too far or descended below 500 meters. It seemed easy to track them across the board, focus the disrupter field and annihilate them with the press of a button. Yet as the hours wore on, the schools grew smaller and more remote. We did a lot of chasing around to find them all, and sometimes shadows and cold currents deceived the sensors, sending us on wild chases that wasted precious time.

By early morning, I felt cross-eyed with fatigue. Our final sweeps of the area indicated few Earthly readings, and we quickly took care of those. One of the last appeared to be a sizable school of anchovies that, frightened by the noise above, had descended below the 1500 meter mark, the crush depth of the submersibles.

Sandra's voice sounded tired as she spoke over the comchannel. "We're going after that last bunch of sardines, Meg, how deep are they?"

"1700 meters, Sandy," I replied. "You'd better wait 'til they come up. Crush depth is 1500, and you won't get a good focus on your beam from that far above."

"Negative, kiddo," she answered. "That school's pretty flinchy and looks like it might break up. If that happens, they're all so small we'll never chase 'em down. I think we can push a little deeper than crush depth. These educated beer kegs are good for more than the engineers say anyhow!"

There's no arguing with Sandra, but tension knotted my stomach as I watched the little yellow cylinder move down the display, past the red line that marks extreme danger. The sub moved slowly and cautiously into position. Sandra took a few extra moments focusing the sound beam to make sure she didn't miss part of the school and scatter them. She fired, and the cloud on my display that symbolized the tiny fish faded.

"Kill confirmed, Meg," she reported matter-of-factly. "We're on our way up!" Seconds later a crunching pop punctuated her last remark, and the sub began to sink.

"Sandra! Are you all right? What's happening?" I screamed hysterically into the microphone, but the comchannel alarm winked back at me, indicating no receiver had picked up my voice. At once I knew what had happened; the sub had imploded. Word of the tragedy spread quickly, and a deathly silence fell over the ship, broken only by the muffled sobbing of one lonely, young girl.

Sandra's death stunned and horrified me. During our first voyage, we had grown much closer than supervisor and subordinate. I loved her as a friend as well. For several weeks after my return to Malthus, I considered quitting my job. I didn't know how I could live with the ship, the crew, and the sea again. With time, my grief and hurt began to subside, and I started to think clearly again. The experiment of Genesis was not just an intellectual exercise for Sandra, but a commitment as important as life itself. My failure to continue my important field work would not be a tribute but an insult to the ideal for which my friend gave her life.

I made eleven more voyages on the *Genesis Ex-*

plorer. The uniqueness of each trip lay not so much in its outstanding adventures as in its fine details: the new life forms I discovered, the interesting rock formations along uncharted shores, the tropical sunsets bursting with rainbows. On nights as black as only nights at sea can be, I loved to lie on the deck watching the moon belt revolve overhead.

As wondrous as Saturn's magic rings, the moon belt of Genesis appears as a ribbon of bright lights moving against the backdrop of stars. The belt changes constantly, due to the different speeds at which the moons orbit. Clusters and thin spots pass overhead in dazzling display that astronomers say will not repeat itself in 20 million years!

On a later expedition to southern Maiden Spring I saw the magnificent aurorae of Genesis. One thousand times brighter than Earth's, it blankets all but the brightest stars. Though the aurorae exists thousand of miles out in space, it seems to touch the ground, coloring objects there with a fairy blanket of light.

Even though I enjoyed my field years immensely and believe they did much to sharpen my powers of observation, after six years I yearned to settle down and build something permanent. When I applied for transfer, the Planning Commission assigned me to the Port Aurora region. The creation of Port Aurora had begun ten years earlier, but now the Commission wanted to develop a hardwood forest to the south, the first such project attempted on Genesis. The forest was to be planted with deciduous trees, like oak, maple, elm, and beech—typical of the middle latitudes of Earth's northern hemisphere.

Success of the experimental project would yield three principal benefits: Since forests retain more groundwater than grasslands, the ability to cultivate them would provide Genesis with a more effective means of flood control. The forest would also serve as a popular recreational area, far more picturesque than the coniferous forests planted to date. Lastly, hardwood from the forest would have valuable decorative uses.

The area to be forested lay in the foothills of a chain of low mountains southeast of the existing developed region. The hilly terrain presented a few technical problems, but the project's principal difficulties were biological. Although Humanity has cut down many forests on a grand scale, it hadn't created any new ones. Reforestation techniques dating back to the 19th century had been perfected for smaller-scale projects in established forest areas, but an endeavor of this magnitude, on barren land, lacked historical precedent.

I lived on the edge of the Port Aurora development in a small prefab house which I owned cooperatively with four other project scientists. We had arrived at the forest site before any other workers to study the terrain, analyze drainage patterns, and develop a day-by-day schedule for soil preparation and planting.

After six months, soil preparation crews and equipment began to work. Giant combination soil preparation machines, larger than any previously built, performed the task. (See figure 3.13.) In one operation, these mechanical monsters broke rock into coarse chunks using laser drills, then ground it fine with ultrasonic grinders and mixed it with nutrients and mulch from giant hoppers they dragged behind them. In past projects, the pulverization depth reached about a meter below the surface, but the depth to which these large trees' roots would grow dictated that rock be broken up three meters deep and pulverized to a depth of 1.5 meters.

Earth's natural soil is very fine-grained, but organic mulch within it inhibits its natural tendency to compact and congeal into a solid mass. On Genesis, soil technicians must add a synthetic mulch prepared from seaweed. Since the mass of the synthetic humus cannot equal the mass of natural humus, special chemical treatments must augment it. The addition of fixed nitrogen and other necessary plant minerals follows the mulch, and the soil is chemically balanced. As a final step, technicians seed the new soil with a hardy grass, specially bred from Earth grasses, called "prep grass." After two years the prep grass is plowed under to provide further mulch and natural nutrients to the soil. The addition of bacteria and worms of appropriate species complete the formation of synthetic soil.

While the prep grass grew we began sprouting seedling trees in the mature soil of the developed area not far from my home. Since no natural barriers existed to blunt the ferocious gales that blew off the south Genesean Ocean, protecting the young trees from wind damage became our most pressing problem. Solid shelter of any kind would have been prohibitively expensive, but fortunately a wind-breaking field had been developed a few years earlier. The field employed a special configuration of the g-field to slow inrushing air to a

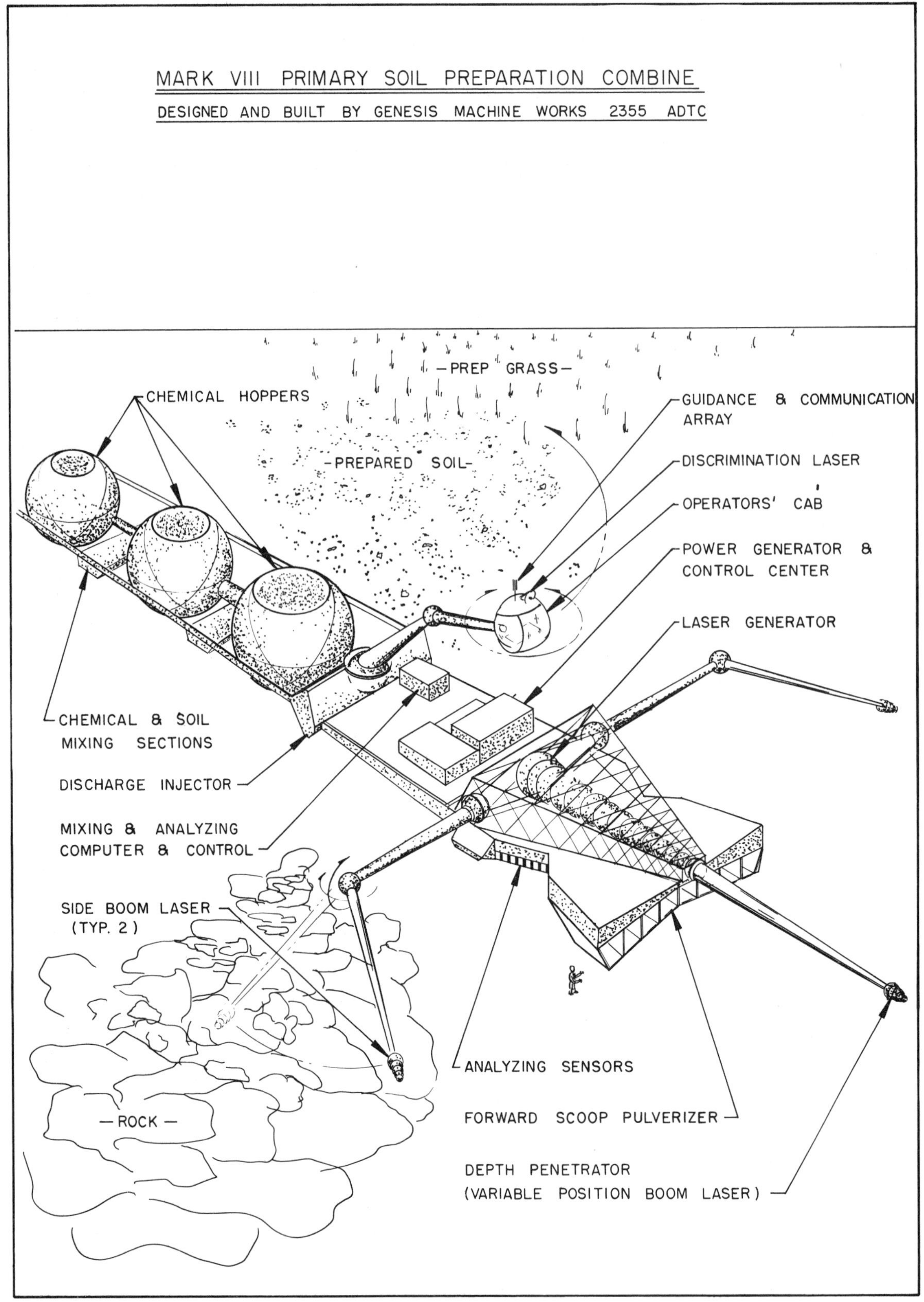
MARK VIII PRIMARY SOIL PREPARATION COMBINE
DESIGNED AND BUILT BY GENESIS MACHINE WORKS 2355 ADTC
-PREP GRASS-
CHEMICAL HOPPERS
-PREPARED SOIL-
GUIDANCE & COMMUNICATION ARRAY
DISCRIMINATION LASER
OPERATORS' CAB
POWER GENERATOR & CONTROL CENTER
LASER GENERATOR
CHEMICAL & SOIL MIXING SECTIONS
DISCHARGE INJECTOR
MIXING & ANALYZING COMPUTER & CONTROL
SIDE BOOM LASER (TYP. 2)
-ROCK-
ANALYZING SENSORS
FORWARD SCOOP PULVERIZER
DEPTH PENETRATOR (VARIABLE POSITION BOOM LASER)

FIG. 3.13

standstill in the space of two decimeters. Unfortunately, the field would kill anyone who accidently walked through it, so its perimeter had to be guarded by sensors, coupled to visual, sound, and telepathic alarms to warn stray children or absent-minded scientists who might not heed posted warnings.

After two years huge transplanting machines began to move the seedlings from the nursery beds to the forest site. The machines passed over the seedling beds, picking up the young trees together with their roots and a clump of soil, then travelled out to the newly plowed forest and deposited the seedlings in holes dug by the machine itself. Each machine carried about 1000 seedlings and planted at the rate of two per minute.

The machines planted the trees in a predesigned pattern that allowed for optimum tree growth and insured that mature trees would someday provide natural shelter for future seedlings, thus avoiding the need to raise the young trees in a sheltered nursery. A major conflict arose among the forest planning staff over the pattern in which the trees would be planted. Some wanted to plant the trees in regular rows resembling a European garden, while others wanted an irregular, pseudo-random pattern that would resemble a natural forest. The regular pattern made the trees easier to care for and would have cost less, but fortunately the "naturalists" won out. The forest, now quite mature, is one of my favorite vacation spots on Genesis. It gives me great satisfaction to walk among the towering trees and feel that I helped place them there. Despite the fact I aligned with the ordered pattern camp, I am glad that the trees now grow in a naturalistic pattern, for I have come to appreciate the marvelous "order" in nature's randomness.

During my years as field biologist, the theory of the origin of species, first advanced by Darwin four centuries ago, continued to fascinate me. I kept notes of my observations and hypotheses on the subject and hoped someday to evaluate them rigorously. That chance came with my transfer to the Advanced Studies Group of the Genesis Scientific Research Institute (GSRO) in Malthus. The ASG became interested in a summary I had written proposing an hypothesis for the mechanism of species creation and offered me a position as a research associate. For the next seven years I poured through computer files containing field data and tried to substantiate my theory. The result appeared in a 100-frame report that shaped the rest of my life.

Though biologists throughout GAILE heralded my first work as a "breakthrough," it formed only a rude beginning for the research I have continued to this day. In response to it, the GSRO created the Darwin Institute for Evolutionary Studies and made me a permanent associate. I did not become its director until five years later, that position going to a more experienced administrator, Dr. Paul Chin. In retrospect, I am glad I did not become the first director, for I was quite unschooled in the ways of the GSRO bureaucracy. Paul's stewardship taught me a great deal about it, and his politic handling of the early years earned the Institute the funds so critical to a new organization.

The Darwin Institute provides important input to the Planning Commission's decision making. Many of our projections look forward thousands of years, although in cases where the introduction of a new Earth species might devastate a native species, the effects can be much more immediate. The Earth species brought to Genesis have embarked on their own evolutionary paths and will become quite different within 50,000 years or so.

My life today combines the best of all the lifestyles I have known on Genesis. I still direct the Darwin Institute and shall probably hold the Director's position for several more years. My work allows me to travel all over the planet and to watch my world grow and progress. My large modern home sits on a hectare of land near Malthus City, now three times the size it was when I was born. I have become more active in local government, for I believe we must begin to develop governmental policies to cope with our increasing industrialization. As we emerge from a biological experiment to become a self-sufficient, industrialized world we must continue to deal with the fact that our natural environment is still young and fragile.

OPPORTUNITIES FOR IMMIGRANTS

The Central Committee of Genesis strictly limits immigrants to those with skills in short supply, yet opportunities exist for a wide variety of skilled tradespeople and professionals. Quotas of needed skills are revised quarterly and prospective applicants are encouraged to review the special publication GAI-SP-8-8336, "Open Skill Classifications for Genesis."

Developed land is parceled out to all residents of Genesis who apply for an allocation. Waiting lists range from six months to three years for plots ranging from one to four hectares, depending on the location. Though no charge is made for developed land, high taxes are levied to pay the substantial costs of preparing it. Land may not be transferred via sale, but improvements such as landscaping or immovable structures may be sold.

GAILE provides transportation for 6000 immigrants annually in a starship of the double hex configuration. Travel time requires 64 days of ship time and 92 days of planet time.

Mammon—Galactic Treasurehouse

IMPORTANT STATISTICS

Planet name: Mammon

Equatorial diameter: 9880 kilometers

Mass (Earth = 1) : 0.41

Surface gravity: 0.69 g

Escape velocity: 8.16 km/s

Albedo: 0.32

Atmospheric pressure at mean sea level: 0.34 bars

Fraction of surface covered by land: 48%

Maximum elevation
above sea level: 7642 meters

Length of day: 22 hours 8 minutes

Length of Year: 349 Earth standard days

Obliquity: 5 degrees

Von Roenstadt habitability factor (Earth = 1): 0.79

Current population: 1.5 million

Number of mining/industrial centers: 9

Year settled: 2318 adtc

Number of moons: zero

Star name: Zeta Triangulum Australe

Star type: GOV

Distance from Earth: 39.3 light years

Distance from planet: 1.02 AU (Earth-Sol = 1 AU)

Star Mass: 1.16 (Sol = 1)

Travel time from Earth

Ship's time: 63 days

Planet time: 92 days

Atmospheric Composition

Oxygen: 51%

Nitrogen: 46%

Carbon dioxide and inert gasses: 3%

Table 3.6

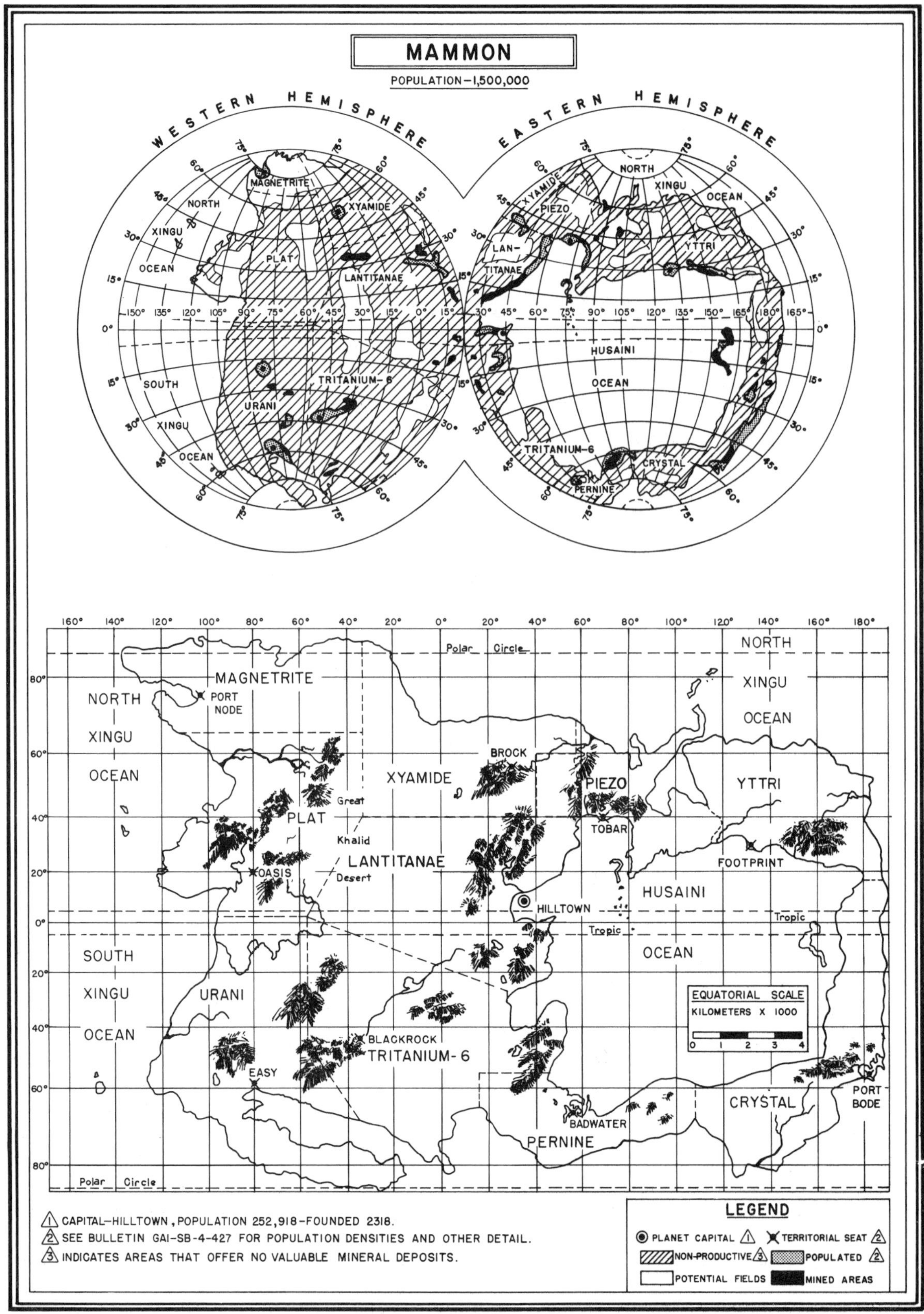
MAMMON
POPULATION—1,500,000
WESTERN HEMISPHERE
EASTERN HEMISPHERE
MAGNETRITE
NORTH
XINGU
OCEAN
XYAMIDE
PLAT
LANTITANAE
TRITANIUM-6
URANI
SOUTH
XINGU
OCEAN
PIEZO
LAN-
TITANAE
YTTRI
HUSAINI
OCEAN
CRYSTAL
PERNINE
Polar Circle
PORT NODE
BROCK
Great
Khalid
Desert
OASIS
TOBAR
FOOTPRINT
HILLTOWN
Tropic
BLACKROCK
EASY
BADWATER
PORT BODE
EQUATORIAL SCALE
KILOMETERS X 1000
CAPITAL—HILLTOWN, POPULATION 252,918—FOUNDED 2318.
SEE BULLETIN GAI-SB-4-427 FOR POPULATION DENSITIES AND OTHER DETAIL.
INDICATES AREAS THAT OFFER NO VALUABLE MINERAL DEPOSITS.
LEGEND
PLANET CAPITAL
TERRITORIAL SEAT
NON-PRODUCTIVE
POPULATED
POTENTIAL FIELDS
MINED AREAS

FIG. 3.14

Mammon—Galactic Treasurehouse

PHYSICAL ENVIRONMENT

Mammon, the fourth planet of Zeta Triangulum Australe, a type GOV star slightly larger than Earth's sun, looks out upon an undistinguished sky. The planet lacks any moons, and its primary is a single star system. Mammon lies far enough from Earth (approximately 39.3 light years) that the constellations familiar to Earth people appear quite distorted when viewed from Mammon's surface.

Mammon's physical characteristics present a series of extremes. The smallest of the colony planets, its diameter measures 78 percent of Earth's, giving the planet about 60 percent of Earth's surface area and only 41 percent of Earth's mass. Gravitational attraction at the surface measures only 69 percent of Earth's, the lowest of any habitable planet. Oceans cover only 52 percent of Mammon's surface, making it the driest among the new worlds. The small oceans allow Mammon's land area to almost equal Earth's, despite the colony's smaller size.

Mammon's two separate oceans create the most distinctive feature of its geography. On most planets, individual oceans form part of one continuous system which separates the continents. On Mammon, the reverse is true. One can travel exclusively on land to 99 percent of the planet's dry surface, while the planet's oceans lie separated by expanses of land ranging from 700 to 11,000 kilometers across. Since the two oceans seek two different levels, the concept of "mean sea level" has no physical significance. Mammon's maps show a reference elevation known as "mean sea level," computed as the weighted average of the areas and elevations of the two oceans, but not equal to the level of either of them. The planet's sun exerts the only significant tidal forces on Mammon's seas, but its effect seems small compared with the effect Earth's moon has on Earth's oceans.

Though scientists believe Mammon to be geologically younger than Earth, its surface cooled and solidified much earlier in its development. The rapid cooling apparently trapped a far greater percentage of heavy elements in its crust; so Mammon's surface is rich in heavy metals such as gold, silver, mercury, uranium, plutonium, and the rare earths.

Climate

Weather on Mammon varies highly, depending on the geographical location. In general, temperatures range higher over most of the planet than on Earth. The planet's inclination toward its ecliptic plane of just five degrees and the low eccentricity of its orbit make seasonal temperature variations almost nonexistent. The planet spins on its axis once every 22 hours, a fact that mitigates daily variations in temperature. As on Earth, milder climates occur near the ocean's shores. The planet's Great Khalid desert records the highest daily temperatures of any Human planet. Mid-day extremes range up to 80° C., and regularly reach over 60°. Relative humidity usually measures a parching fraction of one percent. In such conditions, an unprotected Human being would perish in less than five minutes!

The planet's atmospheric pressure, the lowest among the Human colonies, averages less than 34 percent of Earth's at mean sea level. The atmosphere would be unbreathable, were it not for its high oxygen concentration, more than twice that found in Earth's air. Because atmospheric pressure is low, winds of high velocity have less destructive force than on other planets. Ocean storms tend to be mild, yet winds on the Great Khalid regularly reach 400 kilometers per hour, driving tremendous sandstorms more dangerous than hurricanes.

LIFE FORMS

Marine Life

Life abounds on Mammon in a variety of highly developed forms.

Biologists take greatest interest in its ocean life, because Mammon is the only planet on which evolution in the oceans proceeded along two distinct paths. This separate evolution created life forms that vary almost as much from ocean to ocean as they do from planet to planet. In fact, the origin of life in Mammon's oceans occurred at different times. Biologists estimate that life in the Husaini Ocean may have begun as much as 200 million years later than life in the larger Xingu Ocean. As life evolved under similar physical conditions in both seas, the differences between the two ocean's life forms furnish biologists with an important comparison by indicating which physiological characteristics have been caused by functions of the planet's physical environment and which result from chance.

Life forms in the two oceans differ markedly. In the Xingu Ocean, well-developed vertebrates, similar to Earth's bony fishes, flourish. These creatures take a variety of shapes, colors, and sizes, from large predators ranging up to 2.5 meters long and weighing more than a ton to fingerlings whose mature length is less than four centimeters. The Xingu also harbors less-developed forms that superficially resemble Earth's mollusks and crustaceans.

The Husaini, in contrast, contains no fish-like animals. Creatures resembling Earth's arthropods fill all higher niches in the food chain. Great lobster-like carnivores weighing up to 500 kilograms and measuring three meters from end to end attack their prey with claw-like pincers which can snap a Human limb or puncture a Human body like a skewer. One blow from their fast-moving tails also can be fatal. Observers have noted these monsters moving at up to eight kilometers per hour and swimming in the sea at more than 30 kilometers per hour! Their decentralized nervous system makes them difficult to kill, and they can regenerate lost limbs like Earth's crustaceans. Fortunately, most species act quite timid, and, if given a chance, will flee from Humans. Very large monovalve and bivalve shellfish, resembling Earth's oysters and abalone, also live in the sea. These creatures have been found up to two meters across and weighing in excess of a ton. Both the large shellfish and the smaller varieties taste delicious. Though they resemble Earthly varieties in texture and consistency, their flavors belong uniquely to Mammon.

Both oceans share some common life forms, all of which are warm-blooded, water-born animals that breathe air. In form they resemble the seals and sea otters of Earth, but they do not appear to be as intelligent or playful.

Vegetation

Biochemical evidence indicates that all land life on Mammon descended from creatures in the Xingu Ocean. Scientists believe this occurred because life started first in the Xingu, and thus Xingu forms first learned to adapt to the land. Once established on land, these species prevented the less-developed species in the Husaini from intruding. Land plants exhibit a wide variety of vascular forms, varying in size from short grasses to large trees. None of them resemble the angiosperms of Earth. Most reproduce by spores or seeds contained in cone-like structures. All land plants use green chlorophyl, so the foliage of the landscape appears green. Because no severe winters occur anywhere, virtually all known species remain green year round.

Animals

A class of warm-blooded four-limbed creatures comprise the most evolved animal forms on the planet. They bear more resemblance to Earth's birds than to mammals, since most species lay eggs and no species nurse their young after birth. These creatures fill the ecological niches of carnivorous and non-carnivorous mammals, but no flying species exist. Most species have hair, though some hairless varieties wear a covering of reptile-like scales. They range in size from herbivores as small as field mice to ferocious predators—called "mashers"—the size of a horse.

As on Earth, the largest animals don't field the greatest numbers. Small animals with external skeletons predominate in a variety of four- and six-legged forms.

A BRIEF HISTORY

Baq Husaini and the crew of the Xingu discovered Mammon in 2252. They spent three years studying this tiny world that appeared to be on the lower limits of Human habitability. During this period, they noted enormous deposits of valuable minerals, but in those days, no economic method for transporting these minerals across interstellar space existed. When Baq filed his report with the International Council for Space Exploration, the Council classified Mammon marginally desirable for colonization and indefinitely postponed its development.

The invention of the matter/antimatter reactor had tremendous implications for Mammon's future for two reasons: first, it greatly reduced the cost of transporting payloads across interstellar space, and secondly, it greatly increased the demand for a particular class of minerals known as the "rare earths."

Rare earths consist of a series of naturally occurring elements with very similar chemical properties. All occupy the third column of the periodic table. The control devices of matter/antimatter reactors require rare earths. They are not used as fuel themselves, but form an integral part of the matter/antimatter buffer in which matter and antimatter mix to produce pure energy. As the reactor is operated, the rare earths transform into other elements, rendering them unsuitable for further use. They must then be replaced with fresh rare earths. Human technology has been capable of transmuting one element into another for several centuries, but even today this process remains extremely costly. The early use of matter/antimatter reactors rapidly depleted the supply of rare earths on Earth; so more had to be obtained by transmutation of other metals. Unfortunately, this very expensive process produced large quantities of the wrong rare earth isotopes which had to be separated by even more costly processes.

In 2305, Jocelyn Hill, a 60-year-old independent operator of titanium and aluminum mines, persuaded the Intra Ocean Minerals Company and the World Wide Rare Metals Company to finance rare earth mining on Mammon. GAIL had just been formed, but the new organization still had no plans for developing the planet. The financial backers formed the Mammon Mining Company with Hill as president, and MMC began preparing at once to establish mining operations.

MMC obtained an option to lease the largest of GAIL's starships, and began assembling the required equipment. The miners needed not only mining equipment, but also refining equipment to reduce the rare earths to pure form for transport. Everything required to sustain the workers, including housing, farm machinery, food, clothing, and personal conveniences, had to be purchased and shipped. Planning and preparation took eleven years from the time MMC secured GAIL's approval. Midway through the preparation stage, MMC exhausted the allotted funds, and for a while it appeared that the project wouldn't get off the ground. Yet Jocelyn Hill, always the enthusiastic salesman, secured additional financing from other mining and financial institutions, and in the process, solidified her control over MMC.

Hill didn't plan to establish a permanent colony on Mammon. She originally induced employees to sign a five or ten year contract to work on the planet, then return to Earth and retire on their earnings. Almost from the beginning, this plan broke down. Most of the workers enjoyed life on Mammon so much that they stayed on permanently, using their wages to buy consumer goods from Earth. In fact, only 13 percent of the workers that have ever come to Mammon returned to their home planet!

In 2318, the first party of pioneers arrived on Mammon and established a base camp at a place called Hilltown. Of the 2,000 men and women in the group, only 200 engaged in actual mining operations. The balance worked at constructing shops, factories, and the refinery, or in providing essential services to the workers, such as food, power, and transportation. Every six months, another party of 2,000 arrived with more equipment and supplies. Mining of earths began in 2320. The miners stockpiled ore for a year until the refinery was complete. Export of rare earths began in 2321. Since that date, production and revenues have far exceeded expectations. The pioneers began food production in 2320 and importation of food ceased five years later.

Jocelyn Hill wisely realized that she could exercise little coercion over a group of people on another planet almost 40 light years away. She knew that to assure long term success of the venture, she had to motivate the workers on Mammon to be productive. She also realized that individual entrepreneurs worked harder and more efficiently than employees, and she therefore began very early to get MMC out of the business of providing domestic goods and services to Mammon's residents. She did this by selling the means of production to individual Mammonites in exchange for promissory notes. In this way, she could get a farmer, for example, to purchase farm machinery and establish a farm. With his profits from farming, he would pay off the note in money which MMC could then use to pay its workers. Eventually MMC sold most of its mining operations on Mammon and became primarily a marketing agent for the exported elements. Mammon had evolved from a "company planet" to a fully independent, free enterprise colony.

During the course of this evolution, people on Mammon found it necessary to form a government to settle disputes, protect people's rights and to enforce claims to property. They adopted a constitution similar to Wyzdom's which strictly limits its power and authority.

As Mammon's output of rare earths grew, other members of GAIL wished to obtain some. In 2348, the Ardotians began to barter for rare earths with MMC, offering completed starships suitable for Humans, in exchange for the precious commodity. Ardotian ships call regularly at Mammon to pick up the elements and transport them to their own manufacturing centers. Later, Chlorzi established a direct claim to certain clearly defined mineral deposits on Mammon. They did this by exchanging the right to colonize Athena, which they discovered but found unsuitable for their species. GAIL negotiated the transaction with the Chlorzi, then compensated MMC for the claim in the form of goods from Earth. Today, independent Human operators work the claim for the Chlorzi in exchange for a percentage of the output. Chlorzi supply all mining and refining euipment built to Human specifications.

Throughout Mammon's history, Jocelyn Hill managed the affairs of Mammon Mining Company from Earth. In 60 years, she didn't once visit Mammon because her doctors said she was medically unfit to withstand the "rigors of space travel." In 2365, knowing she had little time to live, Jocelyn took her life in her hands and made the trip to Mammon. By this time, she had arranged for new management of MMC to succeed her, and she intended to live out her life on the planet she helped to found. When she arrived at Hilltown spaceport, she found a million people, virtually the entire population, waiting to greet her. More than any other single person, "The Old Lady," as the pioneers affectionately called her, was responsible for establishing the Mammon colony. Her years of wise leadership had brought unparalleled prosperity to its residents, and their final tribute to her showed how much they appreciated what she had done.

Jocelyn Hill lived two years on Mammon, but didn't participate in MMC operations or government politics during that time. Instead she spent it traveling over the planet, seeing its natural wonders and Human development, and talking with people about their lives and their plans for the future. She lies buried in a small crypt located in the Hilltown central square. It bears a quotation by her which epitomized her life, "Don't tell me this job's too big for one person; no job's too big for one person! Behind every major achievement in history stands one individual with drive and a vision of what could be."

CURRENT STATE OF DEVELOPMENT

Population

Approximately 1.5 million people occupy Mammon, more than 1.1 million of them immigrants. The population is widely distributed over most of the planet, but the majority live on the shores of the Husaini Ocean. Most people do not live in or around cities, preferring to locate within reasonable commuting distance of their work. Industrial and commercial complexes tend to be small and isolated, rather than concentrated in specific geographic areas.

Consumer Goods

Because of its unique natural resources, Mammon boasts the highest physical standard of living of any Human planet. Though its sparse population makes industrial self-sufficiency impossible, virtually any consumer good found on Earth can be purchased on Mammon. In fact, Mammon is the only Human planet on which consumer goods from other colonial planets or from other races in the Galactic Association can be found. Naturally, Chlorzi and Ardotians don't make too many things that a Human might want, but art objects, such as Ardotian geoforms, decorate both homes and public places. Some very chic Human fashions have been styled from materials like aquasheer, produced in highest quality by the Ardotians.

Industry and Technology

Mammon's industry tends to be highly specialized. The giant starships that take away rare earth bring both customer goods and capital equipment with them. Few consumer goods, save clothing, food, and shelter, are manufactured on the planet, though Mammon's industries assemble large items, like transportation vehicles. Mammon now manufactures many specific items that use rare earths for export. The first factory to manufacture matter/antimatter buffer plates opened more than fifteen years ago, and today manufactures plates to Chlorzi and Ardotian, as well as Human Specifications.

Minerals

Manufactured items on Mammon often appear quite distinctive. Mammon has more deposits of gold and silver than the Earth has of metals like chromium and nickel. Therefore Mammonites use gold plating on a variety of common objects, from bathroom fixtures to kitchen utensils.

Mammon's unique contribution to the Galactic Association stems from the essential minerals it provides. The rare earths of Mammon have reduced the cost of space ships and enhanced the interstellar exploration and development programs of all GAIL's members. In addition, Mammon's rather central location in the explored space of GAIL, and the fact that GAIL's members must go there for rare earths has made it an important meeting place and transfer point for the entire Galactic Association. All information exchanges between the Chlorzi, Ardotian, and Human races now take place on Mammon, and diplomats now travel via Mammon to the home planets of GAIL's member species. As interstellar space travel becomes less expensive and more common, Mammon's importance as an intergalactic port may someday outweigh its importance as a producer of elements.

Research and Development

Besides its commercial value, scientific study of Mammon has proven important to Humankind. Mammon's dual ocean system, unique among the known worlds, provides data essential to understanding the beginnings of life. Because Mammon lies on the lower limits of Human habitability, study of its physical characteristics and life forms has enabled scientists to develop better theories for predicting which stars will possess homes for Humans.

AN ACCOUNT OF LIFE ON MAMMON

Editor's Note: Ezra Lilly, 62, general manager of the Mammon Intergalactic Hotel in Hilltown, gives an account of life on his planet. Ezra immigrated to Mammon with his mother, Rose, a mining engineer, at age four. He grew up in remote mining camps all over the planet and during this time developed a lasting appreciation for the planet's natural beauty. He set out for Hilltown at age eighteen to make his fortune, and, after taking a stab at a variety of consumer service businesses, settled on hotel management as a career. As manager of Mammon's largest hotel, Ezra has developed a side business as an art dealer, handling both Human and non-Human art forms. He has edited several books on art and written an entertaining volume about backpacking on Mammon.

My mother is one of those workaholics who can't live with a career and a husband. Before I turned four, she terminated her marriage contract, took custody of me, and signed up with Mammon Mining Company for a stint as a mining engineer. Her divorce had left deep wounds, and she thought a change of scene might make her forget the unpleasantness. Like most immigrants of the day, Mom intended to return to Earth after making a comfortable nest egg, but like most immigrants, she remains on Mammon to this day. She still works 40 or 50 hours per week and has no more desire to live on Earth than on Ignati.

My childhood memories of the original Hilltown recall a haphazard cluster of structures on grassy hills overlooking the sea. Back then, MMC had no time for landscaping the camp, and the great bustle of activity trampled the grass into dust that flew in all directions as the levitrucks took off and landed.

Mother spent eight to twelve hours a day, six or seven days a week, at the mine. I passed the time in an ad hoc day care center run by the spouses of some miners who also had small children. My days were an informal mixture of play, games, and lessons. What our teachers lacked in skill and polish, they more than compensated for in hard work and enthusiasm. Most of the lessons came off computer files, but there weren't enough terminals for every child. Undaunted, the teachers found ingenious ways for all of us to participate, like having us write answers to questions in the dirt with pointed sticks.

Despite her grueling schedule, Mother always took time at the end of the day to sit alone with me and talk. She always asked me what I had done and what I had learned. If I said I learned nothing, she would scold me firmly, but kindly, and tell me that all of my life I should learn at least one new thing every day. On her days off, Mother usually managed to comandeer a truck to take me away from Hilltown. Some days we went to the shore of Fortune Bay and poked among the tidepools or walked along the great sandy beaches. On others we headed for the nearby ridge of low mountains to fish in the tiny streams and lakes or to hike amidst the thick forests. Some days when Mom didn't feel like walking, we glided silently in the levitruck, just above the treetops, looking for wild animals. I liked this best, for it seemed as exciting as a hunt, yet I knew that with the flick of the control lever we could whisk into the air, out of danger. Once we stalked within five meters of a great masher beast, a large, hoofed creature with a mouthful of terrible teeth, as it stood gorging itself on its prey.

By the time I turned ten, the success of the Hilltown mine had proven that mining on Mammon was profitable. From then on, Mammon Mining Company wasted little time in expanding its operations to other mine sites, and they called upon my mother to help. Mother's technical specialty lay in the startup of laser-fusion excited, trispectrum, automining equipment, which employs elements from the most sophisticated of Human technologies. This equipment effectively strips the soil and rock of the oxides of desired elements, then returns the remainder to a nearly natural state. Starting these systems up takes as much art as science, since numerous factors influence the stability and efficiency of the process, including the hardness and composition of the areas being mined, the mix of rare earths being separated, the climate and the construction of the equipment itself. Getting a new mine operational took anywhere from eighteen months to three years after the equipment arrived. Once the equipment began operating satisfactorily, Mother moved on to a new site and a new nest of technical problems.

Hilltown, the home of my childhood, didn't seem like much until we moved away. The town formed the nucleus of the planet. What little farming, manufacturing, and commerce existed on Mammon happened there. Other mine sites looked like remote outposts in comparison, and my lifestyle became more solitary after leaving Hilltown. During the next eight years, Mother and I lived at four different mining camps. Though each had its unique features, many aspects of life in the camps were the same.

When we arrived at each mine site, it swarmed with hundreds of people. Unfortunately for me, most served as temporary contract laborers who brought no families and left within six months of our arrival. Only the permanent operating personnel and a few construction supervisors had their children with them, so the entire teenage population of each mine usually numbered less than thirty. Parents who didn't work at the mines took turns caring for little children, just as they had done in Hilltown's early years. After I turned twelve, Mother no longer felt that I needed such care and let me shift for myself during the day while she worked.

Smaller mine communities didn't have schools, so most parents had to rely solely on computer education for their children. Mother values education highly, and so she got me the best education terminal money could buy. Before it arrived, she began selecting a curriculum of study that included heavy doses of science and business, as well as language, Earth history, and mathematics.

In my years at the camps, my computer teacher became the most important possession I owned. It had an enormous holographic display, nearly a meter square and accepted both voice, fast-tran, and long-form keyboard input as well as telepathic feedback. The lessons came via satellite from Hilltown's library, and though I know each one was just a complex computer program, they seemed more real than that. As I sat before the screen each day, I left the real world of the camp behind me. Geography lessons took me soaring over the face of Mammon, peering down with eagle's vision. When studying astronomy I zoomed into space past brilliant suns and spinning planets to look into the awful maw of black holes that gobble up all but the most intense forms of energy. Studying Earth's history carried me across light years to look upon the battlefields, castles, and pyramids of our mother planet's once great nations.

As I traveled, the teacher's soothing voice explained the things I saw, then asked me questions about them. If I gave the right answer, it praised me and we went on to something new, but when I made mistakes, we reviewed to see where I went wrong. At times I saw reenactments of great events in history. Though somewhere in the back of my mind I knew I wasn't really watching Bonaparte inspire his troops before Waterloo or Armstrong stepping on to the moon, I felt as though I lived through these events and knew *why* people had struggled so hard to do the things that made our history.

Even the abstractions of mathematics came to life on the screen as I manipulated equations and geometric forms against the green backdrop. The teacher always seemed to know when I had missed some point and went back to explain and illustrate again. When studying binary mathematics, I loved to watch the teacher break down the most complex sentence I could concoct to the elemental ones and zeroes of all knowledge right before my eyes.

I rarely desired to play hookey from my lessons, but if I had, Mother would have known. Each evening she looked at my progress score and, on rare occasions when it seemed low, she talked to me about things the teacher couldn't explain.

The teacher wouldn't continue my lessons more than eight hours a day, and Mother didn't let me watch more than two hours of shows after that. She felt that I should spend some time outside each day, getting exercise and associating with people my own age. During these hours I did a lot of hiking and exploring and learned to appreciate the wondrous natural beauty of our planet.

Most outsiders think of Mammon as a desolate, desert world, but though the terrible wastes of the Great Khalid desert dominate its surface, Mammon possesses much territory as beautiful as can be found on any world. Our second home at Blackrock mine lay at the northern edge of a rich, rainswept plain of tall forest. Enormous trees rising up to sixty meters formed a canopy over thousands of interesting plant and animal life forms. Tiny climbing creatures with hand-like claws scamper among the treetops and toss bits of bark and seed cones at intruders walking below.

Our third home at the Oasis mine of the western Plat region nestled in a fertile river valley of temperate climate. The wind sweeps the Plat night and day, every day of the year. When we first arrived, we thought it would drive us mad, but after a while we learned to tolerate it. The region's trees all bend eastward before the howling wind that feeds the great furnace of the Khalid to the west. Mother didn't like the Plat and worked furiously to get the Oasis project on line so we could leave the infernal wind.

The very day the superintendent accepted the automining equipment, Mother arranged for us to hitch a ride back to Hilltown on a returning freight caravan crossing the Khalid desert. In those days, heavy duty antigrav trucks had a top speed of just 300 kilometers per hour. Ideally, vehicles crossing the Khalid prefer to do so at night because a breakdown during the midday heat would be fatal to the passengers. Unfortunately the desert measures 9,000 kilometers wide and can't be crossed in one night.

We left Oasis at dusk riding in the lead truck's cab at the head of a long string of automated vehicles, which followed us. Overhead the sky glowed with the last rays of twilight, while below an inky expanse swept to a distant horizon. I slept most of the night, and as dawn broke I looked out at an ocean of sand extending as far as I could see in all directions. The mountains of the Plat had passed out of view behind us, and nothing lay ahead but white expanse. As the temperature climbed, so did we until by midday we flew at almost 15,000 meters, breathing supplemental oxygen from masks. Despite the high altitude, the cab's temperature climbed to 30° C. The dry air parched our throats and skin, and though we sipped water constantly, it brought little relief. As we traveled, some geographical features came into view: an occasional ridge of mountains or a grey patch that Mom said might be gravel instead of sand. Yet nowhere did we see a drop of water, a speck of vegetation, or any living thing. Mother said no geological evidence of water had ever been found.

Throughout the day the sky never turned blue, but remained dawn pink. Though no water clouds appeared, we could see great dust storms rising thousands of meters into the air, some as high as we were flying. They appeared as opaque, shapeless masses, blocking all light and turning the sky near them a deeper pink. We continued to travel throughout the night, gradually descending to eliminate the need for oxygen. As dawn broke, blue sky appeared, and we saw vegetation below, indicating the desert lay behind us.

We spent the next month in Hilltown while the company's planning staff briefed Mother for her new assignment at the Tobar mine. Hilltown had more than doubled in size since we left for Blackrock four years before. The residents bubbled with ambitious plans for moving the city center so that future growth could be planned instead of haphazard. Everyone we met exuded enthusiasm for the project. They felt that someday Hilltown would become the cultural center for an important, wealthy planet instead of just a staging ground for a gold rush expedition.

After we left for Tobar, I soon forgot the future of Hilltown. Tucked into the foothills of a stunning range of snowcapped mountains, the mine camp overlooks a sky blue gulf, lined with broad beaches of sugar-white sand. The idyllic period at Tobar seemed the happiest in my short life. I spent most days at the beach, had my first torrid romance with another miner's daughter, and began to think of the world as my oyster when Mother got promoted.

The promotion offered a tremendous boost to Mom's career, and I couldn't bring myself to say anything bad about it. She jumped from field engineering specialist to production superintendent. Unfortunately, the job was at the Port Node mine in the perennially frozen northern wastes. My mood reached a low point as we landed on that bleak plain littered with scruffy mounds of snow extending to an infinite horizon. The timing, a month before my eighteenth birthday, worked out well since I was preparing for my comprehensive exams, and the forbidding landscape made staying inside and studying even more attractive.

I turned eighteen, the legal age on Mammon, and finished my formal education. Mother offered to get me a job at the mine or to send me to the University at Hilltown. In those days, the U of H offered little except courses that prepared people for mine-related industry, and I had long ago decided that the mining life was not for me. I wanted to live in a city, enjoying the benefits of luxury. I couldn't stand the thought of spending my days around a dusty, open pit mine, getting rich, but having nothing to spend it on. Indeed, it seemed to me that Mammonites would soon insist on having comforts and luxuries in exchange for the vast sums they were accumulating in banks. I determined then to get into the business of supplying these needs and packed up my belongings for a move to Hilltown. Mother took this news in her usual gruff but good-natured way.

"It's your life, son," she said as we parted. "I can't see what sort of *career* you're going to find in that carnival town! But, nobody could tell me how to run my life, so I can't expect to tell you. Whatever you do, do it well."

In the three years since my last visit, Hilltown had changed a great deal. The citizens had approved the development plan for the new city and several major buildings had sprung up. All of the major spires from the old city had been relocated in the new area with more space and landscaping around them. The plan called for the original town to be left largely as it was, a collection of low structures containing bars, restaurants, cafes, brothels, small stores, and an assortment of small hotels. A high-speed guideway then under construction would soon connect the new town with the old.

Since so many people had headed for higher paying jobs in the new town center, a great many jobs remained to be filled in the old city. I had some experience as a helper in Tobar's camp kitchen; so I started out as an assistant cook in a tiny, classy restaurant. An irascible old Frenchman named Henri Legasse runs the Rive Gauche, and it remains one of Hilltown's finest gourmet eating places to this day. Though foodmasters, autoselectors, and kitchen robots alleviate much of the drudgery of preparing meals at home, Henri taught me the importance of Human dexterity and skill in serving truly fine food. Only Human hands can properly prepare a salad, a paté, or a pastry as pleasing to the eye as to the palate. Only the Human eye can tell when an omelette has reached the instant of perfection, or when a roue has thickened to ideal consistency. Only the Human nose and tongue can correct the seasoning just the right amount to turn a good sauce into a great one.

In his abrupt, arrogant way, Henri must have liked me, for whenever I mastered one task and asked to learn another, he agreed. In the ten years I worked at the Rive Gauche, I learned to prepare dishes from the first course through dessert. Though Henri ranks as the greatest chef I have ever met, he was an awful businessman. Often I had to help out with ordering supplies or computing the payroll. Once a year I made a dreaded trip to the accountant's office to try to sort out Henri's hopelessly scrambled financial records. I enjoyed learning as much as I could about the business, for my ambition then was to own a restaurant, not just to work in one.

After nine years in the kitchen, I made the unprecedented request to work up front as a waiter. To everyone's surprise, Henri agreed, and I was fitted out in formal attire to serve the public. The kitchen and dining room stood in such striking contrast to each other that they hardly seemed part of the same place. The kitchen gleamed with pure white steroglass, stainless alloy, and gold. At both ends, a wall of ovens cycled food continuously from beginning microwave cooking to final browning with long-wave heat. At work stations along each wall and around a center island stood the intricate chopping, grating, and puréeing machines, and the cookers for steaming, sautéeing, and boiling. The clatter of dishes, the whoosh of ventilators, and the constant chatter of the cooks filled the air.

Stepping through the short, sound-blocking corridor from the kitchen to the dining room brought a change as abrupt as a warp through space. Natural hardwood paneling covered the walls and thick carpet lay underfoot. Padded leather chairs surrounded linen-covered tables and formally dressed waiters moved like ghosts between them. White noise absorbers soaked up sound to create church-like silence. Here gourmets from all over the planet came to relax and worship the art of fine cooking. Though not as demanding as the kitchen, serving revealed to me a completely different dimension of the restaurant. Unlike the robot servants in cheaper restaurants, subtle changes in the conduct of a Human waiter influenced the guest's perceptions of the meal.

One of our regular customers was a big, jovial woman whose good-natured manner thinly masked an iron will. I liked her because she reminded me of my mother. One night, after a little too much wine, she remarked that she'd better keep me away from her girls or they'd be giving away the merchandise. After that I inquired about her and discovered she was Bertha Moynihan, owner of the Pleasure Palace, the largest brothel in the old town. The next time she came in I called her by name. She laughed and made a joke about my checking up on her, then asked me to stop by to see her about a job.

I had never been to the Pleasure Palace—my income wasn't in that category. The structure, a thick cylinder five stories high, reminded me of Bertha herself. Inside, amidst rich brightly-colored surroundings, it offered wealthy patrons of both sexes not just the services of a partner, but an amusement park of sexual fantasy augmented by holographs and somafields. All manner of drugs, from the benign to the dangerous, could be bought to dull or stimulate the senses. The Palace also served food in private dining rooms so that customers could dine romantically with their "host" or "hostess" before retiring for the evening. Though the dining rooms remained booked solid, Bertha claimed they lost money and that the food itself was awful. (It was.) Though she realized the customers didn't care, it made her feel cheap to sell a lousy product.

I remember her blunt offer, "I want you to get this food business straightened out! I don't expect you to make it into a place like Henri's. I don't even want you to, but I can't stand the *smell* of that garbage they're serving now!"

Despite my love of Henri, I couldn't refuse the money Bertha offered. So I started work at the Palace three weeks later. The kitchen was a disaster, and straight off I fired the chef and two cooks. Nothing that could be called inventory control existed, and employees had been walking off with enormous amounts of food. I found no rational system of pricing, and the junior cooking staff seemed poorly trained and in very low spirits. After I took over, things began picking up almost at once. Indeed, it would have been hard to make them worse, but more than a year passed before I had the kitchen running to my satisfaction.

Pleased with the results I achieved, Bertha began to give me other problems to tackle. Even in the automated 24th century, prostitutes of both sexes generally have more than their share of personal problems, and such problems often interfere with their work. Most are spendthrift; all worry about getting old and being lonely. They need someone to listen to their problems, and that was often me. If the problem seemed too serious, I recommended a psychiatrist, and Bertha usually paid the bill.

The Pleasure Palace hummed with sophisticated equipment from somafield and holograph projectors to automatic surveillance and neural neutralizers that prevented the "boys and girls" from becoming occupational statistics when customer's sadistic fantasies came on too strongly. In those days, few qualified mechanics could be found on Mammon to service such equipment, and spare parts created nightmares.

The administrative challenges of the job at the Palace kept me interested for several years. Bertha had grown tired of the hassles; I welcomed them as a chance to learn and to prove myself. Yet when I'd mastered most of them, I grew uncomfortable. Work at the brothel began to depress me. Not that I'm a prude or disapprove of sex, but the notion of people paying for something that ought to be freely shared seemed very sad. I realize that all people, even the old and not-so-nice ones, need to be touched and to satisfy their sexual desires, but people at the brothel seemed emotionally crippled. Brothels may be necessary, just as hospitals are necessary for the sick and injured, yet I'm not the sort of person who can work around either.

My years at Henri's and Bertha's convinced me that I enjoyed serving people, and that I like to see people enjoy themselves. I liked the restaurant business, but I appreciated the added challenges of a more complex organization. After some thought, I decided that hotel management might offer a satisfying, lucrative career and began looking for a hotel job.

I started my new career, at a substantial cut in pay, as assistant manager in a small hotel owned by a retired transport pilot named Andy Werner. I'm sure Andy was a better pilot than hotel manager, or he wouldn't have lived to his 100 years. His accounting records were a mess. He had a hard time keeping any staff and a harder time getting them to work when he did. Despite my inexperience with the hotel business, I began to tackle Andy's problems in a businesslike manner, employing the management skills my other jobs had taught me. Cash flow increased throughout the first year, and by the next year Andy showed a handsome profit and rewarded me with a bonus.

With the worst of Andy's problems behind me, I again grew restless. I wanted to work for a really large hotel in the new city, such as the Executive or the Husaini. Only these huge establishments could provide me with a wide variety of management opportunities, and ultimately allow me to work my way up to a salary that equaled my mother's.

Landing even a junior manager's position at one of Hilltown's major hotels took me the better part of a year and a great deal of politicking, including joining the Hilltown Chamber of Commerce and the Mammon Society of Hoteliers and Restaurateurs. However, my work at Andy's had not gone unnoticed, so at long last I was offered a position as assistant night shift manager at the Hilltown Executive Hotel. The job at the Executive opened a variety of doors. At that time, it reigned as the undisputed king of hotels on Mammon, offering the most elegant appointments, the finest service, two of Hilltown's best restaurants, and exotic, live entertainment from as far away as Earth.

During the next ten years I worked in every aspect of the business from staff management to building maintenance. As I learned more, my superiors thrust more responsibility on me, until I rose at last to general manager. By then I had become familiar with many of Mammon's movers and shakers, and the connections I had made would serve me well in the future. Hilltown had grown enormously in that interval. As Mammon's industrial infrastructure had grown and flourished, Mammon Mining Company had been eclipsed as the dominant economic force. The importance of rare earths in the galactic economy was turning Mammon into an important trade center not only for Humans but for every species in the Galactic Association.

My dream was to open a fully-integrated interplanetary center serving the needs of travellers not only from all of Mammon, but from all the known galaxy. The concept would provide not just a luxury hotel but a center in which travellers to Hilltown could satisfy every need. Meeting rooms, computer access, offices, restaurants, and entertainment would, of course, be included, but I also wanted a lower level with shops and service business of every kind. I began to make quiet inquiries, and it didn't take long for me to interest one of

Mammon's wealthy builders, Anwar Shimi, in the idea. We soon negotiated a contract whereby I would join his staff and manage the design and construction of the hotel. After completion, I would remain as general manager and hold a five percent equity interest as well.

We set to work at once, securing a site near the center of town and meeting with architects. Design and construction took three years—a long time by Mammon's standards—because of endless compromises between what I desired and what our expected revenues let us afford. Though I had many doubts, when the hotel opened, the result was everything I had hoped.

The Mammon Intergalactic Hotel, shown in figures 3.15 and 3.16, occupies a 45 hectare site just eight kilometers from Hilltown central square. We chose a slender tower structure instead of a shorter, wider building because the tower offers more window space for its area and allows a feeling of spaciousness between the hotel and the adjacent city structures. The fourteen-tiered base structure and the subterranean spaces beneath contain about two-thirds of the building's usable floor space and house those functions that derive no advantage from window openings. The roof above is landscaped and contains gardens, swimming pools, game courts, and a tiny zoological garden displaying interesting species of plants and animals from the other colony planets. An electrokinetic field surrounds the zoo to prevent these alien life forms from escaping into Mammon's environment.

The lower four tiers and two subterranean levels below them contain a wide variety of shops and service businesses. Though they form an integral part of the hotel complex, I don't attempt to manage them directly, but simply lease space to an appropriate mix of entrepreneurs that will complement the hotel's services. An assortment of shops offers goods imported from throughout the Galactic Association. Complex appliances, specialized furniture, medical devices, and other items not manufactured on Mammon can be purchased along with forms of art from every GAIL planet. Nightclubs and theaters feature live or holographically-simulated entertainment to suit a variety of tastes, and a variety of service business from real estate brokers to travel agents serve the needs of newcomers to Hilltown. A color-coded system divides the hotel into eight sectors to prevent shoppers and guests from getting lost. Though other colors exist within each sector, the color shown in the plan (figure 3.16) predominates. A low-speed guideway system transports people quickly around the shop levels, although most customers prefer to stroll and window shop.

Hotel functions begin on the fifth tier and include bars, outdoor and indoor restaurants, a convention hall, and special entertainment rooms ranging from Chlorzi armball courts to a unique exercise gym. The gym contains computer-controlled equipment to exercise selectively any muscle in the Human body. Such machines heop Mammon residents condition themselves for trips to GAIL planets with higher gravity. A large casino fills the top tier of the base structure and offers a wide selection of computer-controlled games of chance.

The eight-sided, irregular column rising above the casino contains a variety of public rooms, including offices, conference rooms, seven restaurants, bars and lounges and entertainment centers. A small convention hall surrounded by transparent walls occupies one entire level.

One restaurant in the column, the Amonde Room, offers diners an experience unique to all of Mammon. The Amonde serves gourmet banquets of up to 30 courses, yet diners need consume no more calories than a light snack. Meals combine real dishes mixed with simulated ones, created in the mind of the diners by a type of selective somafield. The field allows chefs to tailor diets so foods that aren't healthful to the diners can be tasted, but need not be eaten. The process actually costs more than serving real food, but imagine what a corpulent gourmet, too long restricted to a low-fat diet will pay for an illusionary feast of Beef Wellington followed by strawberry shortcake topped with a mountain of whipped cream!

The Hotel's lobbies lie above the public rooms, immediately below the residence towers. A unique feature of the Mammon Interglactic is its ability to cater to all member species of the Galactic Association. Of course Chlorzi, Ardotians, and Minutae comprise a small percentage of our guests, but for Mammon to continue serving as an interspecies trade and communications center, such accommodations must be provided. The largest lobby caters to Human guests, but we have arranged others to suit the need of alien species. Chlorine gas fills the Chlorzi lobby so they may breathe without respirators. The Ardotian lobby, maintained at 70° contains supplemental oxygen

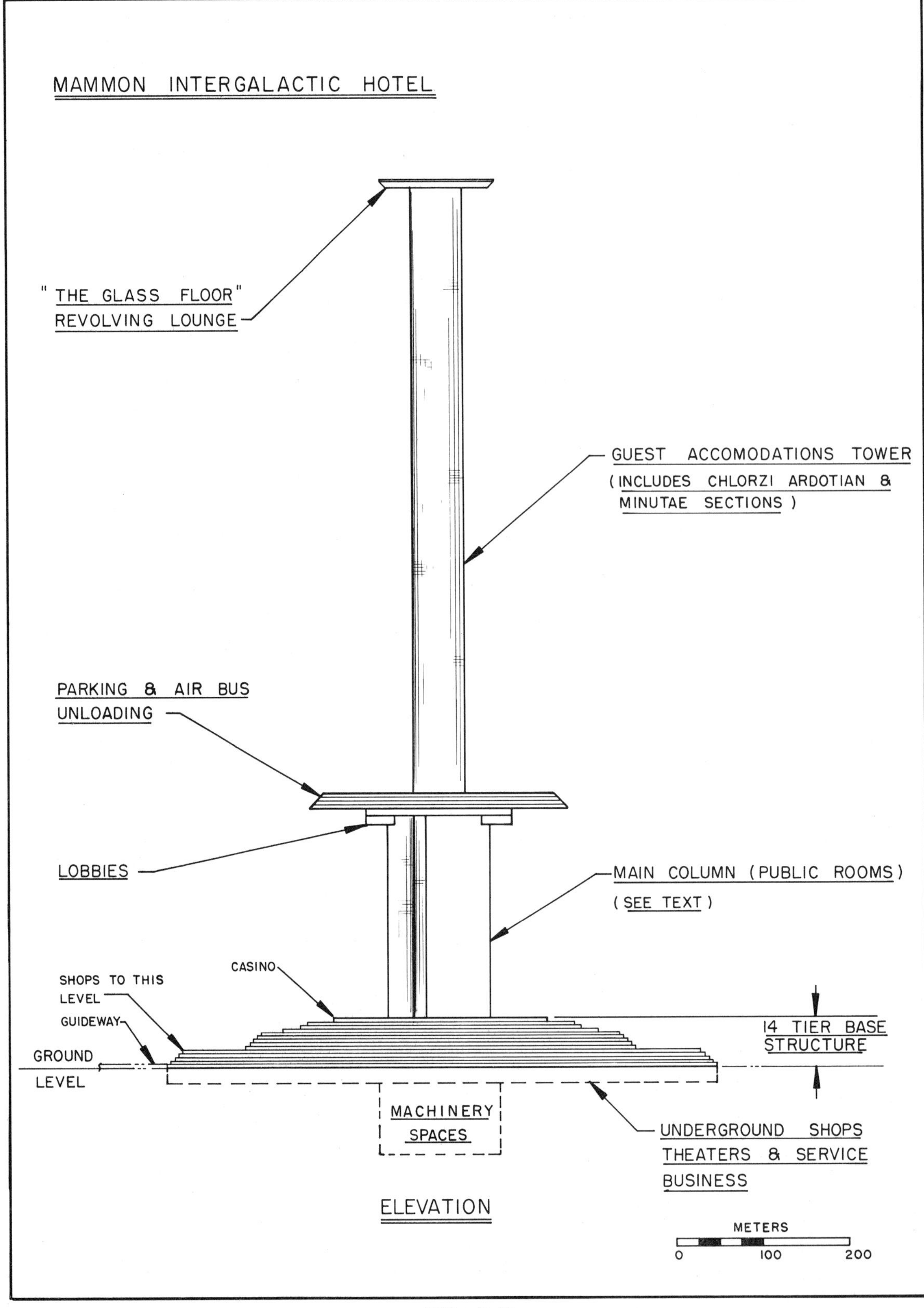
MAMMON INTERGALACTIC HOTEL
" THE GLASS FLOOR "
REVOLVING LOUNGE
GUEST ACCOMODATIONS TOWER
(INCLUDES CHLORZI ARDOTIAN &
MINUTAE SECTIONS)
PARKING & AIR BUS
UNLOADING
LOBBIES
MAIN COLUMN (PUBLIC ROOMS)
(SEE TEXT)
CASINO
SHOPS TO THIS
LEVEL
GUIDEWAY
GROUND
LEVEL
14 TIER BASE
STRUCTURE
MACHINERY
SPACES
UNDERGROUND SHOPS
THEATERS & SERVICE
BUSINESS
ELEVATION
METERS
0
100
200

FIG. 3.15

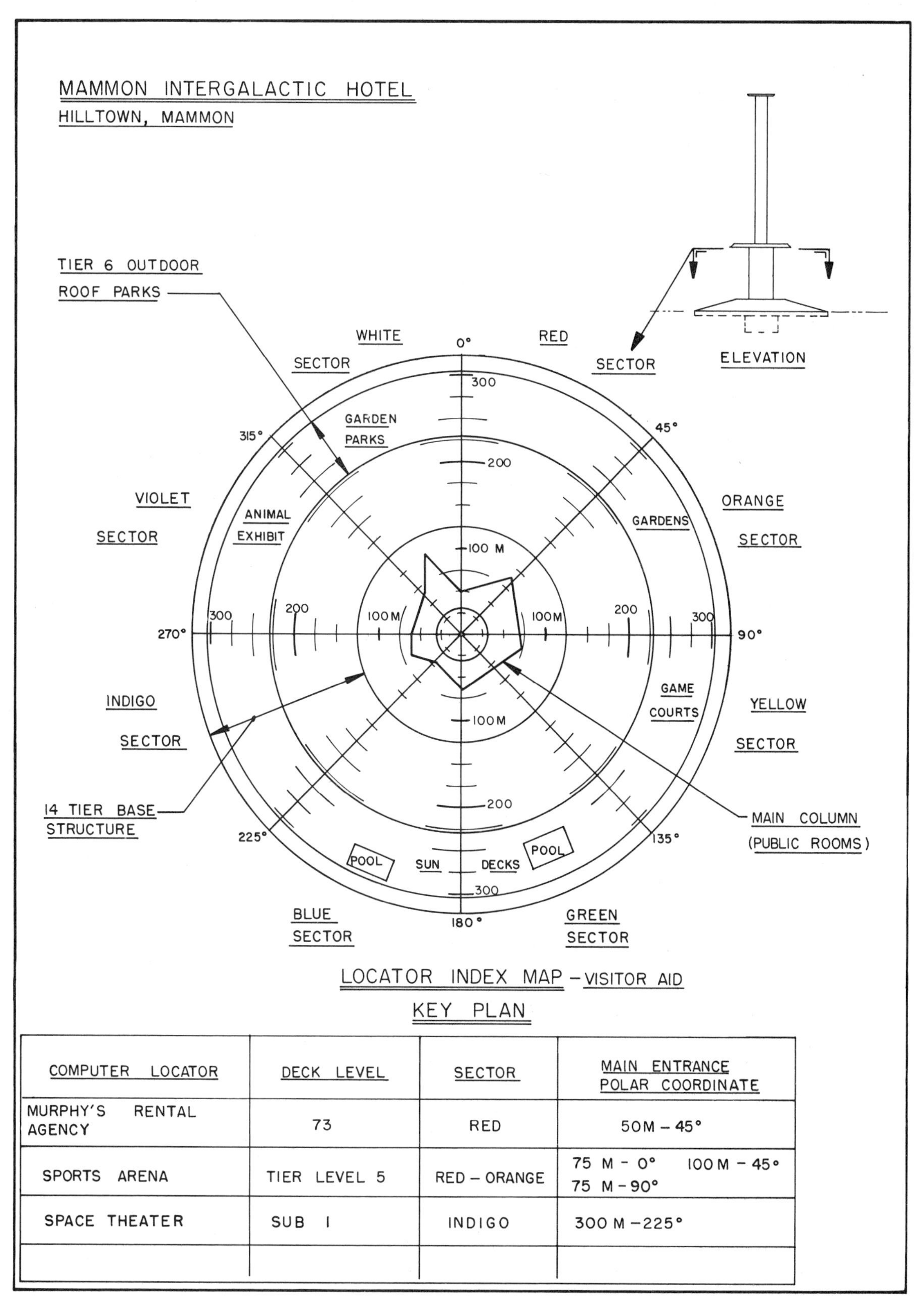

COMPUTER LOCATOR	DECK LEVEL	SECTOR	MAIN ENTRANCE POLAR COORDINATE
MURPHY'S RENTAL AGENCY	73	RED	50M – 45°
SPORTS ARENA	TIER LEVEL 5	RED – ORANGE	75 M – 0° 100M – 45° 75 M – 90°
SPACE THEATER	SUB I	INDIGO	300 M – 225°

FIG. 3.16

to provide a partial pressure of 0.9 atmospheres. The Minutae lobby fills only two cubic meters, yet can accommodate 500 Minutae!

Communications equipment forms the biggest investment in the alien lobby. A Chlorzi, for example, doesn't just walk in and say, "I want a room." Chlorzi communicate telepathically, not by sound. Their ideas do not resemble Human thoughts in any way. They have no concept of "I" or "want" or "room." They cannot sense the feeble emissions of the Human brain. Special amplifiers can sense Chlorzi thoughts and computers can analyze them into elemental binary code and rebuild them into something like our language. The Chlorzi equivalent of "I want a room" appears on the desk clerk's computer screen as "Make us of your environment," and the clerk still needs a phrase book!

The other aliens pose different problems. Minutae voices speak in the 40 to 180 kilohertz range, yet most Human ears can't hear above 17 kilohertz. Ardotian language consists of a complex mixture of spoken sound and physical gestures that can't be intelligibly separated. Most Ardotians prefer to communicate with us using a keyboard of the 173 "letters" that form their written language.

Above the lobbies rise the accommodations towers containing 4,798 rooms for Humans and suitable quarters for 120 Chlorzi, 80 Ardotians, and 400 Minutae. The Human quarters, the most sumptuous in the galaxy, contain full antigrav sleep fields, five-sense full-spectrum somafields, holovision, and infinite source sound fields. Any meal, laundry, or package delivery can be accomplished instantly from each room's transfer cabinet, and full-range data terminals link guests with the planet's central computer. Throughout the Human lodgings, decorators chose the finest natural and synthetic materials. Real wood and natural fibers, such as Agip's wool, have more warmth and character than any synthetics could possess.

Unfortunately the accommodations for other GAIL species cannot meet the high standards we have set for our Human guests. We have done the best job we could to create a comfortable environment for Ardotians, Chlorzi, and Minutae, one in which they can breathe their natural atmospheres, see by natural light, and relax in the company of their fellows. Yet we cannot hope to grasp the subtleties of an alien environment whose characteristics would kill a Human being within seconds. How, for example, are we to know if the "chairs" are "uncomfortable" when we're not certain if they need to sit down or if they have a concept of "comfort"?

Even so, most aliens seem to be good sports about their first visit to Mammon and don their environmental suits for a stroll through town. Of course, the Minutae would get stepped on in the rush of Mammon's traffic, so we arrange guided tours for them. No being in the galaxy ever saw a zoo to equal central Hilltown on a busy day!

My ramblings notwithstanding, it's not the physical structure or the decoration but the people I meet that make the hotel business such fun for me. I enjoy sitting at the crossroads of Human history's most dynamic society and meeting the fascinating range of beings that make it happen. Not just mining executives and engineers, but artists, builders, entertainers, lawyers, doctors, diplomats, prostitutes, scientists, con men, and religious saviors all pass through the portals of the Intergalactic. I have met hundreds of each and become friends with many, including some interesting non-Humans as well. The Chlorzi and Ardotian equivalents of "engineers" and "ship's captains" that come to Mammon for rare earths and the diplomats who await passage to their embassies on Earth have given me unique insights into Human consciousness. Our conversations through the computer-interpreter have made me appreciate that thought can take more forms than Human minds can comprehend.

Though the hotel remains my principal interest, the visual arts run a close second. My art avocation began long ago and has grown from a hobby into a second business. Modern techniques for recording, transmitting, and reproducing information enable people on colonial worlds to appreciate Human art developed over thousands of years. But for these techniques, much of Earth's great art would remain in musty museums on the mother planet. Yet we on the colonies have copies of paintings by Da Vinci, Rembrandt, and Van Gogh, exact in every shade and brush stroke, and statues by Michaelangelo, copied to the nearest micron. Copies of such classics form a solid base of business for my gallery, but the most meaningful art must be contemporary.

Pioneering societies, like Mammon, have traditionally had little time or resources to produce artists. Not one emerged from America to rival the masters of Europe until the twentieth century. Despite our few fine local artists, Mammonites have had to reach back to Earth and to the older colonies for most of their contemporary art. Though copies of Earth's most famous contemporary artists sell well at modest prices, people still pay extraordinary amounts to own the one artist's original. I'm still amazed to see discriminating art lovers pay 100 times the cost of a first-rate copy of a sculpture by a great artist, like Falcone, to have the original work of a much lesser artist.

Perspective and photofilm have vanished as contemporary media. Sculpture and holographs now convey most three-dimensional concepts. Artists still make two-dimensional "wall hangers" in abstract, often textured forms. The expansion to the new worlds has greatly influenced Earth's 24th century artists. Works inspired by the colonies seem most popular on Mammon. My favorite holograph depicts a sinister representation of Mammon as a Dante-like inferno by an artist who has obviously never been here.

Two-dimensional, original media can be shipped from as far as Poseidous, but even the lightest of sculpture remains prohibitively expensive to transport. Yet "Mammon First Copies," computer-produced duplicates of original Earth art, bring as high a price here as the original would on the Old Planet. Like most dealers, I make my purchases from catalogs since I can't afford the time or expense of shopping trips to Earth. All original artworks and first copies I receive bear a unique radiosignature that dates the art and prevents its exact duplication.

Recently art works from other GAIL cultures have become fashionable, particularly Ardotian geoforms and Minutian tapestries. In keeping with the intergalactic style of our hotel, I have large geoforms and several tapestries displayed prominently in the lobby. Yet I must confess, as a sincere art patron, if not the most discriminating, that I can't imagine how any Human being can appreciate alien art. One can marvel at the intricacies of Minutian tapestries or find something pleasing in the regular patterns of the geoforms, but this is no more art to us than the pleasure we might get from enlarging a drawing of a starship's lacy, web-like structure and hanging it on the wall! Art ought to convey the artist's feelings to the viewer, but we can never expect to fathom the emotions of our alien associates in space exploration.

The artifacts of alien culture, the import of goods from other worlds, and the foreign travelers who come and go give evidence of the fact that Mammonites have become part of a transgalactic society. As such we are sensitive to the criticisms levied against us on other planets. We have been accused of being too materialistic and hedonistic, of being too interested in physical pleasures and unconcerned with spiritual ones. Doomsayers prophesy that such ways will lead to social decay, crime, and the ultimate destruction of our society.

I acknowledge that Mammonites do concern themselves with material well-being, that we spend a great deal of money gratifying our physical pleasures. Yet throughout Human history, people have sought relief from toil and the wherewithal to appreciate the sensuous aspects of life. On Mammon, we are simply fortunate enough to have these things. Not since the 20th century, when the tiny oil-rich nations of Arabia controlled half the world's petroleum, has so small a group of people been endowed with such wealth from the ground. Yet like the ancient Arabs, the wealth of Mammonites is vulnerable too. Just as breakthroughs in nuclear, solar, and coal technology greatly decreased the Earth's dependence on petroleum, so could a technical breakthrough in matter/antimatter technology eliminate the need for rare earths.

I, for one, do not fault people here for enjoying their wealth. We have earned it; literally everyone on Mammon works hard for a living. Little inherited wealth exists, certainly not the sort that characterized ancient monarchies and the early industrial revolution. Mammonites have not grown complacent. Within a century, our industry will satisfy all our material needs. Far from degenerating into a crime-infested society, most people on Mammon value work and personal independence. If we lack enough concern for religion and philosophy, perhaps that's because we are a young society. The opportunities to become rich attracted most of our immigrant population. Now that we have achieved our physical desires, I fully expect the next century to bring accelerated intellectual development. Few of Humanity's great philosophers worked fourteen hours per day. Most had ample leisure in which to indulge their thoughts.

I see parallels between the development of Mammon's society and my own personal life. As a young man, I was interested mainly in money and my narrow technical career. As I matured, I began to appreciate other aspects of life. Personal relationships became important to me. As I rose through the management hierarchy, I began to take a broader view of its purposes. I no longer thought just in terms of how to serve food or fix machinery, but of what motivates people, and how to establish a "permanent" institution that would benefit not only shareholders and employees, but our world's society. I began to view business problems in the context of Human desires, instead of the reverse.

Today I reflect a great deal about the purposes of life. What ought we to be doing with our lives? Human history has been a continual struggle over scant resources and an endless battle with nature. In the process, Humans did great harm to each other, more than natural disasters ever did. Now we stand at the beginning of an age in which our struggle with physical need has been won. We have food, shelter, leisure, and diversions in abundance for all. We have learned to live in peace with not just each other but with alien beings from other star systems as well.

Where do we go from here? What shall be our purpose in living? I can't answer these questions, but when I was young I wouldn't have asked them. I hope to spend more time in the future reading the works of other Humans and aliens who have addressed these questions. I hope I shall find answers, if not for all Humankind, then for myself.

* * * *

OPPORTUNITIES FOR IMMIGRANTS

Mammon Mining Company finances immigration to Mammon and selects immigrants. GAIL acts only as a data-gathering agency, and insures that MMC does not, in any way, misrepresent its contracts. Opportunities exist for most skilled workers, especially engineers, geologists, physicists, and technicians with specialized knowledge of rare earth or matter/antimatter technology. A few unskilled workers, of sound moral character and with demonstrated ability and willingness to learn a trade, also are selected.

MMC requires all workers who do not sign a labor contract to sign a promissory note for payment of their passage. Unlike other colonies, return passage from Mammon may be purchased once outbound obligations are fulfilled. Most immigrants pay off passage in full and acquire substantial personal assets within ten years of their arrival. A fixed exchange rate between Earth Interdollars and Mammon credits is established at 0.43 to one.

Land claims, administered by regional governments, resemble those on other planets. Personal claims vary from five to 20 hectares, depending on location, and may remain unused for ten years. Industrial and mining claims must be developed within two years of filing a claim. Legal claims may be sold and may be purchased in Earth money.

MMC operates four starships of the quad-hex configuration, carrying up to 15,000 immigrants per year. Travel time requires 63 days of ship time and 92 days of planet's time. Emigration currently amounts to less than 1,500 per year.

Yom—Wintery World of Three-Sided Life

IMPORTANT STATISTICS

Planet name: Yom

Equatorial diameter: 11,220 kilometers

Mass (Earth = 1): 0.64

Surface gravity: 0.82 g

Escape velocity: 9.55 km/s

Albedo: 0.40

Atmospheric pressure at mean sea level: 0.71 bars

Fraction of surface covered by land: 35%

Maximum elevation above sea level: 6147 meters

Length of day: 26 hours 52 minutes

Length of year: 345 Earth standard days

Obliquity: 36 degrees

Von Roenstadt habitability factor (Earth = 1): 0.91

Current population: 0.83 million

Number of population centers: 3

Year settled: 2331 adtc

Number of moons: one

Star name: Pi Ursa Major

Star type: GOV

Distance from Earth: 50.2 light years

Distance from planet: 0.92 AU (Earth-Sol = 1 AU)

Star Mass: 0.87 (Sol = 1)

Travel time from Earth

Ship's time: 66 days

Planet time: 97 days

Atmospheric Composition

Oxygen: 29.5%

Nitrogen: 69.5%

Carbon dioxide and inert gasses: 1%

Distance (km)	*Period* (Earth days)	*Mass* (Earth's moon - 1)
180,400	11	0.58

Table 3.7

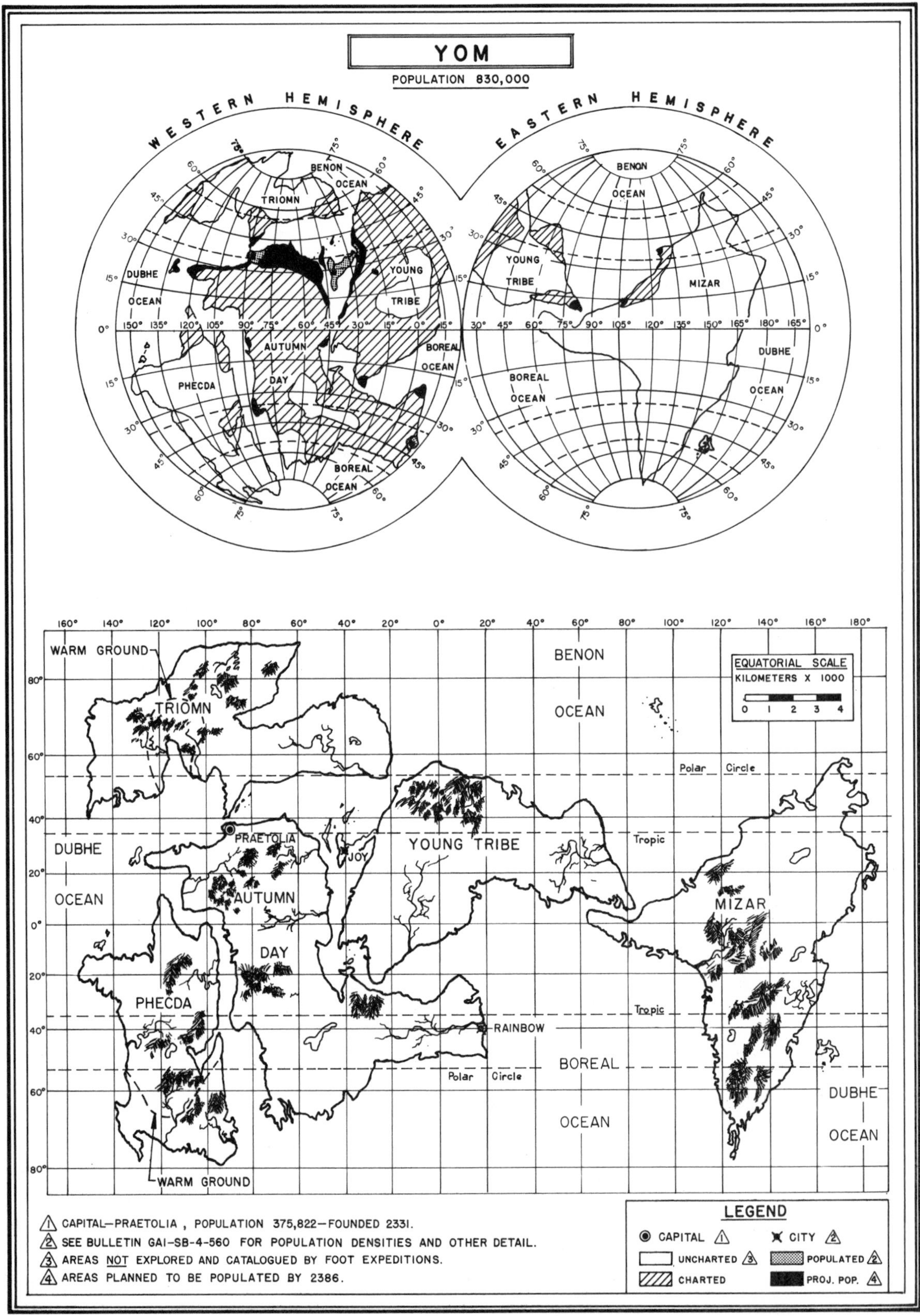
YOM
POPULATION 830,000
WESTERN HEMISPHERE
EASTERN HEMISPHERE
TRIOMN
BENON OCEAN
DUBHE OCEAN
YOUNG TRIBE
AUTUMN DAY
PHECDA
BOREAL OCEAN
MIZAR
WARM GROUND
PRAETOLIA
JOY
RAINBOW
Polar Circle
Tropic
EQUATORIAL SCALE
KILOMETERS X 1000
CAPITAL—PRAETOLIA, POPULATION 375,822—FOUNDED 2331.
SEE BULLETIN GAI-SB-4-560 FOR POPULATION DENSITIES AND OTHER DETAIL.
AREAS NOT EXPLORED AND CATALOGUED BY FOOT EXPEDITIONS.
AREAS PLANNED TO BE POPULATED BY 2386.
LEGEND
CAPITAL
CITY
UNCHARTED
POPULATED
CHARTED
PROJ. POP.

FIG. 3.17

Yom—Wintery World of Three-Sided Life

PHYSICAL ENVIRONMENT

Yom orbits Pi Ursa Major, a type GOV star slightly smaller than Earth's sun, once every 345 standard Earth days in the second position among seven planets. Its nearest astral neighbor, a single moon, circles the planet once every eleven days at a distance of 180,000 kilometers. Since Yom lies 50.2 light years from Earth, the constellations in its sky bear scant resemblance to those seen from Earth, but the brightest stars in Earth's heavens remain clearly visible.

A smaller world than Earth, Yom measures only 88 percent of Earth's diameter and possesses only 64 percent of its mass. Gravitational attraction at the planet's surface pulls with approximately 0.82 g. Dry land makes up approximately 35 percent of Yom's surface, a higher percentage than Earth's. This gives Yom 93 percent of Earth's dry land area despite its smaller size. Oceans divide Yom's land mass into five rather large continents and numerous smaller islands..

Climate

Yom's atmospheric pressure at sea level measures about 70 percent of Earth's, although it has a slightly higher percentage of oxygen in its air. Yom inclines 36° toward its ecliptic plane, considerably more than the 23 1/2° of the Earth. This inclination causes more extreme seasonal temperature variations and causes its tropical and arctic regions to be proportionately much larger than Earth's. Yom's average temperature approximates Earth's, though its polar regions turn colder in winter and its equatorial regions remain unbearably hot. As on Earth, local conditions greatly influence local climates. For example, the high altitude of some equatorial regions makes them comfortably cool.

Yom's most unusual climatological features are the frost-free areas of the Phecda and Triomn continents. The high volcanic activity of these areas has created subterranean geothermal "oceans" which warm the ground. Although not really hot enough to affect air temperatures, the warm ground does prevent the buildup of snow during the winter months, allowing both animals and plants which could not otherwise live in arctic regions to survive.

Yom's day, from which it derives its name, is the second longest among the Human colonies, measuring 27 standard hours. The longer day causes daily temperature extremes slightly higher than Earth's, but the difference creates no significant discomfort for Humans.

LIFE FORMS

Vegetation

Yom's unique vegetation paints a landscape of perennial autumn colors. Green chlorophyl colors Earth's land plants, while on Yom, the chemical that turns light and air into plant material is red, orange, or yellow. Yom's plants exhibit very advanced structures that resemble the grasses, bushes, trees, and ferns of Earth. All but tropical varities are deciduous, losing their foliage and becoming dormant in the winter months. Colors change little in the autumn; leaves just turn dark brown and fall from the plants.

Yom's most advanced plant life rivals Earth's angiosperms in the sophistication of its design. Seeds develop inside an ovary, protected from damage. Unfortunately, none of Yom's ovarian plants display the beautiful flowers found on Earth. Nothing resembling the complex symbiosis between flowering plants and pollen-carrying insects has developed in Yom's ecology.

Despite the fact that Yom's plants bear some fruit that Humans can eat, early pioneers brought native crops with them from Earth. These crops remain the staple food source of Yom's colonists today, though intensive cultivation of native varieties has begun. Earth's plants thrive in Yom's environment, for few native weeds can crowd out the more efficient green chlorophyl plants.

Animals

Yom hosts the most unusual animal life among the Human colonies. On all other planets, advanced animal forms generally have one plane of symmetry, regardless of their phylum. In other words, one half of the creature approximately mirrors the other half. On Yom, however, all advanced species have three planes of symmetry, each at 120° angles to each other. These planes divide each animal into three sections that resemble each other and meet along a central axis. Land creatures on Yom generally have three or six limbs. Six-limbed varieties may have three specialized limbs that aid in eating, or they may use all six for locomotion, resembling self-propelled, two-wheeled chariots.

Marine species, both warm- and cold-blooded, also possess the triplanar symmetry. Some of these animals swim by pushing water back with their fins in fashion similar to Earth's fish and cetaceans; others swim with a corkscrew motion, rotating constantly and looking ahead with one unblinking eye. Symmetry may not be preserved in the internal organs of Yom's animals, just as the internal organs of Earth's animals lack truly symmetrical design.

Most scientists believe the triplanar symmetry to be, in some sense, inferior to the planar symmetry common to most other planets. Many speculate that if large numbers of planar species from another planet were introduced they would supersede the native forms. A minority of biologists dissent, arguing that the triplanar form often offers advantages, and that it occurs so rarely simply because it requires a statistically less probable evolution. The minority viewpoint may be correct for many of Yom's triplanar animals possess extremely well-developed instincts and rudimentary intelligence. In fact, one species, *triangulus dexteralis,* commonly referred to as the "trup," may be the most fascinating lovable animal found on any Human world. Its unique capabilities have made it extremely important to the economic and social development of the Yom colony.

The trup, shown in figures 3.18 and 3.19, is a warm-blooded animal which stands between 110 and 150 centimeters in height and measures between 30 and 45 centimeters in diameter. It has three eyes, three arms, and three legs. On each of its three arms, the trup carries a three-fingered hand. One finger, set at right angles to the other two, serves as a thumb. Trups breathe air through three "noses" located in the sides of their heads. Each nose contains a single nostril and the trup's hearing organs. The trup's mouth opens on top of its head, with its throat passing through the center of a torroidal-shaped brain. A light coating of fur, ranging from reddish-brown to pale gold in color, covers the trup's body.

Like most warm-blooded species on Yom, trups bear their young alive. Unlike Earth's mammal they do not nurse their young from mammary glands. Males and females appear quite similar, though females outnumber males by two to one. Though biologists classify the trup as a single species, several races, or "breeds," have been observed. These vary somewhat in size, shade and shape from one geographic area to the next, similar to the variation among Human races in Earth's early history and prehistory.

Tryps live in small tribal units consisting of fifteen to thirty individuals. When a unit grows larger than this, it usually divides into two smaller groups with each going its separate way. Trups employ a basic language of approximately 500 verbal words and signs. The language contains no abstract concepts and very few adjectives or adverbs. Trups fashion simple tools, such as stone axes, clubs, and shovels. They build small, lean-to shelters to store food and protect themselves from the snow. Trups do not use fire, have no written

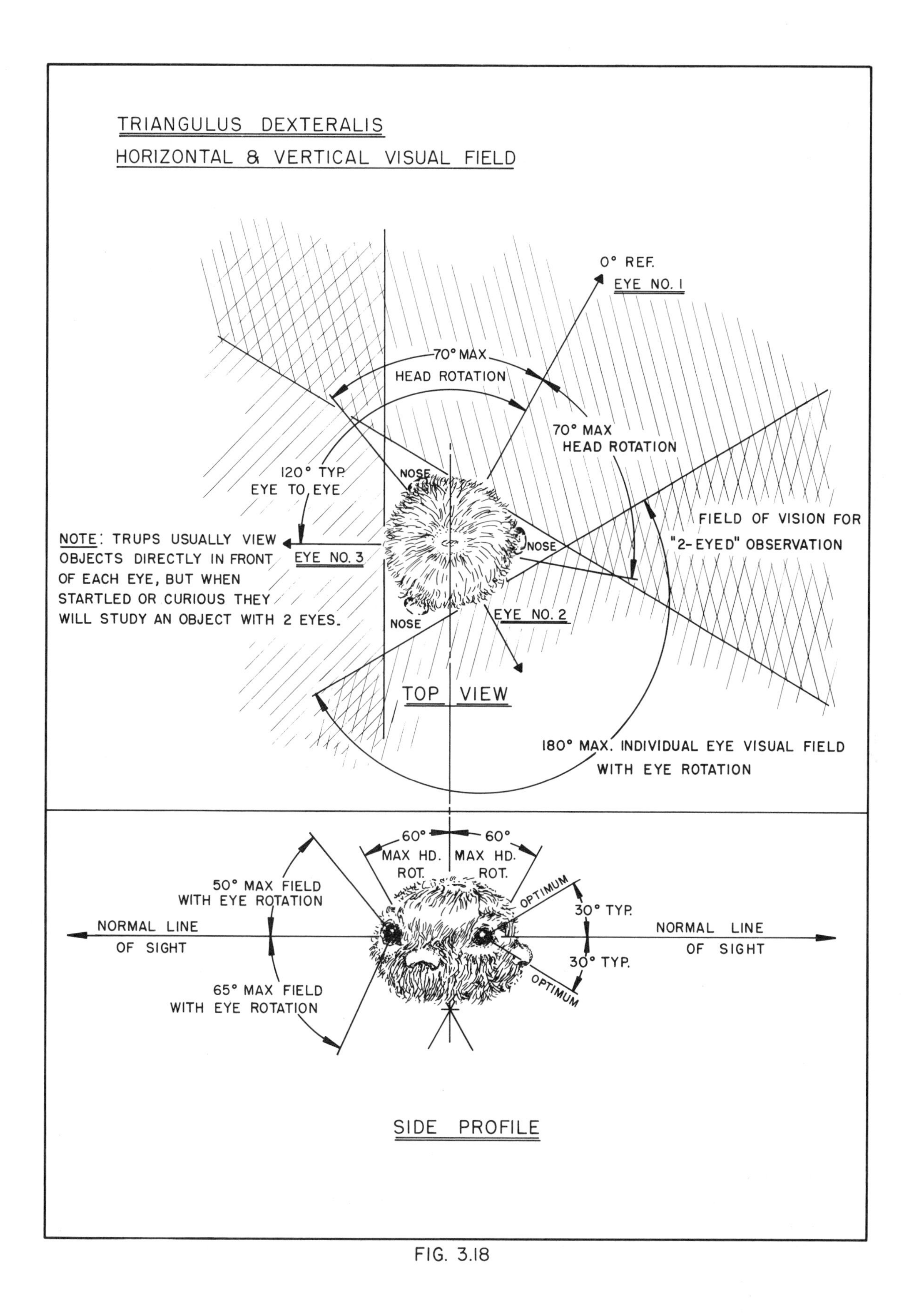
TRIANGULUS DEXTERALIS
HORIZONTAL & VERTICAL VISUAL FIELD
0° REF.
EYE NO. 1
70° MAX
HEAD ROTATION
70° MAX
HEAD ROTATION
120° TYP.
EYE TO EYE
NOSE
NOSE
NOSE
FIELD OF VISION FOR
"2-EYED" OBSERVATION
NOTE: TRUPS USUALLY VIEW
OBJECTS DIRECTLY IN FRONT
OF EACH EYE, BUT WHEN
STARTLED OR CURIOUS THEY
WILL STUDY AN OBJECT WITH 2 EYES.
EYE NO. 3
EYE NO. 2
TOP VIEW
180° MAX. INDIVIDUAL EYE VISUAL FIELD
WITH EYE ROTATION
60°
MAX HD.
ROT.
60°
MAX HD.
ROT.
50° MAX FIELD
WITH EYE ROTATION
OPTIMUM
30° TYP.
NORMAL LINE
OF SIGHT
NORMAL LINE
OF SIGHT
30° TYP.
OPTIMUM
65° MAX FIELD
WITH EYE ROTATION
SIDE PROFILE

FIG. 3.18

TRIANGULUS DEXTERALIS
(COMMON NAME : TRUP)
HEIGHT: 130 - 150 CM
WEIGHT: 40 - 80 KG
(RANGES OF 95 % OF POPULATION)
POPULATION ESTIMATE: 20-40 MILLION

FIG. 3.19

language, and do not paint pictures, make designs, or create art of any kind.

Though omnivorous, the bulk of the trup's diet is vegetarian. Their staple food consists of roots of certain plants dug from the ground with simple shovels. They often store roots in lean-to shelters during the winter months when the ground is frozen. Trups eat some local fruits and seeds when they can and will occasionally eat inquads, small creatures that resemble four-legged versions of Earth's insects. Trups generally eat non-flying inquads and will sit for hours picking them one at a time from the ground. If given cooked or processed meat, trups will eat it, but they have never been observed to eat raw flesh of other advanced animal life.

Before Humans came, trups dominated their planet. Somewhere between 20 and 40 million trups wander Yom's surface, living a foraging lifestyle. Few natural enemies threaten them. Though gentle and passive by nature, the trup's long arms can hurl stones with such impressive force that even the most ferocious of Yom's large carnivores refrain from attacking a group of them.

A curious phenomenon besets most of Yom's animal life. During one hour of each day, all native species become extremely quiet and appear to hide. No sounds can be heard; no cries or movements disturb the silence. The phenomenon has been compared to the silence that proceeds a tornado on Earth, yet it has no apparent physical cause. Silent hour often occurs in late afternoon, but has no straightforward relationship to sunset. Time at which the hour occurs appears to be random from day to day, but always occurs during daypight hours. Length of the "hour" varies between 53 and 71 minutes. Domestic animals brought from Earth seem oblivious to it and continue to act normally. Trups also ignore the silent hour, carrying on their simple tasks as if it did not occur. Silent hour has been observed in all explored areas of the planet.

A BRIEF HISTORY

When Zenon Benon and the crew of the *Boreal* discovered Yom in 2310 adtc, they named the slowly turning planet after the Hebrew expression for "day." The *Boreal's* crew spent the next nine months studying the smallish world and its strange life forms, including the trup. Because the trup had well-developed hands and a rudimentary culture, Zenon feared violating the non-interference policy of the Galactic Association. The crew therefore made most of their observations from space and exercised extreme caution when making occasional trips to the planet's surface. Consequently, though the *Boreal* gathered much useful data about the planet, they learned very little about the living habits or the true capabilities of the trups. When Zenon returned to Earth, he reported Yom to be a planet inhabited by an "advanceable species," ineligible for colonization.

Because such an inhabited world had never been discovered by Humans before, GAIL petitioned the High Council of the Association for permission to send an anthropological expedition to Yom to study the trups. In 2314, Dr. Joy McGillicuddy lead a team of scientists on a three-year expedition, the first of many that would take place over the next sixteen years. They took a great deal of specialized equipment that allowed them to observe the trups undetected at close range in their natural environment. McGillicuddy's team catalogued the habits of the trup in great detail and managed to decipher much of the trups' rudimentary language. Their findings repudiated Benon's conclusion that the trups would eventually develop a civilization. They argued, quite cogently, that though trups indeed possessed a highly-developed hand, they lacked the mental capability to advance beyond their primitive, foraging culture.

The conclusions of the McGillicuddy expedition report ignited a firestorm of debate. In those days, the discipline of predictive sociology was in its infancy, and the most advanced work to date had been done by the Chlorzi. To settle the question, the GAIL High Council authorized a second study, to be participated in by scientists and sociologists from all three GAIL members. The authorization also allowed a non-destructive sampling of 500 individuals from the trup population to be taken aboard GAIL spaceships and subjected to detailed brain scan and analysis.

The interspecies study project launched in 2319 combined all known techniques of the day to analyze the thought processes of the trups into fundamental bits of data. Then researchers, aided by powerful computers, spent the next four years trying to determine if such thought patterns might allow the trups to develop a civilization. Their conclusions ultimately paralleled those of

McGillicuddy's research and determined, with a probability of better than 98 percent, that the trup species, as it now existed, would never develop an advanced civilization capable of increasing its population above the current level or capable of leaving the planet to explore space.

The GAIL High Council authorized the colonization of Yom by Human beings in 2326, with the specific provision that trups must not be abused, mistreated, or indiscriminantly slaughtered. Final planning and preparation for launching the Yom colony proceeded in record time, since Earth's scientists made good use of the time spent studying the trups to determine if and how Humans might live on Yom. The first pioneers from Earth set foot on Yom in 2331, just five years after GAIL gave the authorization to proceed. They established their first settlement at a place they called Praetolia on the northern coast of the Autumn Day continent. Despite the speed with which they conceived and built it, the Yom colony benefitted from the advance

Praetolia's early settlers paid little attention to the trups, for the problems of establishing their tiny settlement took 110 percent of their waking hours. Yet within a few years of their arrival, observant pioneers found that the highly imitative trups could be trained to perform simple tasks. By the end of the second decade, trup labor had become an important factor in the colony's industrial equations and began markedly to affect Human lifestyles on Yom. Now in its 46th year, the Yom colony stands out among the Human worlds for its lack of dependence on automated machinery. An important symbiosis has developed between trups and Humans that has helped both species.

CURRENT STATE OF DEVELOPMENT

Population

Immigrants comprise 540,000 of Yom's 830,000 Human inhabitants. The bulk of the native population consists of children under the age of twenty, though some native adults trace their birth to the early years of the colony's existence. Most of the population lives around three major industrial centers: Praetolia and Rainbow on the Autumn Day continent and the newer Joy on the Young Tribe continent.

Transportation

Limited transportation facilities have kept Yom's population concentrated, despite its abundance of land. A young automobile industry now manufactures a basic antigrav vehicle for personal use. Produced in four- and six-passenger configurations, the vehicle derives its main market from farmers who value its truck-like features. Few families can afford more than one of these basic machines, and many country dwellers rely on flexible-route, antigrav buses to transport them to work or to the cities. A good supply of levitrucks insures that no restrictions occur in Yom's distribution of basic goods, and large hypersonic transports carry passengers quickly and efficiently between the main population centers.

Food

Yom produces abundant quantities of inexpensive food. Many part-time farmers supplement their incomes and their larders with fruits and vegetables grown on their own farms. Yom's farmers became the first people to find a practical use for trups. Today, trups perform many important menial tasks on virtually every farm. Consequently, Yom's farming industry requires less of the expensive machinery used on other planets.

Though Yom contains a variety of seed- and fruit-bearing plants that make suitable foods for Humans, the first pioneers feared to depend on them and brought their familiar crops from Earth. Meat, prohibitively expensive in the early years, has not become extremely popular fare and today most people on Yom eat a largely vegetarian diet, supplemented by an abundant supply of dairy products. Native foods have begun to creep into the Yomite diet, and Yom will no doubt develop its own intriguing cuisine someday. Yom's farmers now cultivate "orchards" of "quard" bushes, which produce small fig-like fruits that are usually served dried. Large-scale farming of "Ibo barley," a hearty grain with a slight coffee-like taste, has begun near Joy. No doubt the most interesting of Yom's native grains is the "sweeta." Sweeta forms a small kernel that looks like wheat but is filled with a nearly pure fructose-like sugar. Bakers make cakes directly from sweeta flour, using no artificial sweeteners, and popped sweeta has become a favorite dessert or snack among children from six to 106.

Consumer Goods

Yom lacks a plentiful supply of consumer goods, and the novelties and luxuries commonly found on Earth can't be purchased at any price. Some ready-made clothing can be bought in rather limited and simple styles. Inexpensive yard goods have induced Yomites to sew much of their own clothing, particularly for dressy occasions, and so a basic, semi-automated cutter-sewer machine has become a standard appliance in nearly every home.

Housing

Spire structures imported from Earth still serve as housing for more than half the population and are well equipped with the most modern of Earth's appliances. People moving to single-family dwellings must give up some of these conveniences, though most can afford manually directed clean fields and a basic food prep center in addition to necessities like preservators, clothes treaters, and space conditioning. Yom doesn't produce any domestic robots, but most country dwellers keep a few trups to do basic cleanup chores.

Communications

Yom's pioneers enjoy the same access to entertainment and information available to people on Earth. A universal "communications center" has become as standard as indoor plumbing. This simple device serves as book viewer, holovision projector, and musical reproduction center. Small remote speakers and terminals extend the device's capability to more than one room. Yom's central computer contains all books, movies, music, and recorded shows produced throughout five colonies and on Earth. Since Yom has no entertainment industry of its own, the dependence on Earth helps the largely immigrant population to feel that they maintain a strong link with their former home.

AN ACCOUNT OF LIFE ON YOM

Editor's note: Indira Hodara, 70, a founder and trustee board member of the University of Praetolia and Chairman of its Department of History and Anthropology, presents a brief account of life on Yom. Indira has written two books on the care and use of trups that have become standard references on the subject. Also, she has edited a scholarly history of Yom's first 30 years. At age 28, Indira and her husband immigrated with the third shipload of pioneers to arrive on the planet. Although she held a doctorate in anthropology from the prestigious Oxford University, she and her husband began their life on Yom as farmers. Indira became interested in a colony of trups that lived in the nearby woods and began to use them as domestic helpers around the farm. Within ten years she had learned so much about these useful animals that she wrote a scientific, yet practical book that would help other people to understand them.

By the time she reached age 54, Indira and her husband had succeeded so well at farming that she left the farm and moved to Praetolia to help found the planet's first university. Amidst this busy career, Indira raised five children and taught all of them herself in the local cooperative school. Indira Hodara's involvement with all aspects of life on her burgeoning new world, as well as her accomplishments as a writer and a scholar, make her extremely qualified to write of life on Yom.

We inserted into final orbit during the sleep shift, but from the moment I awoke, I knew we had arrived at last! The ship felt different, strangely quiet. The daily bustling routine of life in space had ended. I shook Satish, my husband, from a deep sleep. "We're here!" I cried like a child on Christmas morning who can't wait to open the presents. "*Please* can we take a look?" I dragged Satish, still groggy with sleep, to the closed viewport and waited a long moment before pressing the button that opened the shutters. As the covers rolled back, we looked out upon what seemed to be a cold forbidding little world. Three-fourths of the planet appeared in full sunlight. Its most prominent feature, an enormous ice cap, seemed to stretch to its equator. Clear weather over most of the planet's surface let me pick out three of the major continents as if they lay on a map, but frozen ocean around the northernmost continent of Triomn masked its shape.

I shivered and instinctively reached for my husband's arm. Was this the promised land for which we had traveled so far? It looked more like a replica of Earth's ice age! As I studied the surface I noticed the warm ground, a large, brown blotch in the snow, looking as if someone had pressed a finger against a frosty window pane and left a thawed spot. As the ship gradually orbited the planet warm areas came into view. I learned later, that the high inclination of our orbit from the equator made the ice cap look much larger than it is, and I saw, as we orbited the southern hemisphere hours later, that the ice cap there had shrunk to a tiny spot. This first fast trip gave me a memorable introduction to the highly seasonal nature of my adopted world.

We spent most of the "day" in orbit, packing and stowing both the ship's equipment and our own personal belongings for the landing. After we departed from Earth, the daily cycle of activities on board had been increased a few minutes each day until now we lived on a 27-hour wake/sleep cycle. (I actually like this better than Earth's 24-hour cycle because I sleep no more than I ever did and I have three extra hours each day!) By this time the crew had disconnected the great passenger spires from the ship and they floated freely in space. The great space tug lowered the spires one by one during our sleep period. The viewports stayed shuttered to protect against accidental damage during descent.

The spire touched down in the early morning, and we opened the shutters for our first view of the surface. The spire stood on a broad plain of yellow grass that sloped gently toward a white, sandy beach. On the inland side, the plain ended at the edge of a forest unlike any I had ever seen. A dense blanket of red, orange, and yellow leaves shrouded the trees, reminiscent of an Earthly forest in autumn, but unmistakably different. It gave the impression of a holovision set whose color adjustment is way off. Yet nature did paint Yom's foliage in these strange colors. Green can only be seen on farms and orchards filled with plants from Earth.

In the year we arrived, Praetolia consisted of nothing but a cluster of spires and a few low, freestanding buildings. One of these served as an office building and community center, the other as a fabrication shop. A small powerhouse supplied power to the cluster using deuterium from a small separating plant set up in a cove a few miles down the shoreline.

Though Satish and I met while pursuing advanced degrees at Oxford, we both came from farming families. A few years after graduating, we grew tired of playing bureaucratic games at the college where we served as instructors. We longed for honest, straightforward work in a world less bound by tradition. We applied to GAILE's colonial program and requested Yom, then the most recently discovered planet, as our first choice. We knew that our degrees in philosophy and anthropology would be of little use on an unsettled planet, so we emphasized our farming experience in our applications. Shortly after our arrival, we found ourselves solely responsible for operating a farm in an unpredictable climate on alien soil with little knowledge of the pests or diseases that might attack our crops. We faced these difficulties with little modern machinery and few of the chemicals that had become an integral part of farming on Earth.

Vehicles cost their weight in gold in those early years, so during the first few weeks of our stay, we went out each day with a group of new farmers to look for farm sites. With so much land available, we had little problem choosing hectarage for all, but we then faced the formidable task of clearing the forest. The colony had just one demolition unit suitable for clearing farmland, so all the new farmers helped each other with the tedious job. Each day, we commuted together in our own levitruck from apartments in the spires to one of the farm sites. The clearing took most of the summer. When winter came we worked in the fabrication plant prefabricating our farmhouses and the other structures needed by the colony's growing industries. During this time we lived mainly on imported food supplemented by the meager crops produced by the farmers who had begun a year before us. At night we attended lectures to learn what little others knew about farming on Yom. After the prepared talk, the experienced farmers held an open discussion of the problems they had encountered and of any solutions they had found to them. This information sharing continued on a more or less weekly basis for twenty years and did much to make Yom's farm industry a rapid success.

The following spring before the snows had melted, we set our farmhouse and tool shed and began to make preparations for the spring planting. Our buildings had been completely preassembled in one- or two-room sections during the previous winter, and a large, specially-designed levitruck lowered them in place. The rooms rested on special self-leveling jacks that allowed us to compensate for settling each year and let us add more rooms one at a time. Our original farmhouse contained just three tiny rooms. Satish called it a glorified lunch box. Over the years it has grown with our family until now it is quite spacious and comfortable.

Immediately after the spring thaw we plowed our fields and began to plant. Because farm machinery cost so much, we shared a single cultivating machine with a cooperative of nine other farmers. During the plowing season it worked around the clock. Our first year we planted wheat, corn, soybeans, and a truck garden of a dozen vegetables. Farming on Yom took more hard work than I ever imagined; yet I felt immense satisfaction when the plants began to grow.

After the planting, the hectic pace slackened. We still had tools to fix, bins and sheds to erect, and we had to maintain constant surveillance over the crops so that we would be aware of attacks by pests, diseases, or predators in time to take action. No fences or protective fields surrounded our land, and we planted to within three meters of the forest's edge.

One day as I walked along the perimeter strip checking the crops, I was startled to see a creature watching me from beneath the trees. It stood erect, a little shorter than I, and looked at me through two large green eyes. Its head seemed much larger in proportion to its body than a Human's, and a turned-up nose with a single nostril in the center gave its "face" an appealingly cute appearance. A soft coating of reddish brown hair covered its body from head to foot. Except for the action of its three hands, which picked over its fur for flea-like parasites, it stood motionless on three round, flat feet. This was my first encounter with a live trup, and I would have felt quite scared if I hadn't seen pictures of them during shipboard lectures.

For several minutes the trup and I stood still, watching each other. Then I slowly approached it, one hand outstretched in front of me. The trup let me get within one or two meters, then began to move off into the forest. It didn't turn around but simply began to walk. As it moved, it turned its head slightly, and I could see that it had a second "face" in the back of its head! In fact, trups have three "faces" formed by their three eyes and noses. Their bodies have no "front" or "back"; they just scuttle in any direction they wish using their leading arm to push brush or branches out of the way. On impulse, I followed the trup into the forest. It appeared to show no apprehension, since it could keep one eye on me and still use the other two to guide itself.

Less than a kilometer from the spot where we entered the forest, we came upon a group of trups. They were engaged in a variety of activities as I approached. Some broke limbs of trees to build a makeshift lean-to; others approached with armloads of grass that would later be used for a crude thatch covering. Still others dug up the fat roots of native marjom bushes using crude stone axes and pointed sticks. A "nursemaid" trup presided over four tiny young trups, feeding them bits of marjom root from time to time and not allowing them to stray too far.

When the other trups saw me, they stopped their work and stood erect and motionless, watching me. I felt more than a little nervous at this point; though a trup stands shorter than a Human it weighs about the same. The group had ten adults in it, and they could have made short work of me if they wished. I worried about how they might react to an alien intruder their size and what they might do to protect their young. After a long moment in which no one moved, the largest of the adults walked toward me and appeared to ask the first trup a short question to which it gave an equally short reply. The trups spoke with strange chirping voices reminiscent of an Earthly songbird trying to talk. The sound they made put me somewhat at ease. After all, I illogically reasoned, how could any creature with such a voice be dangerous? The larger trup continued to look at me for a few minutes, but as I made no threatening gestures, he chirped a word to the others, and all resumed their tasks without paying further attention to me.

I watched the trups for the rest of the afternoon, my attention riveted to their every move. Although most people might not have found their mundane tasks interesting, they seemed to me the most exciting anthropological study ever! Their behavior patterns looked very complex, and, for the first time, I ceased to wonder how Zenon Benon had mistaken the trups for advanceable life forms in an early stage of development. I didn't realize how late it was until it started growing dark. Jumping to my feet I ran most of the way home, knowing Satish would be worried about me. He seemed relieved though still angry when I returned, but I so bubbled with news of my discovery that his temper cooled. He agreed to go with me to see the trup colony the next day.

The trups remained in the nearby forest during the spring and summer, and I spent what little free time I had left after the farm chores observing them. I kept careful notes and filed a Volunteer Observer's application with GAILE so I could get a camera to photograph them. During the summer, the trups continued to build lean-to huts and to store marjom root. They seemed to have little interest in the crops Satish and I were growing. Satish admonished me not to feed them lest they develop a taste for Earthly foods.

As autumn and harvest time approached, the trups stopped storing marjom root and began to eat what they dug on the spot. Their normally industrious behavior deteriorated and they spend most of their days wandering aimlessly about in the forest looking at things. I had little time to study them, as we were in the midst of harvest and working from before sunup to after sundown. Occasionally the younger trups would come to the forest's edge to watch us work. We got the corn, wheat, and soybeans in without incident, but disaster nearly struck as we tried to harvest our largest crop—tomatoes. We planned to cryofreeze these for sale during the winter. The cooperative's one tomato-picking machine, a cantankerous piece of equipment at best, broke down. The machine shop couldn't deliver parts for several weeks and by that time the tomatoes would have spoiled.

Since other farmers had their own harvests to worry about, we could count on no help. With a hectare of tomatoes before us, Satish and I bravely began to salvage what we could by hand picking. As we worked, the trups watched us from the field's edge. We picked tomatoes and put them into chemical drums which we had cut in half and sonically cleaned. The next day, when our turn came to use the truck, we planned to carry them to the freezing plant and empty them. Suddenly, for no apparent reason, one of the trups stepped into the field and began to pick tomatoes and place them in the barrels! This was no instinctive behavior; it had learned by watching us! The others soon joined in and together the four of us picked twice the tomatoes we had planned that day.

As we worked, Satish and I would occasionally clean and eat a tomato. After doing this several times, we offered some to the trups, in hope that, if fed, they might return to help us the next day. The trups appeared to like tomatoes and ate several as they worked, though they made no attempt to stuff themselves. The following day, four trups returned and continued with the harvesting. All of them worked steadily, except for occasional pauses to eat tomatoes. The number of trups working increased gradually each day until we had a total of eight, and within a week we had all of the tomatoes harvested. The other members of the co-op couldn't believe it when they heard our crop was in, and they were incredulous when we told them how we did it.

Even now I'm not sure why the young trups began to imitate us that day. They had never eaten tomatoes before, and they hadn't previously shown any desire to help me or any other Human beings. Regardless of the reasons, I knew the behavior we witnessed would ultimately revolutionize Yom's farming industry. Suddenly the trup studies that I previously regarded as a hobby had gained considerable economic value.

The winter months passed quietly during Yom's early years. We had little to do but repair and service the equipment we used during the growing season, attend Grange meetings, and fix up the house with little personal touches to make it seem like home. Every day I managed to steal a few hours to bundle up in my winter clothes and go out to observe the trup colony. Winter created a difficult time for the trups. They spent most of it huddled together in their huts, living off the roots they stored during the summer months. They had no fire to keep them warm, and their grass and twig shelters made poor defenses against the damp cold and winter winds. The first winter, two older adults died, and the other trups carried their bodies into the forest and buried them in shallow graves they scratched in the frozen soil with sharpened stones. Watching their pathetic labors brought tears to my eyes, for it appeared as great a tragedy as death in any Human family.

During this time I spent many hours musing about how we could encourage the trup colony to stay near our farm. They had dug up most of the marjom root in the area and I suspected that they would move to a new place in the spring. I finally hit upon an idea; perhaps we could encourage the trups to stay by teaching them to eat our crops: carrots, turnips, or potatoes. We could also build them little houses of titanalum sheet, or even wood, which would keep them warmer and drier than their thatch-covered huts. When I told Satish, he seemed very enthusiastic about the idea and began to work on a design for the trup house at once. Nobody could get titanalum sheet or any other essential building material that winter unless they had ordered it months before. We did have a universal w-field cutter able to cut literally any substance with high precision, and we had an abundance of wood in the large trees that surrounded our farm. Using natural materials, Satish fashioned a three by five-meter hut at the edge of the forest near the trups' camp. Meanwhile, I obtained a detailed chemical and nutritional analysis of marjom root and began to compare it with crops from Earth in an attempt to formulate a balanced trup diet.

Convincing the trups to live in their new home took considerably more effort than its construction. Trups have rigid habits and strong instincts, so they weren't about to forsake their painstakingly constructed huts or their piles of marjom root for some unorthodox scheme. Though they will imitate many actions, in this case we could offer no clear example for them to follow. We tried a variety of lures including digging marjom root out of the frozen ground and placing it in the shed. Nothing worked until one very cold night when Satish went to the trup camp carrying a small, portable space heater. The trups came out of their huts to crowd around the warmth, and when they all had gathered about him, Satish slowly walked to the new shelter. The trups followed him all the way there and stayed clustered inside with the heater even after he left them. Satish and I then returned to the trup camp, loaded our antigrav sled with marjom root and carried it back to the shelter. We mixed this food with carefully selected proportions of many root vegetables: carrots, potatoes, sun chokes, turnips, and radishes. With warm shelter and a supply of food, the trups no longer desired to return to their huts. Throughout the winter, we gradually thinned all the marjom root out of their food mixture as the trups became used to the imported vegetables.

Since that day, the trup colony has remained a permanent part of our farm and now numbers about 40 individuals. Although we have built them modern houses with titanalum-sandwich walls, concrete floors, radiant heating, and windows, we have earned this modest investment back a hundredfold by the work the trups have done for us. The year after their arrival, we expanded the vegetable hectarage under cultivation and began an apple orchard. Trups have been helpful not only in picking, but in sorting, weeding, and cultivating these labor-intensive crops as well. Trups have also reduced our need for expensive farm machinery by harvesting our fruits and vegetables by hand. During the first decade that the trups lived with us I learned much about their dietary, mating, and territorial habits, as well as some basic trup medicine. Though I can't really speak their language I learned to understand it very well. We have developed a sort of pidgin-trup dialect that allows us to communicate with them very effectively.

When our fellow farmers saw the trups at work, they too wanted to use them. I dispensed a lot of free advice about trups in exchange for a good deal of help with our horticultural problems. While I was pregnant, Satish and the trups took over a lot of the physical labor that I wasn't able to do so production continued to rise. In our fifth year, an infestation of macaws, small reptile-like flying creatures, threatened to decimate all crops. Satish and I fashioned great broom-like flyswatters. With these in hand, the trups guarded the fields night and day to keep the macaws away.

After ten years of living and working with the trups I decided that my knowledge of these amazing animals might be valuable to fellow pioneers. Pregnant with my last child and with lots of time on my hands, I began to write my first book about trups. Though it is a serious work, filled with many well-documented, vital facts, in a lighthearted moment I gave it a rather flippant title, *Raising Trups for Fun and Profit*. The book became more popular than I had ever imagined. Today I am preparing the sixth edition with added emphasis on new medical discoveries and the use of trups in industrial as well as farming economies.

The use of trups as workers raised many sticky legal and ethical problems for Yom's pioneers. Our claim to the planet, authorized by GAIL, prohibits mistreatment of trups. As their use became widespread Yomites had to devise laws for their protection. Though trups may be kept by people they may not be owned in a legal sense, nor may they be restrained from leaving when they wish. In practice, trups tend to remain in the same place when given proper care. We regard our forty trups almost as family members, and we have given each of them names to which they answer. Less than ten percent of all the trups we have kept ever wandered off.

Some pioneers make a business of rounding up trups for use by others, but they cannot legally be "sold," nor can their "keepers" legally manacle or incarcerate them. Physical or mental abuse of trups threatens the rights of the entire Human population. Yom law dispenses harsh judgments for mistreatment of trups, including stiff fines, loss of the privilege of keeping trups, and ultimately deportation.

My years on the farm watching our children and our enterprise grow to maturity seemed the happiest of my life. During that time we saw our planet grow too and felt great satisfaction at having been part of the tremendous venture that took a primitive world and made it a good place to live. Yet by the time I reached 54, my children had grown into adults or teenagers with activities and interests of their own. The farm had grown as large as we wanted it to, and it ran so smoothly that the challenges it offered in the early years had disappeared.

My scholarly work with the trups revived my interest in history and anthropology again. Because of my education in the humanities and because I took an active interest in my children's education, I worked as a teacher in our regional cooperative school. At the school, I met and became close friends with Mary Roberts, a teacher of chemistry and general science. At that time, my youngest and her eldest children were reaching maturity, and we thought they and the generations that followed them should be able to continue their educations beyond the basics offered by the current system.

Neither of us pursued this idea until we met Harold Bartholome, a Doctor of Physical Science, at a teaching workshop in Praetolia. He also wished for a university on Yom but didn't feel up to the task of starting one alone. The three of us formed, in Harry's words, "a critical mass." We began our campaign to continue the tradition of higher education on Yom. None of us knew exactly how to begin to found a university. On Earth such institutions had existed for centuries, having amassed huge pools of capital which supported them in perpetuity. We found numerous books in the planetary computer library on Earth's history of education, but most of these dreary things gave little practical advice on how universities actually begin.

After mulling the problem over, we decided we needed three ingredients: students, teachers, and a place to teach them. Getting the three together at once became a "chiken or the egg" type question. Without teachers we couldn't attract students, and without both we certainly couldn't afford a campus. We decided to go after the faculty first and began by contacting every person we could find who possessed a doctoral degree from a university on Earth. We had concocted a list of specialties for which we needed faculty members, and our plan consisted of trying to obtain teaching commitments from one or two qualified people in each field. These people were to remain in their current jobs until the college opened, then assume their teaching duties as we needed them.

While looking for faculty, we filed a special land claim for a university campus of 150 hectares near Praetolia. We chose this site because it would enable the university to avail itself of Praetolia's facilities, such as housing, stores, and services. It would also provide us with a pool of part-time students whose tuitions would provide much-needed revenue. We asked for a five-year extension of our claim's development provisions because we didn't know how long it might take to get our project off the ground. Though this required a vote of the Council of Governors, we secured approval without much debate.

After claiming the land and getting commitments for faculty, we began to solicit both students and money. Neither seemed plentiful in Yom's thirty-second year. Most young people preferred to become independent, claim land of their own, and begin their careers. High school graduates could obtain specialized education in a variety of fields from computer programming and electrokinetics to farming. Few desired to pursue several more years of intensive study to find themselves equipped with no specific skill, trade, or prefession. Many highly intelligent young people openly questioned the purpose of studying literature, history, art, anthropology, music, or philosophy. Such things seemed irrelevant in a world so new and so full of need. We persisted in our efforts to attract students precisely *because* no one seemed interested. We all felt it would be terrible for Yom if all its people lost contact with its past, if they forgot how to appreciate the art and philosophy developed during 6,000 years of civilization. Perhaps reading Homer will never butter anyone's bread or build anyone's levicar, but it will uplift the souls of those people who read and appreciate works like Homer's. Without a living, continuing appreciation of Human culture among a significant percentage of the population, our links to Human heritage will soon rust and break despite the fact that all the works of art, literature, and music lie at our fingertips in computer files.

Soliciting money for the University discouraged us even more than soliciting students. Few people on Yom, even wealthy people, had capital to spare in the early years. Most of those we approached questioned the value of the University, calling it an obsolete institution in an age when computers allow instant access to all information. Many told us that we should teach only "practical" subjects like engineering, law and chemistry. Still, we managed to obtain a few credits here and there for our endowment fund. We also managed to secure loans against our future tuition receipts. Our most important contribution came from a building fabricator, John Weiskoff, who donated a small office building containing six classrooms and a dozen tiny offices.

Despite these obstacles, the first classes commenced five years from the date that Harold, Mary, and I first conceived the project. It had been a harrowing five years, often separating me from my family, and had taken me over most of Yom's inhabited areas. Had my family not wholeheartedly supported my efforts and cheerfully picked up my workload on the farm, I would have abandoned the project.

The first University of Praetolia curriculum covered eight major subjects taught through six departments: Art and Human Literature, Philosophy, Chemistry, Physics, Biology, and History and Anthropology. The first year, each had only one faculty member and a handful of students, but we attempted to set standards as high as those of any university on Earth in the hope that the value of what we did would manifest itself in the quality of our graduates. Each department justified its own budgets in proportion to the number of students it had. Though we encouraged professors to do research and write, we never coerced them to do so, and all funding in support of research had to be solicited by the professors themselves.

The early years tested the dedication of the entire faculty, but the quality of the service we offered proved itself, and our student body grew. As our graduates passed on to other careers, they remembered and appreciated the value of their experience with us. Contributions, both small and substantial, began to flow in. During most of its existence, the University walked one short step ahead of the bill collectors. We channeled any spare money we had into more classrooms, laboratories, and dorms. The departments of biology, chemistry, and physics secured important research contracts from local industry, and Harry Bartholome founded a department of Astronomy and Astrophysics to augment the research undertaken by the GAILE space station.

Today, the hungry years have passed. Our University has established itslef as a major and permanent educational and research organization on Yom. My History and Anthropology department, now with seven full-time professors and more than sixty students, is growing by 12 percent each year. We can't do much original research in the field of anthropology on Yom since the Human society here is just a small transplanted version of 24th century Earth society. Largely because of my role as department chairman, however, study of trups fell to the Anthropology department and became an important source of support. Grants from the Yom Farm Bureau and the Manufacturing Institute of Yom paid for important research into the nature of trups. I am particularly pleased by our interdisciplinary study program which combines sociological, biological, and veterinary studies of these fascinating and important animals.

Today I have more time to devote to my own research, partly because I need not spend so much time out hustling students and funds, and partly because my children, now fully grown, live lives of their own. I spend weekends with my husband, who remains on the farm, and I consider that I enjoy the best of both worlds—the academic and the practical. I look forward to the next twenty years and hope to help the University grow. I enjoy working with my students as a teacher, counselor, and director of their research projects. The University now maintains a good balance between teaching and research. I hope to maintain this balance so we can serve our planet most effectively. The University must remain a lean, efficient organization, and as it grows I shall strive to prevent it from becoming the sort of bloated bureaucracy that universities became on Earth.

No Human being can hope to influence society much after her life has ended. As Yom grows and changes from a frontier society to a mature one, the values and ethics of its people will change. I can only hope to impress my values upon the generation that succeeds me and hope these values will withstand their scrutiny and be passed on by them. This personal goal is, in a sense, the goal of everyone on Yom. We build not just for ourselves today, but to create a future world better than the one from which we came.

OPPORTUNITIES FOR IMMIGRANTS

Yom offers practically unlimited opportunities for all types of applicants. Lack of a specific skill poses no obstacle to those who wish to emigrate to Yom, though skilled trades people and professionals in every field are urgently needed. Medical doctors, electronics manufacturing technicians, cooks and engineers and designers of all kinds remain in greatest demand.

Land may be claimed in every major populated area as well as in uninhabited areas. No limit exists on farm or industrial land claims, although use provisions similar to those of the other colonies are stipulated. Personal housing claims of 20 to 30 hectares, divided into several parcels, may be made. Immigrants may continue to occupy the starship spires in which they traveled for unlimited periods, paying only a nominal maintenance fee.

GAILE currently operates a starship of the quad hex configuration which makes two trips annually between Earth and Yom carrying 6,000 pioneers per trip. Travel time requires 99 days of planet's time and 68 days of ship's time.

Romulus—Fairer of the Twin Worlds

IMPORTANT STATISTICS

Planet name: Romulus
Equatorial diameter: 11,600 kilometers
Mass (Earth = 1): 0.72
Surface gravity: 0.87 g
Escape velocity: 10.0 km/s
Albedo: 0.38
Atmospheric pressure at mean sea level: 0.77 bars
Fraction of surface covered by land: 29%
Maximum elevation above sea level: 5928 meters
Length of day: 43 hours 58 minutes
Length of year: 360 Earth standard days
Obliquity: 19 degrees
Von Roenstadt habitability factor (Earth = 1) 0.98
Current population: 0.44 million
Number of population centers: 2
Year settled: 2360 adtc

Number of moons: one (Remus, a twin planet system)
Star name: Upsilon Lupus
Star type: G2V
Distance from Earth: 58.2 light years
Distance from planet: 1.05 AU (Earth-Sol = 1 AU)
Star Mass: 1.19 (Sol = 1)

Travel time from Earth

Ship's time: 68 days
Planet time: 99 days

Atmospheric Composition

Oxygen: 25.6%
Nitrogen: 72.9%
Carbon dioxide and inert gasses: 1.5%

Distance (km)	*Period (Earth Days)*	*Mass (Earth's Moon = 1)*
71,436	*1.83*	*64.2*

Table 3.8

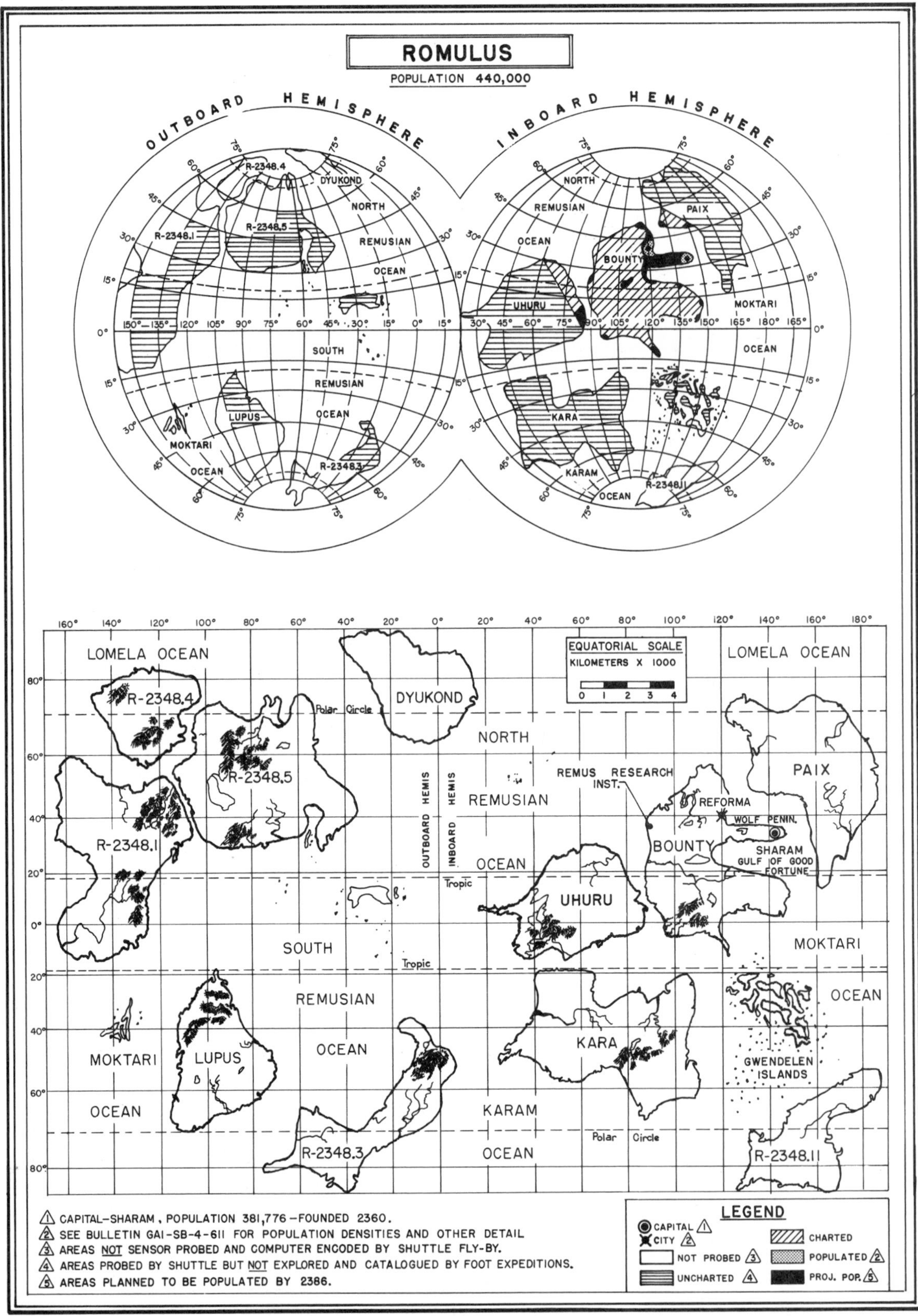
ROMULUS
POPULATION 440,000
OUTBOARD HEMISPHERE
INBOARD HEMISPHERE
R-2348.4
DYUKOND
NORTH
REMUSIAN
OCEAN
R-2348.1
R-2348.5
SOUTH
REMUSIAN
OCEAN
LUPUS
MOKTARI
OCEAN
R-2348.3
PAIX
BOUNTY
UHURU
MOKTARI
OCEAN
KARA
KARAM
OCEAN
R-2348.11
LOMELA OCEAN
LOMELA OCEAN
EQUATORIAL SCALE
KILOMETERS X 1000
0 1 2 3 4
Polar Circle
Tropic
OUTBOARD HEMIS
INBOARD HEMIS
REMUS RESEARCH INST.
REFORMA
WOLF PENIN.
SHARAM
GULF OF GOOD FORTUNE
GWENDELEN ISLANDS
1 CAPITAL-SHARAM, POPULATION 381,776-FOUNDED 2360.
2 SEE BULLETIN GAI-SB-4-611 FOR POPULATION DENSITIES AND OTHER DETAIL
3 AREAS NOT SENSOR PROBED AND COMPUTER ENCODED BY SHUTTLE FLY-BY.
4 AREAS PROBED BY SHUTTLE BUT NOT EXPLORED AND CATALOGUED BY FOOT EXPEDITIONS.
5 AREAS PLANNED TO BE POPULATED BY 2386.
LEGEND
CAPITAL 1
CITY 2
NOT PROBED 3
UNCHARTED 4
CHARTED
POPULATED 2
PROJ. POP. 5

FIG. 3.20

Romulus—Fairer of the Twin Worlds

PHYSICAL ENVIRONMENT

The twin planets, Romulus and Remus, orbit type G2V star Upsilon Lupus located about fifty-eight light years from Earth. As figure 3.21 illustrates, the planets orbit each other about their common center of mass occupying the third position in the planet system of Upsilon Lupus (refer back to figure 3.1). The mass of Romulus measures less than 0.3 percent greater than the mass of Remus and both planets move in a nearly circular orbit about their mass center, forming one of the great natural wonders of the known galaxy.

Because just 71,000 kilometers separate the two planets, their gravitational attraction has stopped their spin with respect to each other, just as the attraction of the Earth has halted the rotation of the moon with respect to it. Since Romulus and Remus keep the same side toward each other, their day nearly equals the period of their orbits, about 44 Earth standard hours. Neighboring Remus can be seen only from the "inboard" hemisphere of Romulus and remains in the same position of the sky both night and day. When full, Remus appears about 18 times the diameter that Earth's moon appears from Earth, but because it possesses a higher albedo Remus shines more than 1700 times brighter than Earth's moon. Because it is so bright, clear nights are never dark on the inboard hemisphere. Remus appears as a half circle at dusk each day, grows to fullness at midnight, and diminishes to a half disc again at dawn. The absence of total darkness on the inboard half of Romulus has had interesting effects on the development of the planet's lifeforms which will be explained in the following section.

Smaller planets than Earth, Romulus and Remus measure 91 percent of Earth's diameter and 75 percent of its mass. Gravitational attraction at the surfaces pull with 87 percent of the Earth's force. The twin planets are similar in many respects. An oxygen-nitrogen atmosphere surrounds both planets and oceans cover a majority of their surfaces. Both planets teem with highly complex forms of life, though colonization by Humans is permitted on Romulus only.

Dry land covers 29 percent of the surface of Romulus, giving the planet a total land area 83 percent as large as Earth's. Oceans divide the land into eleven small continents distributed on the planet's surface in three groups. Because Humans have lived on Romulus for a relatively short time, several of the continents remain as yet unnamed.

Climate

The peculiar orbit of Romulus has important consequences for its weather. Figure 3.21 illustrates how the plane of the planet's orbit tilts approximately 19 degrees from the plane of the primary orbit. This causes the planets to have seasons that are somewhat less extreme than Earth's and also prevents the planets from eclipsing each other except at the equinoxes. Romulus' very long day causes higher daily temperature extremes than Earth's, and thermal winds generated by these temperature extremes reach high velocities in coastal regions. Fortunately, the planet's lower at-

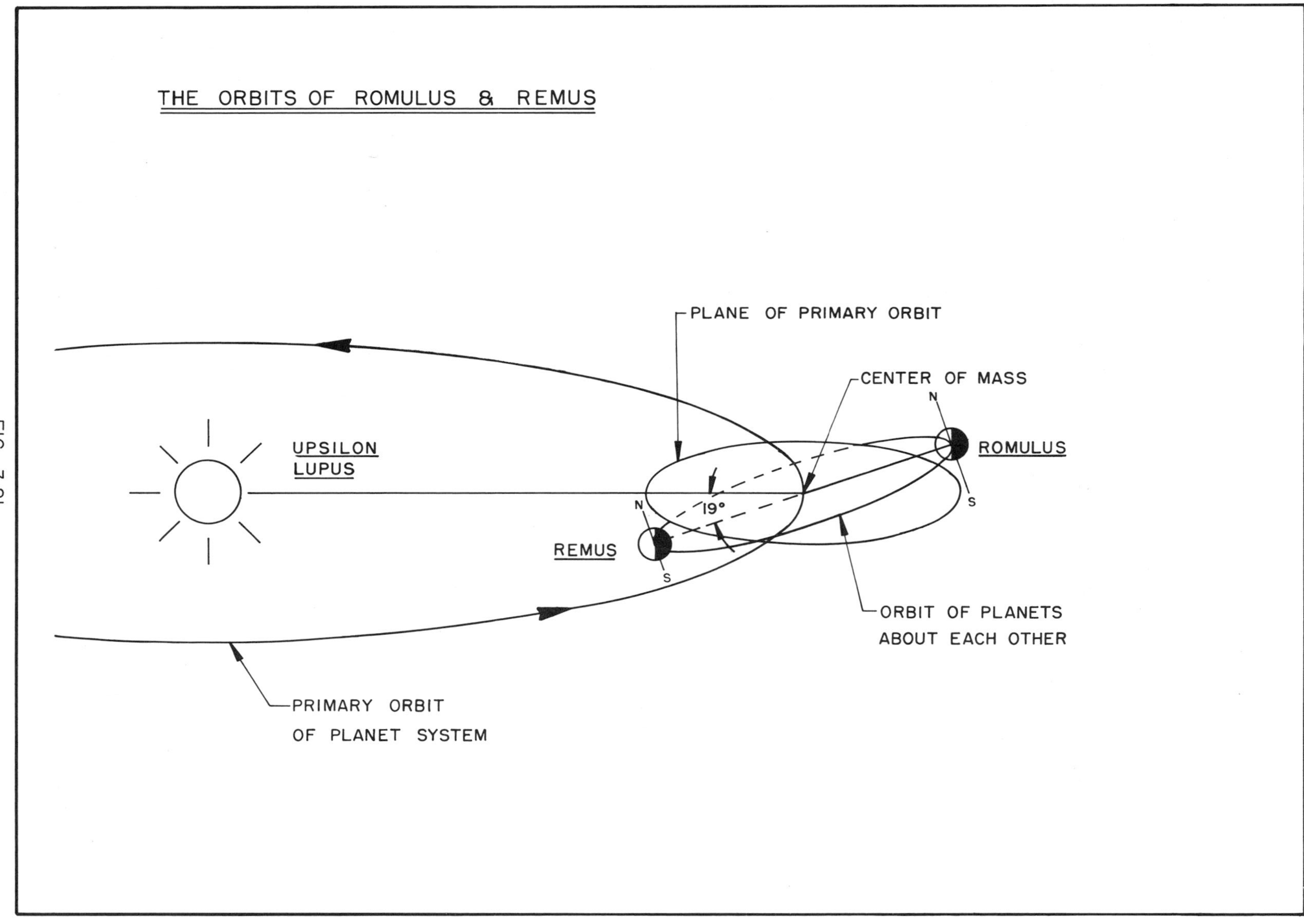
THE ORBITS OF ROMULUS & REMUS
PLANE OF PRIMARY ORBIT
CENTER OF MASS
N
ROMULUS
S
UPSILON
LUPUS
N
19°
REMUS
S
ORBIT OF PLANETS
ABOUT EACH OTHER
PRIMARY ORBIT
OF PLANET SYSTEM

FIG. 3.21

mospheric pressure causes them to have less destructive force than comparable winds on Earth.

LIFE FORMS

Both Romulus and Remus offer physically and biologically suitable homes for Human beings, but GAIL has approved only Romulus for Human settlement because Remus is inhabited by a primitive, yet highly intelligent and advancable species. As such, Remus is excluded from colonization by any Association member, so that its race may develop naturally without external interference. More details will be given about this fascinating life form in the description of the Remusan study project given by Cheryl Cooper in the following section.

Although it is nearly identical to Remus in age and in the state of evolution of its life forms, Romulus has brought forth no intelligent life. Life apparently evolved distinctly on both planets, since all life on each planet shares certain biochemical properties unique to the planet. Species and classes of animals on Romulus and Remus are, of course wholly different. The differences between life forms of the two planets give evolutionary scientists important evidence to substantiate theories of the development of life.

Vegetation

Plant life on Romulus is highly developed in comparison with that of the older colonial worlds. The most numerous species resemble primitive forerunners of Earth's angiosperms, the flowering plants. Some of these plants produce grains and fruits that are suitable food for native animals, but most of these have proven unsuitable for Human palates. Plants on Romulus grow in a variety of sizes from large trees to microscopic aquatic species.

Animals

Romulus hosts a variety of complex animal forms ranging from small creatures with external skeletons, resembling the insects and crustaceans of Earth, to larger warm-blooded animals which fill most of the ecological niches of Earth's mammals, save that of the man. Taxonomists classify the warm-blooded species as "ovarals." They appear in flying varieties resembling birds, four-legged land species and marine varieties similar to Earth's porpoises and whales. The offspring of ovarals are born alive, but unlike Earth's mammals, overals do not nurse their young through mammary glands. Instead they lay protein-rich pellets, which the newborn offspring eat. Just as cows' milk has become a food for Humans, Romulans consider the food pellets of ovarals to be a delicacy. Pioneers raise large ovarals, called "cows," primarily for the "ovals" they produce.

The ovarals of the sea are among the most intelligent life on the planet. The most intelligent land life appears to be an order of large, cat-like predators, some of which are potentially dangerous to Humans. Fortunately, Human technology has provided colonists with defenses against these animals in the form of force fields, laser guns, chemical repellants, and sonic "stun guns." Because of these defenses, people can allow the predators to remain in their natural habitats without fearing their dangerous behavior.

A BRIEF HISTORY

The twin worlds discovered by Captain Tshuapa Lomela and the crew of the *Kara* in 2342 remained uncolonized for eighteen years. The need to further refine the Galactic Association's noninterference policy caused much of the delay. No primitive but advancable life form had ever been discovered in a star system that also contained a potential colony planet. The debate over Romulus settled this question, the outcome being that an intelligent, but still planet-bound race is entitled to freedom on its planet, but it does not ipso facto lay claim to all habitable planets in its star system. The Ergints' opinion carried the most weight. Their arguments implied that they too had once inhabited planets in the same star system with primitive, but intelligent life, perhaps even the systems of the other Association members.

To ensure that Humans are not tempted to interfere with the development of the Remusans, other Association members make periodic surprise visits to the twin worlds. Interference with the Remusans by Humans will cause forfeit of the colonial rights on Romulus. GAIL has made a partial exception to this strict policy. The development of any intelligent life form from the primitive state to an interstellar technology is, of course, of great interest to all GAIL members. For this reason, GAIL maintains a permanent surveillance project on a space station orbiting Remus. Scholars from all of the Association worlds participate in this project, but Humans naturally provide most of the workers. Should the presence of this investigation, through accident or carelessness of some individual, cause the Remusans to become aware of the Association members, this, of itself, shall not be deemed a violation of the noninterference treaty.

The first pioneer on Romulus grounded in 2360 adtc, making Romulus the youngest of the worlds presently settled by Humans. The pioneers chose the tip of a long peninsula extending off the continent of Bounty at 35°50′ north latitude for their first settlement, Sharam. The climate at this spot is mild year-around with moderate rainfall. The shorelines border on a large sound formed between the continents of Paix and Bounty and are therefore protected from the full force of ocean storms.

CURRENT STATE OF DEVELOPMENT

The initial shipload of colonists numbered 4000, but before a year passed more than 12,000 had arrived. Development of the colony proceeded peacefully and efficiently, for the colonists had the experience of six new worlds before them on which to base their plan. Rather than pursue a purely free enterprise economy, the Romulan plan called for the establishment of community-owned industry. Thus telecommunication, power generation, primary metal production, housing, appliance manufacture, production of transport vehicles, and medicine are the province of large monopolies in which all colonists own an interest. The planet-wide constitution reserves farming, clothing manufacture, and all other small-scale industries and industries not yet developed to individual entrepreneurs. The Constitution also guarantees the property rights of all citizens and no laws prohibit competition with the community-owned industry. It is too early to say how this form of economic organization will perform over the long term. To date, economic progress on Romulus has been as fast as can be expected on any new world.

Research and Development

At this writing, the lifestyle of people on Romulus is changing so quickly that detailed description of the state of development would be out of date by the time this bulletin reaches the reader. The map of Romulus shows how much of the planet remains unexplored and detailed aerial reconnaissance has not even taken place over large portions of the planet's surface. At this writing, pioneers have explored only the continent of Bounty, the only one inhabited at this time. On the outboard hemisphere, little more than orbital reconnaissance has taken place over several of the unnamed continents. Within the next ten years, Romulans hope to complete shuttle fly-by reconnaissance of these regions. For years to come, though, much territory will remain to be explored on the ground by adventurous pioneers.

The Bounty population is spreading rapidly into the remote areas of the Wolf Peninsula thanks largely to the availability of crude but rugged antigravity cars. Improved transportation allowed the creation of a second industrial center named Reforma at the base of the Peninsula just 12 years after the colony commenced, and Romulans expect it to be a focal point for even more rapid development.

Industry

The Industrial Planning Board of Romulus has ventured some projections which should give the potential pioneer an idea of what he or she can expect of Romulus in 2386 adtc. Beginning in 2379, all pioneer's spires will ground at Reforma instead of Sharam, to prevent excessive congestion at the original population center. Every ten or fifteen years, Romulans hope to found another city to absorb incoming colonists and to promote development of other areas of the planet. The industrial plan will emphasize increasing production of prefabricated building materials and consumer goods. At this writing, Romulan industry is confined primarily to assembly industries, relying on importation of many critical items such as electronic components, medical devices, biopolymeric materials, and nuclear fusion reactors. During the next decade Romulans hope to begin original manufacture of some of these critical items. First efforts will begin with those components that are needed as replacement parts for existing equipment and expand into items of original manufacture or design.

AN ACCOUNT OF LIFE ON ROMULUS

Editor's Note: Francoise Patreau, 59, one of the 4000 original pioneers on Romulus, gives an account of the beginning of the Romulan colony. Francoise left a successful career as a communications executive on Earth to emigrate with her family. Her industry experience caused her to be elected vice president in charge of operations of the planetary communications company, a position she held for ten years. She succeeded the chief executive of the company, and she holds that position to this day. At this writing, she also serves as a member of the Industrial Planning Board of Romulus. Though Francoise has no experience as a professional writer, she is an articulate spokesperson for the Romulan way of life.

Why does a rising communications executive who's just approaching the prime of life quit her job, forfeit all Earthly assets and start over again on a wild, uncivilized world half a light century away? I made the decision on a sultry afternoon eighteen years ago as I sat in my 152nd-floor office overlooking the Paris region. Outside, thick, yellow haze blurred all detail of the rows of building that seemed to stretch to infinity. Long ago, Paris was a city, and at its edges lay farms, woods, and fields. Though people had manicured and civilized the European continent, other parts of Earth contained virgin forests filled with wild animals. In 2357, all of Europe lay under a blanket of Human structures or chemical farms, stripped of all natural life and cultivated to the limit. This depressed me.

I turned once again to my editing screen and looked at the report I was unable to continue writing, yet another revision of the *Economic Justification for Upgrading Central Infotranscore in Service Region 138*. For each person that actually lifts a driver or a bonder, there had to be ten people to write justifications for him to do it. Why? When I was an engineering student, developing ways to improve communications between people seemed a worthwhile endeavor. Now twenty years later, with any other person on the globe at my fingertips, it seemed pointless. In my youth, I thought it foolish to leave the comforts of Earth and risk one's neck on one of the colony worlds, but now?

I cleared my editing board and entered the code to link me with the Eurodata library. As soon as I connected, I requested the index of GAILE publications. An enormous list began to roll by before me. It contained so many documents, I didn't know what to ask for; so in desperation I requested the latest issue of the bi-weekly *Pioneers Update*. The lead story announced that GAILE had opened applications for people who wished to colonize Romulus, the uninhabited planet of the twin worlds discovered fifteen years before. A new world! The idea resonated in my soul! I turned back to the index and called up all publications on Romulus.

I read all afternoon until well past quitting time when my husband called. His face bore concern as he looked at me from the screen. "Is everything all right, Francoise?" he asked.

"Pierre," I replied, "I am going to Romulus. I wish you and the children would come with me, but if you do not, my decision is unaffected."

His eyes opened very wide, his jaw dropped and for a long moment he said nothing. Finally, he stammered, "You will be home for supper tonight, yes, Francoise?"

"Yes," I answered, "tonight."

The apartment was in an uproar when I walked in that evening. Pierre still seemed dumbfounded by my announcement, but my two teenage daughters, Marie, 16, and Collette, 13, positively bubbled with enthusiasm and bombarded me with nonstop questions. "Will we live in the wilderness? Can we have a dog? A horse? Will we be farmers or tradespeople? Are there wild animals?" It took some time to calm everybody down and to explain some of the serious things about the pioneering program and my serious reasons for wanting to join.

Early the next morning while the others slept, I rose to begin preparing our applications. I knew it wouldn't be easy for a man and a woman, both past 40 with two children, to be selected to live on the newest and therefore most rigorous of the planets. Yet I also knew that a new planet would need mature, experienced people as well as the young to develop a technological society. I felt I could make a good case that Pierre and I had the skills and experience essential to a new colony and that we were also physically fit to go.

Becoming physically fit took the greatest effort. Neither Pierre nor I could be considered athletic and our jobs kept us at desks most of the time. I devised an exercise program for both of us including swimming, running, and yoga which I hoped would get us into reasonable condition in time for our preselection physicals. The first few weeks seemed agonizing, but soon I could feel my body firming up. I ate like a horse while losing five kilograms. Pierre said I never looked better, a good thing for him because my sexual appetite responded to my improved physical and mental attitude.

After extensive review of our applications, personal interviews, and thorough physical exams for the entire family, GAILE accepted us as a group for the first shipload of pioneers to Romulus! We felt jubilant, for less than 20,000 people in the whole of Human history had the honor and the thrill of being the first to settle a virgin world. I took great pleasure in informing my employers that I would no longer be working for them and left my job within a few weeks. Although our ship would not leave Earth for almost two years, Pierre and I wanted to spend time learning about our new world and to participate in the colony planning meetings. Since our Earthly wealth would be of little use to us on our new planet, we could afford to live off our savings until we departed.

As soon as the Galactic Association approves a planet for colonization, GAILE staff planners begin to define the shape of the first colony. Much of the preliminary work consists of scientific studies to identify potential diseases, earthquake zones, weather patterns, dangerous animals, edible plants, and other phenomenon of importance to settlers. The GAILE staff also identifies preliminary settlement sites and selects suitable strains of Earth's food crops for each of these areas. This research takes many years but forms a vital data base that helps pioneers to make their decisions rationally. During the preliminary stage, planners draw up a multitude of wild designs for the planet's development, estimate maximum and minimum immigration rates, set final pioneer selection criteria, and generate dozens of reports (many of which are never read).

About three years before the scheduled settlement date, GAILE staff members begin to select applicants for the first colony. As these people are chosen they join in the planning process and make the final hard decisions about the location of the first settlement and its organization, both political and industrial.

Pierre and I became involved early in this second stage of planning. Because of the enormity of the task, the Pioneers' Planning Committee divides itself into subcommittees along lines of interest. I served on the Communications and Computer Services subcommittee while Pierre served on the Power Distribution subcommittee. Other pioneers formed committees to deal with government, housing, farming, industrial construction, land use, natural resources, and transportation. As with most committees, only interested individuals ever accomplish anything, and so most of the work of the Communications committee fell to three or four people with expertise in that field. GAILE had specified much of the hardware we would be allowed to take with us before we pioneers entered the picture. It then became our task to make best use of what was allotted. We soon found it impossible to plan communications without considering other factors such as the location of the first settlement, its population distribution, and the forms its industry would take. We therefore held numerous meeting with members of other committees to determine what they were doing.

In short order this process bogged down, and it became clear that some people would have to take responsibility for the overall design of Romulus's society and provide guidelines for the detailed planning groups. The GAILE staff called a meeting of the 9500 pioneers thus far selected to elect a Steering Committee for the colony. The meeting wasn't held in one place. Pioneers from around the Earth linked into a private channel that allowed them to listen to each candidate for the committee's seats. The candidates then answered questions and the pioneers voted, recording their choices electronically. As expected, the election produced a panel of moderates who believed they should solicit the detailed opinions of all colonists before drafting a master plan.

Historically, there had been two forms of colonial organization: Hadar's model of total free enterprise and the Genesis model of a centrally planned society. Soliciting the opinions of 9500 people naturally produced a wide spectrum of views ranging from stoic communists to rabid free enterprisers.

Once again, the Steering Committee adopted a middle ground approach. They provided that major industries be owned by all Romulans, but operated as profit-making entities. New industries and work that lends itself to fragmented industry are allowed to operate freely. Upon arrival, each pioneer is given non-transferable shares of stock of major industries in proportion to the amount of capital equipment brought by his colonial vessel. This stock pays dividends and may be transferred only to the colonists' children or designated heirs. Stockholders (which theoretically includes most adults) elect the officers of major firms.

The Steering Committee set up this complex scheme to cope with the problem of allocating the very precious supplies of imported equipment, such as power reactors, computer main frames, comsats, hospital equipment, mining equipment, and machine tools. It ensures that any excess profits generated by the operation of these vital industries will accrue to all citizens and not to a select few.

Shortly after the organization of the major industries, I found myself elected Vice President in Charge of Operations of the Romulus Telecomplex Co-op. Most of my life I had worked as a specialist. I moved from job to job in the communications industry, but in any one job I controlled no more than one small segment of a very large endeavor. For the first time in my life I found myself responsible for all aspects of the communications industry, including erecting and starting up the computer center, training new craftsmen and programmers, designing access terminals for homes and businesses, and placing communications satellites in orbit. On Earth I had a virtually unlimited supply of scientific knowledge and expert opinion to assist me with everything from building contracts to orbital calculations. On Romulus I would have only the aid of a few overworked individuals and the computer's library. I would have to make most of the critical decisions alone.

The months of intense preparation finally passed and the day came for us to leave Earth forever. We climbed aboard an orbital shuttle at the Paris spaceport carrying small bags that held what remained of our Earthly belongings. The antigrav shuttle drive operated so silently that we found ourselves in the blackness of space almost before we knew we had left the ground. The shuttle linked with two orbiting space complexes to drop off and take on passengers before heading toward the orbit of our starship, the *Romulan Provider*. The enormity of the starship overwhelmed me. As we docked, it loomed larger and larger until it seemed to fill all of space with a metallic silver array of long, slender cylinders.

After linking, we climbed up a tight circular staircase into the airlock, and I felt a dizzy sensation as we passed between the g-fields of the spacecraft and the starship. The ship's airlock opened on to an enormous anteroom containing circular rows of built-in benches and a central core of six large elevators. Four crewmembers awaited us and showed us to our quarters. As a family group, we were given a small split-level apartment complete with a mini kitchen and computer access terminal. Simply but tastefully decorated, the apartment felt so cozy that during the weeks of our journey we nearly forgot that we hurtled at near light speeds through the hostile void of space.

The loading of the pioneers took more than a week as they came in small shuttles from all parts of the Earth. Our family arrived early so that I could make final checks of the telecomplex equipment as the ship's crew loaded it. The loss of a single critical item could cause serious problems for the first pioneers. We had no backup, no existing equipment already in place, and the *Romulan Provider* would not return in less than six months!

At last the final pioneers climbed aboard and the buzz of 4000 voices filled the corridors of the ship. Although many of us had worked together for years, this was the first time that most of the pioneers met each other face to face. When the moment of departure came, the captain announced it over the loudspeaker and every person on board crowded to the windows to take one last look at Earth.

I didn't feel the slightest sensation of motion as the ship moved out of orbit. It traveled very fast, for as we watched, the Earth began to shrink visibly. In forty minutes the moon came into view beside the tiny Earth and within two hours the two spheres had shrunk to tiny points of light against a backdrop of billions of stars. During those two hours, 8000 eyes stared at the shrinking globe. No one spoke and a cold somberness descended upon the festive pioneers. I can't express the emotions that filled me as I watched the Earth drop away into the void of space, knowing that I would not see it again. That tiny sphere of my primordial origins had contained my whole life. As the ship pulled me away, I felt the roots of my soul being torn apart as they clung to the soil below.

The busy routine aboard ship made us soon forget the somberness of our departure. The business is planned so that pioneers don't have time to brood about the past and the loss of their lifelong home. Most attended classes about the new planet since they had little time to study it in detail before departing. Because I had spent so much time in preparation, I didn't go to school but spent each day planning the telecomplex's first months of operation after our arrival.

After nine weeks in space, Upsilon Lupus began to grow from a point of light among the billions to a bright sun. Once again all pioneers crowded to the view ports as the Twin Worlds came into view. Without doubt, Romulus and Remus form one of the most beautiful astrophysical sights in the known galaxy; two irridescent spheres, marbled in blue and white, contrasted against the black background of space. Slowly the great ship glided into a low orbit over one of the worlds, and we began to recognize the continents of Romulus we had studied on the maps.

Unloading the first shipful of pioneers on a planet takes much longer than unloading subsequent ones. The ship spent eight weeks in orbit before the last of the pioneers and their living spires grounded. First the great space tug, brought disassembled in one of the cargo spires, had to be put together. The tug then took the first cargo spire to the surface, along with a small crew of workers who prepared the foundations for landing the personnel spires. As soon as the first foundations were ready, the tug brought our spire to the ground. The crew had arranged people in the spires in the approximate order they were needed on the surface. The plans called for Pierre and me to land at once. He had to begin work on the fusion reactor that would supply power to the city, while I had to start supervising the installation of the planetary computer. Our daughters stayed aboard the ship for several weeks until the rest of the spires could land.

Our spire grounded in the early morning of the 44-hour Romulan day, and we opened our view port to a scene that few 24th century Earth people had ever seen. The spire stood on a broad plain of green grass and low shrubbery extending unbroken to the sea. In the distance, a forest of trees climbed up a ridge of low hills on the horizon. From our window, we could see no sign of Human presence. No artificial structures blemished the natural vista of green and blue.

We went outside and deafening silence accosted our ears. Gone was the constant whir of machinery that had become welded to our subconscious on Earth. As I looked on this beautiful, peaceful scene, I felt the tensions of the journey drain from my body leaving a calm lightheartedness I had never felt before. I took off my shoes and ran toward the beach. "Come on, Pierre!" I cried childishly.

"But Francoise, the computer?" he replied.

"The computer can wait a few hours!" I yelled back. "This will only happen to us once!"

I was born again during those first hours on Romulus. My world became filled with the newness of a child's world. Every plant and tiny creature, every rock and grain of sand seemed new and fresh and so very, very fascinating. My senses opened to the natural fragrance of the soil, the plants, the ocean. The rising sun warmed the air, and small flying creatures drifted overhead in the first stirrings of the sea breeze. We walked for hours along the shoreline, stopping to look at each new thing we saw. Sometimes we splashed barefoot through the icy surge at the water's edge. Several kilometers down the beach, we lost sight of the spire. I put my arms around Pierre and whispered, "Make love to me."

"Here!" he exclaimed with astonishment, "Outside?"

"We have the whole world to make love in if we want it." I replied laughing. "All of it belongs to us now." We held each other on the sand, in the warmth of the sun, and it was the best it has ever been for both of us.

Upon touching the planet's surface, the pioneers had scattered in all directions like dandelion seed, but several hours later they began to drift back to begin long days of work preparing for the arrival of the others. Pierre and I were among the first to return, and we began to examine the cargo spires that contained the equipment. They lay on their sides like great cigars, and opened by splitting crosswise into short cylinders, each of which had a cargo door in one end. I found the section containing the computer equipment, opened a small access hatch and climbed inside. Most of the equipment survived the landing intact, but one large crate containing a temporary storage drive had shifted and crushed some access modules. I located the drawing viewers tucked just inside the hatch. With these the construction crews could begin the foundations of the prefab building that would house the computer center.

The next grueling weeks taxed all my strength. The 44-hour day of Romulus is rough for Humans to get used to, but until we had the major buildings up, we worked during as much of the day as possible. Our general plan consisted of rising at dawn and working 16 hours. Then we napped for four hours and worked another 12 hours, with the bright light of Remus overhead. Before the rest of the spires came down, we had to get the power plant operational and the computer center fully checked out and functional. We had to be ready when the *Romulan Provider* departed with its priceless data base of Human knowledge.

All through the summer months, the 4000 original pioneers labored to erect the shops and mining facilities required to make machinery, electronic components, and housing materials. Stores of freeze-dried food supplemented the food grown in hydroponic gardens. In the winter, the second shipload of pioneers arrived with fabricating tools and farm implements. For the rest of the winter, the new arrivals put this equipment to work making trucks and prefab houses for the future farmers. That winter we prepared the first communications satellite that would link farmers and miners in the outlying regions with Sharam, furnishing them with communications, entertainment, and vital access to the computer library.

There were problems placing a satellite in synchronous orbit about Romulus. Ordinary satellites tended to drift off location during the early years of operation and constantly had to be moved back to position. We finally solved the problem by building a self-powered relay station that would automatically reposition itself.

In the second year, after putting the farms and mines in operation, the time came for the pioneers to assign priorities to the kinds of goods they wanted to have. Since most of them had come to Romulus in part to escape the crowding on Earth, the overwhelming majority of pioneers left in Sharam first wanted land of their own. This required the production of housing units and personal transportation in large quantities. In order to minimize waste, pioneers had to place firm orders for prefab housing, power generators, and automobiles before the shops tooled up to produce them. During the third year, pioneers scavenged communications sets and personal appliances from the spires for their homes in the wilderness. The vacant space was converted into offices and larger apartments for the people left in Sharam.

Because our work tied Pierre and me to the original settlement, we were among the last of the original pioneers to secure our own plot of land, some eight years from the day we grounded. We enjoyed the eight years in Sharam, for as people moved out it became a more pleasant place to live. The colonists who remained made attractive additions to the spire buildings and landscaped the city with ornamental native trees. We purchased our own levicar and used it to take us far from the city on weekend camping trips that gave us the chance to compare potential living sites.

In the colony's eighth year we took delivery on a small prefab house and staked our claim. We settled on 25 hectares of a hillside overlooking a small lake on the Wolf Peninsula about 250 kilometers from Sharam. As both Pierre and I still had to work in the city, we retained our spire apartment. By this time Marie had been married for almost two years and Collette was engaged. Pierre and I spent most weekends alone at our retreat, enjoying the natural beauty of our new world. Though I have spent most of my life amidst the bustle of cities, I relished being surrounded by unspoiled wilderness. We had a few neighbors within forty kilometers of us, but we couldn't detect a trace of their presence from the ground. I developed a passion for gardening and cultivated a variety of imported and native plants. In the summer months, the crystal lake became warm enough for swimming, and on summer days we often spent the entire day at its edge.

In the colony's tenth year it became clear that Sharam was reaching its limits of growth. Although the population of the entire planet numbered only 160,000, 70 percent of these worked in or near Sharam. Though still a tiny town by Earth standards, we knew it would not adapt well to the inevitable growth of the future. We had designed it to serve as a first settlement in a world of scarce transportation and building materials. Establishing the Romulan colony's industrial base obsoleted the crowded little city.

The Industrial Planning Board recognized that the lifestyle of Romulus would remain essentially a rural one for many years to come. Cities would serve primarily as focal points for shopping, commerce and socializing. To relieve the strain on Sharam, the Board decided to establish a new industrial region at the base of the Wolf Peninsula. The region would include a city designed to accommodate the rapid future growth of Romulus and to supercede Sharam as the colony's principal commercial center for the rest of the first century.

To this end, Reforma would consist of two principal parts: the first, a densely constructed commercial/entertainment zone where no levicars would be permitted and public transportation would be integral; and the second would be a giant industrial tract designed to accommodate future factories and plants. This larger, low-density region would be equipped with an automatic levicar control system along principal transit corridors, enabling workers to make the daily commute from their rural homes in speed and safety.

The planning of Reforma presented an exciting challenge to all business leaders on Romulus. Architects drew general plans for the new city in less than two years, and within months the ground had been broken for the first industry, including a power plant, a metal forming shop, an automotive factory, and a new telecomplex facility. Soon thereafter, work commenced on the hub transportation system. As soon as the first guideways began to run, small businesses, many of them from Sharam, moved in. To encourage the growth at Reforma, spire sites have been integrated into the design and soon all new pioneers will land there. Because Reforma has tremendous potential, both of my daughters and their husbands now work there. Telecomplex headquarters remain at Sharam, but the new facilities at Reforma will soon overshadow the original computaplex.

Every year on Romulus seems different from the last. Today I serve as Chief Executive of the Telecomplex Co-op, a position I hope to hold for many more years. As such I am less concerned with daily operations and more concerned with planning for the future needs of our world. The difficulties of the early years have passed. They were exciting times, yet I think the most exciting times for Romulus are still to come. In many ways, today's young pioneers from Earth have greater opportunities than the first pioneers. I hope to continue to participate in our planet's development for the rest of my life, to see my grandchildren, the first generation of Romulans, grow and prosper, and to partake of the future comforts that the growth of our planet's new industry will bring us. I know in my heart that in the free and stimulating environment of Romulus, future generations will make outstanding contributions to the knowledge and achievement of all civilizations throughout the galaxy.

THE REMUSAN OBSERVATION PROJECT

Editor's note: Cheryl Cooper, 27, recounts her experiences as a staff member of the Remusan Observation Project. She did not intend to become a pioneer. She joined the Project shortly after receiving her Doctorate in sociology from Northwestern University and today is an enthusiastic resident of Romulus. Cheryl spends six months of each year aboard the Matthew Obo Laboratory, a permanent space station orbiting Remus, engaged in an ongoing study of the neighboring planet's native civilization.

No single historical event has had greater impact on the disciplines of sociology and anthropology than the discovery of Remus. For the first time, Humanity encountered an intelligent, alien, humanoid species, more primitive in its intellectual development than Humans, though every bit as capable as we of developing into an advanced civilization. Here was a living laboratory of our past. Not our historical past, of course; Remusans are a different species living on a different planet. Their development from a fragmented iron and stone technology to an interplanetary society must parallel Humanity's own development in many ways. Observing this advancement, in itself a fascinating study, cannot help but shed light on our own development and the reasons *why*, as well as how, it occurred.

The Remusan Study Project commenced in 2350 adtc, shortly after the discovery of Romulus. Exploratory parties that had traveled to the Upsilon Lupus system to survey Romulus initiated the Study, although the return of the space ships to Earth for reprovisioning and reporting data periodically interrupted it. GAIL could not begin a continuous study of Remus until Humans established the first permanent colony on Romulus. In 2362, just two years after the first pioneers set foot on Romulus, GAIL placed a permanent space station in orbit around Remus to be used exclusively for the study of that planet and its culture. The station, il-

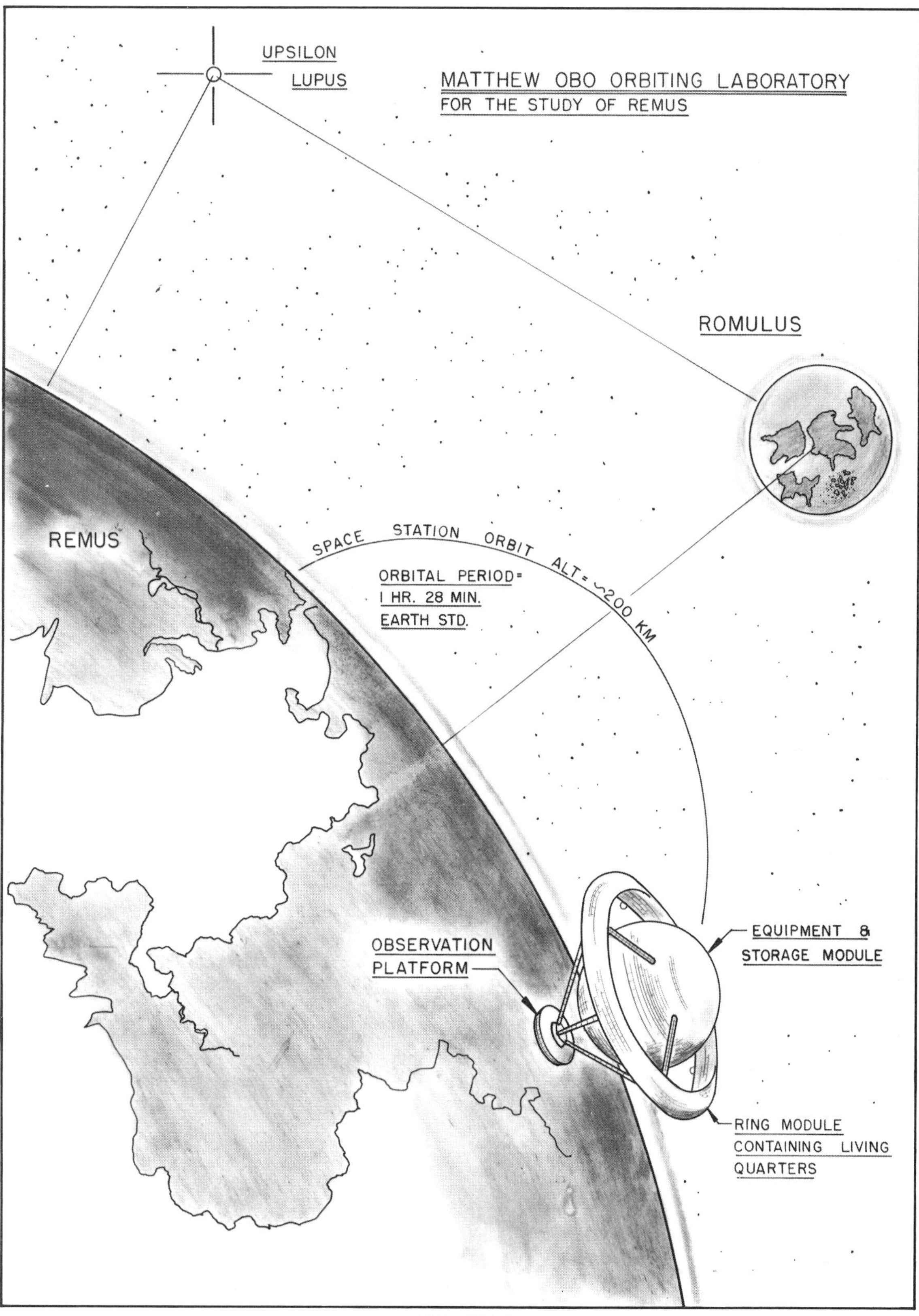
UPSILON
LUPUS
MATTHEW OBO ORBITING LABORATORY
FOR THE STUDY OF REMUS
ROMULUS
REMUS
SPACE STATION ORBIT ALT= ~200 KM
ORBITAL PERIOD=
1 HR. 28 MIN.
EARTH STD.
OBSERVATION
PLATFORM
EQUIPMENT &
STORAGE MODULE
RING MODULE
CONTAINING LIVING
QUARTERS

FIG. 3.22

lustrated in figure 3.22, is known as the Matthew Obo Laboratory, named for the anthropologist of the *Kara* who first observed Remusan civilization. It is resupplied from Romulus where a larger research base for analyzing data is located. From the Romulan base, Project staff workers prepare data for transmission to Earth and on to the other GAIL members.

The provisions of the GAIL noninterference policy place strict limits on the methods used by the Study Project. The Remusans must not become aware that they are being studied by advanced races from other planets. In itself, this knowledge would profoundly alter their natural development. Consequently, the Project Staff takes extreme precautions to preclude discovery. Though the station travels in a low orbit, a special light-absorbing field surrounds it, preventing it from being seen from the surface, even with optical telescopes. Except in very special circumstances, approved in advance by GAIL, the Project Staff must avoid direct contact with the Remusans. Studies of their culture must be carried out by extensive use of sophisticated sensing devices and the occasional photographing of written documents.

The Project would not have had much value if we hadn't studied the physiology of the Remusans. Although Remusans regularly engage in the deliberate slaughter of their own kind, GAIL forbids this practice by its members. Yet taking live Remusans for study and returning them to the surface would have revealed GAIL's existence. We solved this problem by developing techniques for hypnotizing our specimens so they are subsequently unaware that they have been taken and examined. Generally they can recall only the few brief moments before their capture and the fact that they had "lost" several hours, or sometimes days of time.

We make it a practice to take individuals while they walk alone, usually at night, so none of their fellows witness their kidnapping. To date, some 150 Remusans from different parts of the planet have been examined in this way. The practice has been strictly controlled. Two people are never taken from the same village or tribe. In most cases, the specimen's fellows don't believe the story of his capture, discounting it as possession by devils, religious visions, or outright fabrication. We have now obtained a complete physiological data base, so the number of specimens studied in the future will be limited to ten per century.

I joined the Remusan Study Project shortly after graduation from Northwestern University. My doctoral thesis had analyzed certain aspects of Remusan culture, and so I was a natural candidate for a staff position. I had no desire whatever to become a pioneer on an alien world. I enjoyed the soft life on Earth, the luxuries and conveniences. I functioned well in a bureaucratic hierarchy, and the thought of cutting down trees to build a home or encountering wild animals made my skin crawl. When I applied for a position with GAILE, I fully expected to get a desk job on Earth, analyzing results sent in from the field and acting as liaison between Earth Branch and scholars from the other GAIL species. You can imagine my surprise when the Project director chose *me* to go into the field. The offer was one that I couldn't refuse. The chance to observe the Remusans first hand can't be resisted by any scholar, no matter how disposed she is to the comforts of her Earthbound library.

I have served with the study project for almost three years, spending six months aboard the orbiting laboratory, followed by six months analyzing the results of my observations at the Remus Research Institute on Romulus. My duties aboard the space station allow me to make frequent trips to the surface of Remus to plant sensors and to observe the Remusans at close range. Some missions include cloak-and-dagger style attempts to plant bugging devices, copy documents, and steal artifacts, and they have required me to use personal antigrav packs, cloaking fields, and neural neutralizers to avoid detection. To date, no observer has evern been captured by the Remusans, but the danger exists and the consequences of such capture are serious indeed.

Remus' societies are highly fragmented with thousands of different languages and customs. Using the limited methods at our disposal, it is difficult to piece together a coherent picture of their thought, lifestyles, emotions, or even their language. The following is my best attempt to give an up-to-date thumbnail sketch of Remus as we perceive it today.

We call Remusans "humanoid," which means their external shape resembles Humans more than it resembles Chlorzi, Ardotians, or any other animals. Remusans stand erect, have two arms, two legs, and four fingers or toes on each arm or leg. They possess an opposable thumb as one of the four fingers. This gives them manual dexterity equivalent to Humans. Remusans are, on the average, smaller than Humans. Males stand about 116 centimeters in height and have an average mass of 21 kilograms. Females are slightly larger at 122 centimeters tall and 23 kilograms mass. Throughout the different regions of the planet, Remusans vary in size, coloring, and quantities of body and facial hair in the same fashion that the Human race varies.

Just as Humans comprise the dominant species of mammals on Earth, Remusans dominate the class of warm-blooded animals known as "extoplacentals." Extoplacentals have warm blood and bear their young inside a tough membrane which performs many of the functions of the placenta in mammals. The membrane breaks 12 to 48 hours after birth instead of immediately prior to birth as in mammals. Females bear the young and are equipped with mammary glands to nurse them with a milk-like food.

Remusan society is primitive, technologically, politically, and culturally. We cannot say with precision how old it is or compare it with past civilizations on Earth. In some respects, Remusan civilization appears as advanced as twelfth century Human societies, but in other respects it seems centuries more primitive. One can't accurately speak of "Remusan" society or civilization, because the people of Remus are fragmented into literally thousands of cultures. The fastest transportation is on the backs of large animals which serve the function that horses once served on Earth. Most communication is verbal, though the major cultures use written languages. Some written languages employ symbols, while others have alphabets, yet no society has printing presses to reproduce their writings. All written language is carved into clay or stone tablets or transcribed by hand onto coarse parchment. These barriers to communication retard intellectual development and tend to encourage the fragmentation of cultures.

Technologically, the most advanced Remusan society corresponds roughly to the state of Earth in the tenth century. Fire, the wheel, iron, the hand loom, numbers including a zero, and agriculture all have been invented. Remusans build large fortresses of stone and have achieved a fairly high degree of precision in masonry and stonework. Yet no structures yet discovered compare in size, craftsmanship, or precision with the pyramids and temples of ancient Egypt, the Incas, or the Mayans. Remus's most primitive tropical cultures wear no clothing, use no metal tools and cultivate no agriculture. No Remusan society, even the most advanced, has achieved mass production, interchangeable parts, the telescope, gunpowder, the waterwheel, the steam engine, or any form of electric power. No Remusan scientist has uttered scientific principles similar to Newton's or Galileo's, let alone unraveled the mysteries of electromagnetism or chemistry. Astronomy is well developed among the cultures of the outboard hemisphere of Remus. On the inboard hemisphere, the bright light reflected from Romulus masks all but the brightest stars. Some Remusan societies have built sailing ships and used them to cross the narrower oceans. A few societies on the outboard side have apparently developed a crude form of celestial navigation.

Politically, Remusans seem quite primitive and barbaric. To date nothing resembling democracy appears to exist, although such political organization may have existed in the past. Virtually all government takes the form of highly localized dictatorships or monarchs. Some of these accede to the throne by inheritance, others by their knowledge and presumed mastery of theology, but most seize their power by brute force and treachery. A virtually constant state of war prevails over most of the planet, the outcome of which appears to have little effect on the average Remusan. Weapons of war consist primarily of clubs, spears, and axes. The longbow and crossbow have not made the scene, nor, of course, have rifles and cannon. Soldiers commonly wear primitive body armor, often made of chain link or very hard leather.

In general, Remusan societies exhibit little sexist prejudice. Women play as great a role in military and political affairs as men, largely because they are somewhat bigger and stronger. In most societies, males accord pregnant females and females nursing children special consideration, but otherwise women work, fight, and rule their fellows as much as do men. Most societies are too small to have much royalty or upper class surrounding the king or queen. The monarch rules and everyone else obeys in approximately equal misery, save for the ruler's few trusted henchmen. Some Remusan societies practice slavery, but such slavery is seldom along racial lines.

We know less about the philosophies and religions of Remusans than about any other aspect of their societies. A major reason for this lack of understanding is our poor comprehension of their languages. Translation of languages has largely been accomplished by computer analysis of taped conversations and a few written documents. The computers compare words with their context and try to analyze patterns of use. From these patterns, meanings of more common words can be inferred. Once the common words have been filled in, less common words can be deciphered. Words describing common actions or objects, such as "fight," "tree," "food," etc. can be easily understood. Abstract concepts, such as "God," "ought," "right," and "evil" are most difficult to translate precisely. The large number of languages each containing hundreds of dialects further compounds the problem of understanding. Remusans have few existing copies of important philosophical and religious works, and these can rarely be stolen or even photographed.

The information garnered to date indicates that philosophic and religious thought on Remus is not well developed in this period of its history. If thinkers comparable to Aristotle, Plato, or Confucius have lived in Remus's past, we have discovered no evidence of their works. A few important religious books, most of them compendiums of writings like the *Bible,* are currently being analyzed. Religions on Remus vary from polytheistic and anthoropomorphic to monotheistic and abstract. Religion often forms the basis for the divine right of monarchs, and religious doctrine often justifies war between political divisions. On the inboard hemisphere, many tribes attach religious significance to the neighboring planet Romulus, which is sometimes thought of as the "eye of God" or the seat of gods. Few Remusans comprehend that Romulus is a planet nearly identical to their own. On the outboard hemisphere, many tribes do not know Romulus exists or believe it exists only in legends. Some Remusans have taken pilgrimages to see Romulus. These pilgrimages have occasionally caused wars.

We have insufficient data to analyze accurately the intelligence of the Remusans. They appear to have approximately the same mental abilities as Humans; perhaps on average they are slightly more intelligent. Variation of intelligence between individuals appears to be slightly less among Remusans than among Humans. In other words, though Remusans produce fewer morons, they probably produce fewer geniuses as well; yet it is the geniuses that create the great leaps forward in the knowledge of a civilization.

If our assessment of Remusan thought processes seems sketchy, our understanding of their psychic abilities is even more so. Remusans appear to have much stronger telepathic and extrasensory abilities than Humans. Field observers have long noted the ability of Remusans to detect Human presence even though the Human hasn't made any sound or given any other physical sign of being there. Remusans often appear to summon each other over long distances using some sort of telepathy. Their control may even extend to lower animals that they have domesticated.

No researcher doubts that at some time in the future Remusans will begin to develop a technological society. The current scenario for this begins with the invention of the printing press. This would vastly increase the exchange of ideas. This exchange of thought will catalyze understanding of the laws of physics and chemistry, and the development of mathematics. Improved communication will unify Remusans into the larger societies needed to create industry. We can't say precisely when this process will begin. It could start in less than 100 years or it could take more than 5000 years. The beginnings of this process will be of greatest interest to social scientists studying Remus for this murky period in Earth's history is poorly understood. We are certain, however, that once the industrial revolution begins, it will lead inevitably to the development of a space age technology that will take Remusans to neighboring Romulus.

Once there, Remusans will find more than a billion members of a highly developed technological society. Suddenly a society of the technical and political sophistication of Earth's 20th century will find itself face to face with Human society of the 30th century. Future generations of Romulus will find it a great challenge to help Remusans in their rapid transition from a parochial, primitive society to a member of a transgalactic community spanning thousands of worlds. The danger of war causing the destruction of one or both societies is very real. The Remusans are savage and primitive, but we must remember that just 300 years have passed since the last international war on Earth, and only 280 years since the last major civil insurrection. At that time, when Humans began reaching for the stars, they had still not learned to live in peace and cooperation with each other. Let us hope that by the time the first Remusan sets foot on Romulus, Humanity will have matured enough to cope with this crisis wisely.

Until then, we who serve the Remusan Study Project must learn as much as we can about them. Regardless of the ultimate consequences, I find the study of this developing society a fascinating occupation that I hope to pursue for many years to come. Life aboard the space station and on Romulus is unhurried and pleasant. The station houses a staff of seventy-three. Of these, twenty-one serve as station operators responsible for life support, power, and control systems. Four cooks feed the staff and six shuttle pilots ferry scientists and technicians between the station and the planet's surface. Twelve communications technicians support the scientific staff installing and monitoring listening devices, hidden cameras, and biosensors, as well as by maintaining communications with Romulus. Twelve social scientists work in the station at all times. One or two of these may be visiting Chlorzi or Ardotians. The Chlorzi live in a special chlorine filled chamber on the observation platform and do not participate in surface missions. The biological staff consists of ten biologists and eight laboratory technicians. Biological study of Remus is as important to science as the social study. Romulus and Remus provide a unique biological laboratory in which to study the mechanisms of evolution. Both planets have the same size, physical composition, sunlight, and rotation. The differences between life forms on each planet give scientists clues about the relative effects of physical environment and chance upon the evolution of life.

Staff members at the station cooperate very closely. Each day we hold meetings to discuss the day's planned operations. The station houses two antigrav shuttles designed for transport between the station and the surface of Remus. Except for maintenance days, these vehicles fly missions around the clock. Most missions are routine and involve planting sensor arrays and repairing defective units. The sensors have been designed as self-contained units and will self-destruct if disturbed from their original placement or if their power drops to a critically low level. Retrieving these devices would be expensive and possibly hazardous, particularly if the device should be discovered by the natives. Special cloaking fields shield the shuttle craft and cause them to reflect their backgrounds. When viewed from the ground, the shuttlecraft assumes the color of the sky overhead. When viewed against a backdrop of trees, the shuttlecraft appears to be green or grey or whatever color the trees might be that season.

I love the excitement of surface missions and generally go on one each week that I'm in space. Missions may be as simple as filming a battle from the spacecraft hovering just over a nearby forest, or they may involve entering a monarch's walled fortress under cover of darkness to copy records and documents reserved for royal eyes. I find the suspense of nearly being discovered exhilarating. I enjoy being close to discovery but sufficiently in control of a situation that I can escape undetected at the last second.

I had my closest call while escaping from the castle of the Whar (monarch) of Zarobar. I had just taken a furtive picture of the Treaty of Zweig, a pact that would end war among ten tribes covering a third of a continent. I slipped out the door of her private closet moments before the Whar returned and headed for a nearby balcony, intending to launch myself into the air with my antigrav pack. Just before I engaged the drive, I noticed a young female guard staring at me. Her mouth was starting to open, and I knew that in the next second she would call for help. Calmly, with the experience gained by more than seventy missions, I took the neural neutralizer from my belt and fired it at her.

The neural neutralizer casues a temporary loss of consciousness without corresponding loss of motor functions. The target feels as if she has gone into a trance very like a lapse of memoryI n the thirty seconds she remained dazed, I made my escape through the air to the waiting shuttlecraft. Each time a field worker uses the neutralizer, he or she must file a written report. The neutralizer is used only in moments of direct contact with Remusans. Too many contacts would be indicative that observers are getting careless and that security procedures need to be tightened. This was the only time I've had to use my neutralizer, the lowest record for any field sociologist with three years service. I haven't been able to assess the effect of this incident on the culture, but I suspect the guard felt she had seen a spirit rather than an alien being. Belief in such spirits forms part of the religion of Zarobar's natives.

Six months of excitement in space is enough, and I relish my return to the calm life on Romulus. During the six months I spend on the ground, I analyze data gathered during my stay in space. The Remus Research Institute lies nestled on a wooded shoreline of the North Romulan Ocean far from the planet's major population centers. The weather is mild, but cool ocean currents keep it brisk and invigorating. I live in a sort of commune with seven other sociologists. Together we own a large, rambling, modular house on a fifteen-hectare parcel of land. Most days I spend some time in my office at the Institute which overlooks the ocean, but I work at home as well.

I have come to appreciate the natural beauty of the Romulan forest so much that I have no wish to return to Earth again. I find working in our gardens and orchards or making improvements and additions to our house at the commune to be pleasant diversions from my real work, and my friendships within the commune and at the Institute are the warmest I have had in my life. As a reluctant pioneer, I can only say to Earth people who have doubts about leaving home—you don't know what you're missing!

* * * *

OPPORTUNITIES FOR PIONEERS

Romulus offers pioneers from Earth unlimited opportunities. The planet's light atmosphere, mild climate, and low gravity place no physical restrictions on any potential colonists. Romulans desire any immigrants, skilled or not, who wish to work hard and pursue a useful trade.

Personal land claims vary from ten hectares in the industrial region of Sharam to fifty hectares in undeveloped regions. GAILE operates a starship of the quad hex configuration capable of carrying 6,000 pioneers and supplies twice annually. Travel time requires 68 days of ship's time and 99 days of planet's time. GAILE is currently considering the addition of a second starship to increase capacity.

Athena—Wisdom of Earth Born Anew

IMPORTANT STATISTICS

Planet name: Athena

Equatorial diameter: 12,235 kilometers

Mass (Earth = 1): 0.87

Surface gravity: 0.94 g

Escape velocity: 10.7 km/s

Albedo: 0.36

Atmospheric pressure at mean sea level: 1.05 bars

Fraction of surface covered by land: 24%

Maximum elevation above sea level: 7689 meters

Length of day: 25 hours 14 minutes

Length of year: 539 Earth standard days

Obliquity: 21 degrees

Von Roenstadt habitability factor (Earth = 1): 1.09

Current population: zero

Number of population centers: zero

Year to be settled: 2380 adtc

Number of moons: 2

Star name: Sabex 50 (HR 7345)

Star type: G8V

Distance from Earth: 77.6 light years

Distance from planet: 1.45 AU (Earth-Sol = 1 AU)

Star Mass: 1.40 (Sol = 1)

Travel time from Earth

Ship's time: 79 days

Planet time = 123 days

Atmospheric Composition

Oxygen: 21.3%

Nitrogen: 77.6%

Carbon dioxide and inert gasses: 1.1%

Distance (km)	*Period*	*Mass* (Earth's moon = 1)
287,800	18.7	0.72
50,400	1.4	0.0012

Table 3.9

Athena—Wisdom of Earth Born Anew

PHYSICAL ENVIRONMENT

Athena occupies the fourth orbit of the star system Sabex 50 (Chlorzi designation, Earth Catalog HR 7345) approximately 77.6 light years from Earth. Sabex 50 is a type G8V star approximately 1.4 times the mass of Earth's sun. Because it lies so far from Earth, the constellations in the Athenian sky bear little resemblance to the familiar patterns, although some of the giant stars, like Antares and Betelgeuse, can still be seen plainly. Two moons orbit Athena, both smaller than Earth's. The larger one, some 2030 kilometers in diameter, appears as a sphere slightly smaller than Earth's moon, while the smaller moon, the size of a captured asteroid, looks only slightly larger than a point of light. Both moons are barren, lifeless planetoids with relatively low albedo.

Of all the colonial worlds, Athena most resembles Earth. Its diameter measures 96 percent of Earth's creating a surface gravity 94 percent as strong as Earth's. Land covers approximately 24 percent of the planet's surface area and is divided into six continents distributed evenly over the surface. Although Athena possesses less land area than Earth, the absence of a polar continent like Antarctica makes a greater percentage of Athena's land habitable.

Athenian weather, like Earth's, varies greatly, ranging from tropical rain forests to arid deserts and frozen tundra. In general, seasons are less extreme because of the planet's lower obliquity, 21° versus 23° for Earth. The atmospheric pressure, just five percent higher than Earth's, creates higher wind pressures, but the difference is insignificant, being less than the normal variation in atmospheric pressure due to changing weather patterns.

Athena revolves on its axis once every 25 hours, creating a day most nearly compatible with the natural rhythm of Human beings. Because the planet orbits farther from a larger star, it receives nearly the same amount of sunlight as Earth. The spectral radiation of Sabex 50, though of a slightly lower frequency than Earth's sun, is not noticeably different.

LIFE FORMS

Athena hosts an extremely varied and advanced array of life, though, of course, no intelligent life has developed. The first exploratory mission to the planet catalogued more than 200,000 plant species and 75,000 animal species; yet this forms but a small fraction of the estimated one to two million species that may exist there.

Vegetation

Of greatest importance to Humanity is the existence of flowering plants resembling the angiosperms of Earth. Like their Earthly counterparts, these Athenian plants produce seeds in an ovary contained in a flower. Form of the flowers varies tremendously. Some grow large and brightly colored. Others develop edible fruits, and still others turn into edible grains. As on Earth, the

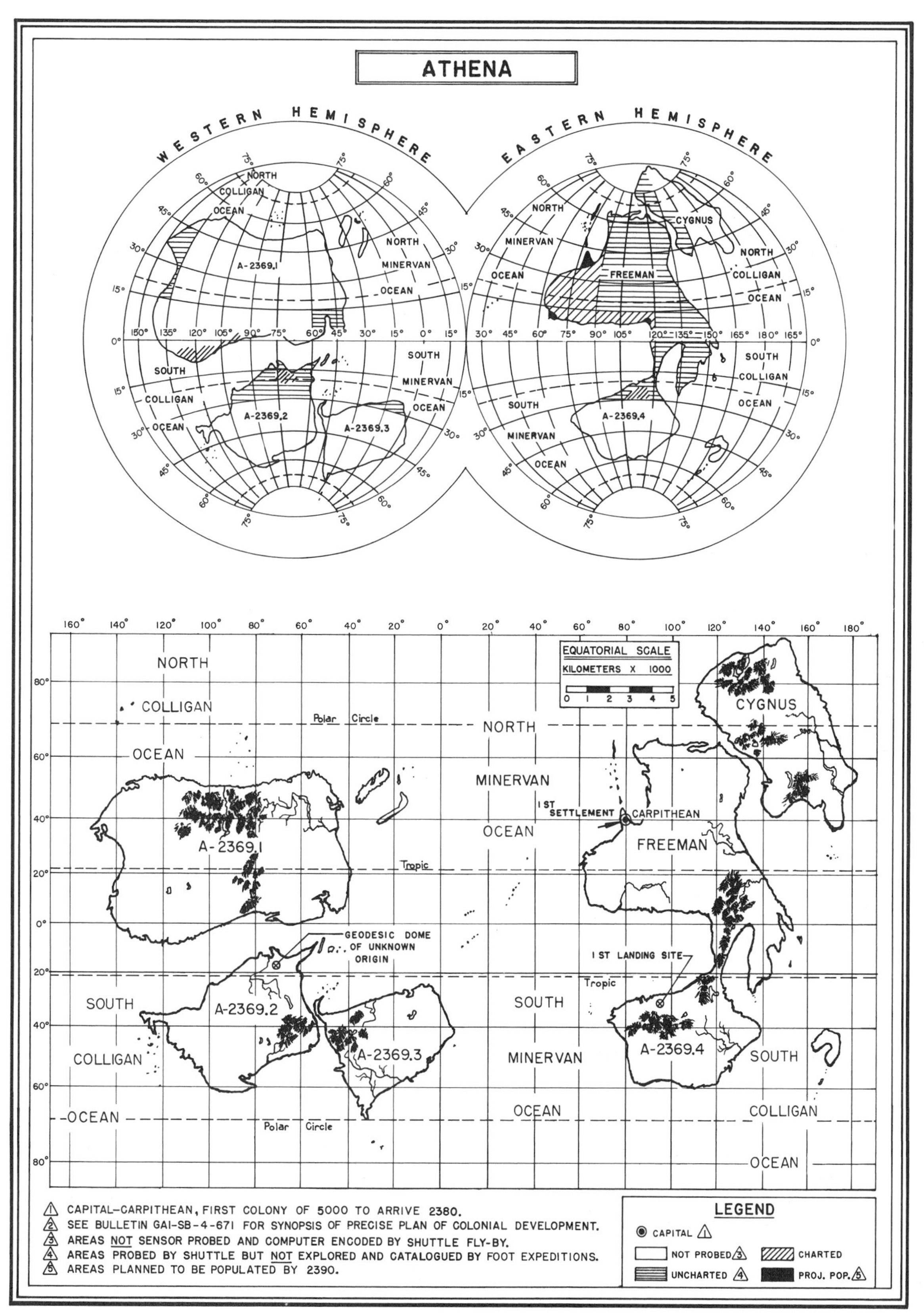
ATHENA
WESTERN HEMISPHERE
EASTERN HEMISPHERE
NORTH COLLIGAN OCEAN
NORTH MINERVAN OCEAN
A-2369.1
SOUTH COLLIGAN OCEAN
SOUTH MINERVAN OCEAN
A-2369.2
A-2369.3
CYGNUS
FREEMAN
A-2369.4
EQUATORIAL SCALE
KILOMETERS X 1000
Polar Circle
Tropic
1 ST SETTLEMENT
CARPITHEAN
GEODESIC DOME OF UNKNOWN ORIGIN
1 ST LANDING SITE
1 CAPITAL–CARPITHEAN, FIRST COLONY OF 5000 TO ARRIVE 2380.
2 SEE BULLETIN GAI-SB-4-671 FOR SYNOPSIS OF PRECISE PLAN OF COLONIAL DEVELOPMENT.
3 AREAS NOT SENSOR PROBED AND COMPUTER ENCODED BY SHUTTLE FLY-BY.
4 AREAS PROBED BY SHUTTLE BUT NOT EXPLORED AND CATALOGUED BY FOOT EXPEDITIONS.
5 AREAS PLANNED TO BE POPULATED BY 2390.
LEGEND
CAPITAL 1
NOT PROBED 3
CHARTED
UNCHARTED 4
PROJ. POP. 5

FIG. 3.23

number of flowering species is enormous, varying from grasses and wildflowers to giant trees.

Animals

Athenian animals boast high variety and development. Numerous insects fill as wide a range of niches as their Earthly counterparts though important physiochemical and structural differences exist between this class of animals on the two worlds. Yet insects of both planets perform such similar functions that the first explorers called them by such names as "bees," "ants," and "caterpillars."

A class of warm-blooded creatures resembling the animals of Earth comprise the most advanced of Athena's animals. These animals with soft, fur-covered skins and internal skeletons fill the spectrum from small gopher-like root eaters to giant predators more ferocious than Earth's Bengal tigers. These "pseudomams," as taxonomists classify them, bear their young in ways that look like a compromise between Earth's marsupials and placentals. Pseudomams birth their young in a more premature state than do Earth's placentals but not as premature as the state of marsupials, who literally cannot live outside their mother's pouch and are physically attached to the mother's milk nipples during early development. Pseudomams do nurse their young and many have pouches similar to marsupials for carrying them about just after birth. Others build nests in trees for housing their offspring, and still others employ built-in "straps" or highly developed prehensile tails for transporting their babies. Some pseudomams, even though they nurse their young, lay eggs. The eggs have soft, pliable shells rather than the brittle shells of Earth's birds.

Most important of the pseudomams are those that resemble Earth's primates. Their forelegs have developed into arms containing six-fingered hands. These hands do not have the opposable thumb, characteristic of the Human hand, but they can grasp tools and perform simple manual tasks. Unlike the primates of Earth, Athenian primates are carnivores. They devour their prey with extremely sharp teeth, and their strength approaches that of Earth's cats. In large groups they form the most dangerous animals of the known colonies, though they still make no match for well-armed and protected Human beings. The primates live strictly in the tropical regions of continent A-2369.1, and therefore will pose little threat to the first pioneers. Other important pseudomams include a large genus of dog-like predators the size of Earth bears and numerous large beasts resembling cattle that can be used as a source of milk and meat.

Marine Life

Athena also supports highly advanced marine life. Biologists believe that, as on Earth, Athenian life originated in the seas. Microscopic plants which produce 73 percent of the planet's breathable atmosphere comprise the most important life in the food chain. These plants begin the marine food chain which leads all the way to great fish similar to the bony fishes of Earth. These fish, resembling Earthly varieties in many ways, range in size from fingerlings to monsters that rival the great sharks of Earth. Pseudomams also live in the sea and these orders contain the largest animals on the planet. Marine pseudomams range in form from amphibious creatures resembling the sea lions of Earth to streamlined versions like Earth's cetaceans. Scientists know little about the habits of these creatures; in fact, most remain uncataloged. Yet they are believed to be highly intelligent, perhaps the most intelligent life on the planet.

THE PLAN FOR ATHENIAN DEVELOPMENT

Editor's note: At this release, Athena is yet unsettled by a permanent colony. Aristotle Chianakis, 35, this year's elected chairman of the Pioneers Steering Committee of Athena, gives his summary of the state of the colonial plan. Despite his youth, Chianakis's record of accomplishments bespeaks his ability and experience as a planner and organizer. He has operated a warehousing business and a commodity trading firm, and he has served as mayor in his home town of Burrwood (population 123,000). He became interested in the pioneering movement, joined the GAILE planning staff and served as director of loadout operations of vessels departing for Yom and Romulus.

Athena shines brighter than any other jewel in Earth's crown of colonies. The Chlorzi starship *Gnar 8*, commanded by Tnuptic Kzorange, reported the discovery of Athena to GAIL in 2365 adtc. Athena's oxygen-nitrogen atmosphere made it an unsuitable home for Chlorzi, so they proposed to swap the planet for certain exclusive mining rights on Mammon. Within six months of receiving the offer, Earth dispatched the starship *Jennifer Freeman* under command of Captain James Colligan to make the first Human observations of the planet. Four years later, the *Freeman* returned and reported Athena to be the most beautiful planet that Human eyes had ever seen.

Since then, a second party has spent another four years on the planet, bringing back enough data to finish planning the first Human settlement. Early in 2380, the first shipload of 5000 pioneers will set out for Athena aboard Earth's largest starship. Our modular vessel of the seven hex configuration will transport more cargo across greater distance than any ship ever built. After this initial settlement, the ship will make round-trip voyages between Earth and Athena every ten months, bringing up to 7,000 pioneers in the fifth year. After that, funds permitting, a second starship may be constructed for the Athenian crossing in order to increase opportunities for pioneers.

GAILE has already begun to choose pioneers to fill the first shipload. The first 500 of them have elected me Chairman of the Pioneers' Steering Committee. Like the other ten members of the committee, I will serve for a one-year term. At that time, and every year thereafter, the pioneers selected to date will hold another election to retain or replace each member of the Pioneers' Steering Committee.

The PSC's members coordinate the overall planning of the Athena colony during its first five years. Planning is crucial for a small group of people attempting to gain a foothold on an alien planet. The amount of supplies and equipment we can take will be strictly limited, yet our survival during the early years depends in large part on the hardware we bring from Earth. Such equipment is, in a sense, priceless. We can not replace it immediately ourselves, for during the early years we will lack Earth's sophisticated manufacturing technology. Yet loss or inefficient use of critical equipment can spell success or failure for our enterprise. Some of America's early colonies disappeared without a trace and others barely survived. Although their technology was crude, in many ways these American pioneers faced problems less severe than those we confront on a new planet.

Eleven people can't undertake this monumental task alone so we rely on the help of the GAILE staff and our fellow pioneers. The GAILE staff provides technical and legal expertise, scientific data, computer analysis, secretarial service, and clerical assistance. GAILE staff planners offer suggested development plans based on the experiences of other colonies; yet we pioneers must make the final hard decisions ourselves. Special planning committees of pioneers create detailed plans and operating methods for specific activities, including the court system, computer center operations, money and banking, farming, housing and building, communications, medicine, power, land use, manufacturing, transportation, and mining and natural resources. The plans of the subcommittees and the PSC must be ratified by 75 percent of the first 20,000 pioneers to provide a check on the abuse of power. During the three years until our departure, the PSC intends to submit two partially-completed plans to straw votes so that they can be modified to suit a three-quarters majority. We don't want ratification of final plans to postpone the settlement date.

At this writing, the PSC has been at work for ten months, and all plans drafted to date may be changed. Yet certain fundamental policies have been formed, and surveys of the pioneers selected so far indicate that they will probably survive final approval.

All Athena's pioneers desire the freedom to live their lives as they wish. This means that Athena's economy must be structured around a system of free enterprise and that its government must be constituted to prevent any person or group from violating the basic Human rights of others. All pioneers are also committed to preserving the physical environment of our new world and preventing the abuses that despoiled the Earth.

Developing plans and social structures for achieving these ends won't be easy. Early in Athena's development, we must face the difficult transition from an economy where all pioneers cooperatively own the means of production to a diversified economy in which no one person or group wields too much power over any one industry. Simultaneously we must guard against starting patterns of development that take the abundance of our natural resources for granted. We must insure that the future generations born on Athena, who have not seen what Earth has become, won't forget what could happen to them.

The PSC hasn't all the answers to these thorny problems and hopes that as the selection of pioneers continues, many will come forward to contribute their ideas. A constitutional subcommittee is currently laboring over a draft of the planetwide constitution. This will, no doubt, see many revisions before its ratification.

We pioneers have little choice about the equipment we will take with us. Limitations of cost and space aboard ship dictate that only the minimum necessary machinery be brought. GAILE selects this equipment carefully, based on the experiences of the older colony worlds and upon the physical characteristics of the landing area.

The GAILE staff has already chosen the landing site for the first settlement on Athena. Powerful computers correlated billions of bits of data gathered by exploratory parties, and analyzed thousands of factors ranging from climate and geologic stability to soil chemistry and native life forms. The final selection maximizes the probability of success of the first settlement. Prospective pioneers may feel disappointed that they didn't get to participate in the final choice, but the ability to choose actually has less significance than most pioneers realize. Within a few years we will spread across the landscape for hundreds of kilometers from the original site. Within two decades, new settlements, perhaps even on different continents, will offer pioneers a range of choices.

For my money, I think the planned first settlement, Carpithean (shown in figure 3.24), will be a very pleasant place to live. Situated on the west coast of the Freeman continent, about 40° north latitude, the landing site overlooks the North Minervan Ocean. Low mountains separated by narrow valleys cover the coastal region, and small rivers offer water for drinking and irrigation year around. The region has a mild and not too humid climate. High temperatures range from 27° in summer to 16° in winter. No snow falls this near the ocean, and rainfall averages 75 to 90 centimeters annually, an amount sufficient to cover the surrounding hills with broad, leafed, flowering trees.

The first spires will ground at the seaward end of a flat-bottomed valley that will provide an excellent site for the first experimental farms and the first small industrial complex. To the west, a forest and a wide beach opening on a protected bay will serve as recreational areas. To preserve the natural beauty of the beach area, the current plan calls for locating the deuterium plant that supplies our fuel further to the south.

So much remains to be decided, that I can only speculate on what the first five years on Athena might be like. During those years, life in the first settlement will resemble life in a small town. We will spend most of our time just setting up the myriad forms of equipment needed to support our modern lifestyle. We must prepare to grow our food, mine raw materials, establish a communications network, and build shelter for ourselves and our business enterprises.

Despite the compactness of our settlement, I expect thoughts of expansion will occupy every person's mind. Each of us will work toward the day when we can move beyond the confines of our narrow valley and spread out on comfortable tracts of our own land. I suspect that even in the first year pioneers will take every opportunity to explore and examine the area around the first settlement, looking for places to live, new life forms, and vistas of natural beauty that no one from Earth has viewed in our lifetimes.

The Athenian colony will attempt an experiment unique among the colonial worlds. Athena possesses an abundance of native fruits, vegetables, and grains, of sufficient variety and food value to be acceptable to a Human population. Consequently, the agricultural committee's current plans call for developing native Athenian food crops instead of importing familiar strains from Earth. This program will require much more work during the early years. Our familiar crops on Earth are products of centuries of selective breeding techniques which have produced hybrid strains with high yields and resistance to disease. No one has ever cultivated Athenian foods, and we are uncertain about the pests and diseases that might beset them.

In the long run, however, Athenian species will unquestionably adapt better to the native environment than Earth species could hope to. When Athena's farm technology is well developed, it will undoubtedly provide a cheaper and more nutritious source of food than imported varieties. In the unlikely event that these plans don't work out, the colony will keep on hand one year's supply of emergency food that can be replenished by starship if necessary. Contingency plans will also be developed to introduce appropriate Earth species should the native crops fail.

The most important and exciting potential lies in the type of society we choose to develop. We, the first pioneers, will shape the lifestyles, social cus-

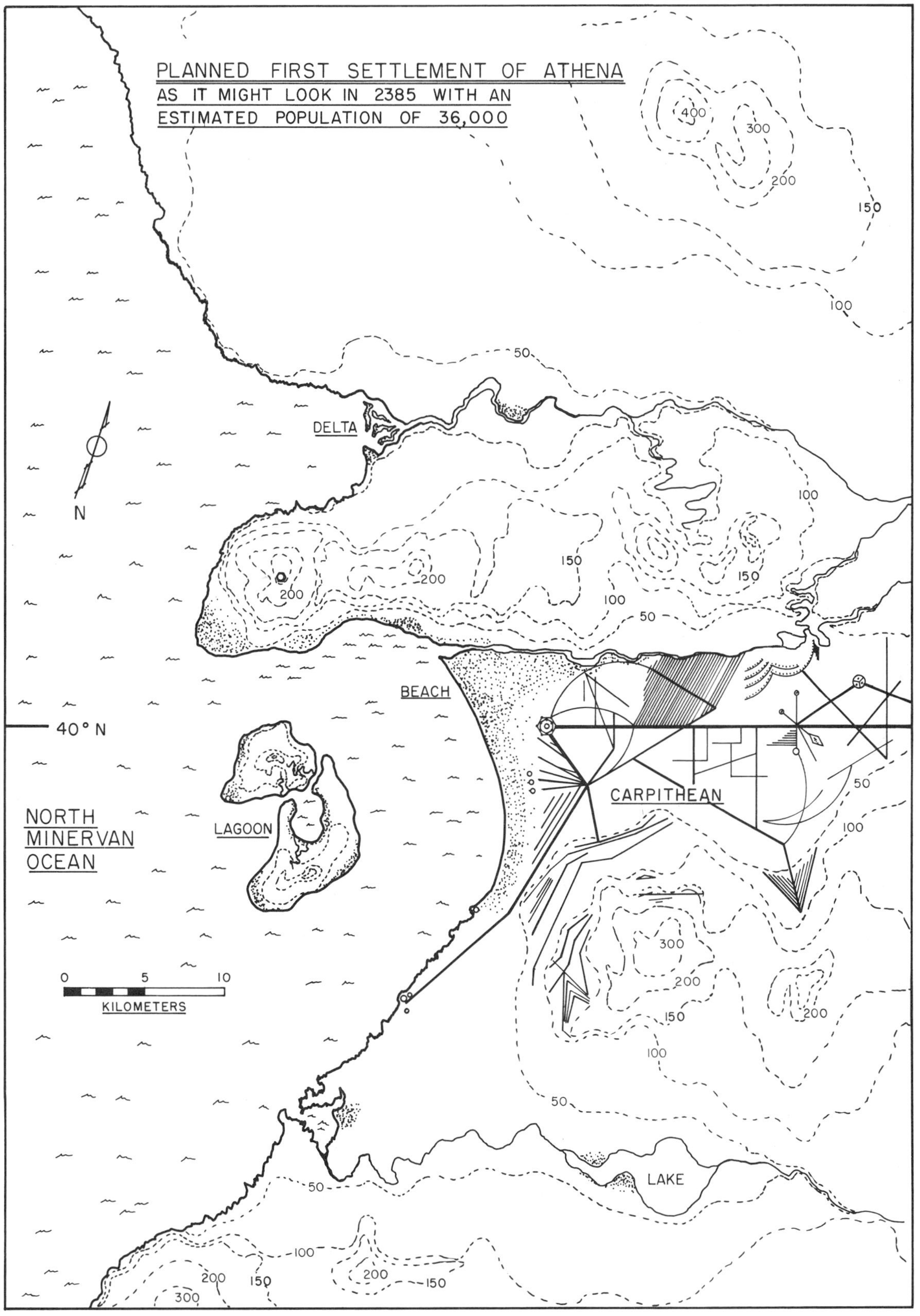
PLANNED FIRST SETTLEMENT OF ATHENA
AS IT MIGHT LOOK IN 2385 WITH AN
ESTIMATED POPULATION OF 36,000
400
300
200
150
100
50
DELTA
N
100
150
200
200
150
100
50
150
BEACH
40° N
CARPITHEAN
50
100
NORTH
MINERVAN
OCEAN
LAGOON
300
200
150
200
100
0
5
10
KILOMETERS
50
LAKE
50
100
200
150
200
150
300

FIG. 3.24

toms, ethics, and institutions of our planet. Whether Athena becomes a good world for people or a bad one, whether our world becomes a cultural backwater or the center of a transgalactic civilization depends on us. If we went to Wyzdom or Brobdingnag, we would have to accept a large part of the cultural norms established by their first pioneers centuries ago. True, we would still have more freedom there than we have on Earth, but it would be a limited freedom. Only on Athena will we have the options to make our world anything we want it to be!

AN ACCOUNT BY THE FIRST PERSON ON ATHENA

Editor's Note: The following is an account of the first Human exploratory party on Athena, written by ship's captain and mission commander James Colligan, 57. Colligan, a graduate of the International Space Academy, started at the position of Junior Astrogator and has worked most of his adult life aboard GAILE's starships. He served aboard the GSS Kara *when it made the historic discovery of the Twin Worlds, Romulus and Remus, in 2342. In 2365, he was acting as executive officer of the exploratory starship* Lewis and Clarke, *then being overhauled at Earth's Bergland shipworks, when GAILE announced the discovery of Athena. Colligan assumed command of the GSS* Jennifer Freeman, *a cargo vessel, and supervised its crash conversion into an exploration vessel for the mission. His promotion at age 42 made him one of the youngest starship captains in GAILE history, but the urgency of the mission, the lack of qualified officers on Earth at that time, and his experience as an officer of an earlier exploratory party made him the most logical choice for the job. Colligan wrote this account shortly after his return from Athena. He currently commands the GSS* Juan Cabrillo *on an exploratory mission into unknown space.*

You have to enjoy chess or reading to like space exploration. It's a job that consists of endless hours of boredom punctuated by brief interludes of fascinating, stimulating discovery. During the months in space, when there is little for anyone save the ship's astronomer to do, most of the crew passes the time by monitoring the instruments and tinkering with the inevitable bugs that occur in the hundreds of systems, from life support to sensor arrays, that fill a starship. When the ship arrives at its destination star, the chances number less than one in twenty that it will hold any planets with life remotely resembling life on Earth. Yet no matter what the find, the ship must make a routine examination of each star system's geological and biological characteristics before abandoning it to continue the search.

Despite the slim odds, excitement always builds when approaching an unknown star system. The day the *Kara* discovered Romulus and Remus, I was standing first shift at the long-range scanner panel. We had spent five years in space and had discovered seven barren systems. Upsilon Lupus was our last try before the long trip home. The ship tingled with excitement. The entire crew, on duty and off, crowded around the main screen monitors throughout the ship as we dropped into a mid-ecosphere orbit. One hundred pairs of eyes probed the field of a billion lights, searching and hoping. Captain Lomela paced the bridge, licking his dry lips, occasionally walking to my panel to peer over my shoulder. For three hours I scanned the emptiness when suddenly the mass point indicator came alive. A planet? I focused the sensor and confirmed that it was a planet, but the reading showed nearly twice the mass of Earth. Another Wyzdom? Or just a barren sphere blanketed in poison gas.

"I've got a mass point, sir!" I cried, "Bearing 354 mark 7!"

"Helm, bring her to bear!" the captain ordered, and a murmur of excitement passed through the bridge. "How big is it, Jimmy?" the captain asked.

"Two times Earth, sir," I replied, "but I can't get a firm focus. It's as if it were a d . . . d . . . double planet!"

A double planet? This would make news even if uninhabitable! The ship accelerated toward the invisible point in black space. The captain briefly announced the facts to the crew while the main screen focused on the point at maximum magnification. All eyes strained.

Then out of the black background studded with a billion points of light, two blue dots began to emerge. Could they be terrestrial? The color looked right. It didn't have the greenish cast of an ammonia giant. All the ship's sensors came to bear, and the science officer began reading over the intercom so all the ship could hear.

"Spectrographic analysis indicates extensive liquid water present; atmospheric nitrogen, 73 percent; oxygen 26 percent; mean surface temperature 26 degrees; mass, point 75 Earth; indications of carbon/nitrogen life. They're terrestrial! They're two Earths!"

Tears welled in my eyes; my hands shook so hard I had to clasp them together. Throughout my body I felt the orgasmic release of tensions that had been building throughout our five-year journey. We had made it! We had found not one world with life, but two!

A roar went up from the entire ship that shook her from citadel to drive beam. Suddenly I felt a slap on my back and a hug around my neck. Everyone was laughing and yelling, stomping their feet and kissing one another like New Year's Eve and Unification Day all rolled into one. We had done it! We had accomplished our mission to find new life in the galaxy.

The discovery of Athena didn't happen like that. I had been on Earth, or more precisely, orbiting Earth for eighteen months during the overhaul of my ship, the *Lewis and Clarke*. The overhaul was the routine denouement of an eight-year exploratory mission. Technicians were giving all the ship's systems a thorough check and replacing obsolescent equipment with the latest advances. I lived at Space Station Seven, a quiet little world, contained in a giant cylinder. Each day, at the end of my shift, I returned to my tiny house to eat, watch the Earth's news and sometimes do a little work in my garden. On one such evening, reports flooded all news stations that the Chlorzi had discovered a planet suitable for Humans. Of course, the news interested me, but I thought little of its consequences. I am a space officer, not a colonist.

Within a week of the announcement I was called to GAILE Command Headquarters on Earth's surface, a rare occurrence for a ship's officer. When I arrived, a staff aide ushered me into the office of the Chief of Starship Operations who informed me that I had been given a command of my own. The transport vessel *Jennifer Freeman,* then in the latter stages of construction, would be outfitted as an exploratory ship to make the first Human survey of the as yet unnamed planet. The *Freeman* would carry a crew of 470, much larger than is typical for exploratory missions. We were charged with determining as quickly as possible if Athena would be suitable for Human colonization. If the answer was yes, we were to remain on the planet for up to four years, taking enough data to let colonial preparations begin. The mission dictated a very tight schedule. The Chlorzi pressed for a quick decision from Earth because they wanted valuable mineral rights on Mammon. Yet Earth could not begin negotiations until Human scientists had verified Athena's habitability. Chlorzi data on such matters is always questionable. They live in a chlorine world at 150 degrees. To them, Earth appears as a frozen wasteland, and it is easy for them to miss little details like a maximum noon temperature of 130 degrees or two percent ammonia content in the atmosphere.

Working around the clock, the builders equipped the *Freeman* for its special mission in less than three months, and we put out to space as soon as she was ready. The mood of the crew seemed upbeat, for this time we knew we would discover something! Chessboards gathered dust, for we had much to do to prepare for our arrival. Unlike other exploratory parties, we operated under time pressure. Also unlike other explorers, we knew about what to expect when we arrived. We studied and analyzed the Chlorzi data to plan our observations in great detail, drafting logistics plans for support of the surface party from the ship.

As Captain, I held responsibility for the entire mission, but Science Director Gayle Edmunsen actually directed the staff of more than 300 scientists and technicians. My executive officer, Fred Shetterly, a meticulous and dedicated spacer, ran the day-to-day operations of the ship. Despite their able help, I still had plenty to worry about, including supplies, surface support logistics, accidents, bacteriological contamination, communications with Earth, and how to maintain a semblance of order among 470 people who, until three months ago, had never seen each other, let alone worked together.

After eleven weeks in space and two separate warp maneuvers, we found ourselves in orbit over Athena. Before setting foot on the planet, however, we had to answer the two critical questions that every exploratory party must ask: First, we had to be sure Athena wasn't inhabited by a civilized species. To this end we scanned all conceivable communications frequencies and adjusted our sensors to detect abnormal energy sources or large masses of refined metals that would indicate the presence of power plants or artificial structures.

Secondly, we had to determine that we

wouldn't encounter any extremely lethal diseases on the planet's surface. The greatest fear of space explorers is that some microorganism to which the Human body has no immunity will be carried from the planet's surface to kill the entire crew within hours, or worse, will be carried back to Earth and there destroy millions of people. To answer this question, we sent dozens of small, unmanned probes to the surface containing a variety of life analyzers and a few tiny creatures called "Armonk's mice." These mice have been specially-bred so that physiological reaction to all known diseases is identical with Humans. We could monitor the life functions of the mice in the probes as they breathed the air and drank the water of the planet.

While we awaited the results of our biological tests, we selected the regions we would explore. Four hundred seventy people are a very small number to explore a world the size of Earth. To fulfill our mission expeditiously, we couldn't afford to waste their energies combing frozen wastes, deserts, and other areas that would never serve as living sites. We had to concentrate our efforts in the not too mountainous, temperate zones of the planet. If Humans couldn't survive in these regions, they couldn't survive at all.

After three weeks in orbit, all the Armonk's mice remained alive, and we began preparing to go to the surface. For several more weeks we would wear bio-barrier suits to prevent biological contamination. We chose a site at 28° 37′ south latitude and 92° 22′ east longitude for our first landing. I made the trip with the Research Director, two biologists, a geologist, and the shuttle pilot. As the shuttlecraft dropped silently toward the surface, the details of the landscape came slowly into focus.

We hovered about 100 meters above the ground, searching for a clear spot to land. The Chlorzi had told us about the atmospheric composition, the size and shape of the land masses, and the planet's magnetic field. They neglected to mention the flowers. Below us lay the most beautiful spring forest I have ever seen, filled with big, bushy broad-leafed trees, bedecked with flowers so gorgeous they put to shame the dogwood, wysteria, and cherry blossoms of Earth. Each of these great flowers rivaled the Earthly rose and orchid in its color and irridescence, and there were *millions* of them!

We put down in a clearing on a low hilltop. As I stepped from the craft, I felt the soft spring of natural earth beneath my boots. At first, the forest seemed quiet, yet as we stood in silence our ears opened to the rustle of a light breeze in the trees and the buzzing and chirping of tiny, unseen creatures. No one spoke for perhaps fifteen minutes as we stood transfixed by the beauty of the scene. At last the biologists switched on their life analyzers and began to take data.

In those first hours on Athena, Human knowledge of the planet more than doubled. In that tiny clearing one found thousands of species of plants and animals, all alien in type and origin, yet strangely familiar. We found no toxic substances, and the Armonk's mice seemed unperturbed after eating a few seeds of the native grass. The pilot and I set up a small titanalum monument commemorating our landing and a data station that would monitor weather, seismic shocks, and life forms in the clearing after we departed.

At last the time came for us to return to the ship. I turned to the biologists and asked, "Have you detected any questionable microorganisms or toxic substances in the air?"

"No sir," they replied, "nothing that seems the least bit harmful."

"Well then," I replied, "I've got to know . . . even if it breaks every regulation in the book," and I removed my helmet. At once my nostrils filled with the soft fragrance of a million blossoms; the sweetness of lotus, the pungent tang of orange, and the richness of roses flooded my senses. The warm sunlight caressed my face like gentle fingers, and the soft breeze sifted through my hair. At that moment, I knew in my heart as well as my mind that God made this world for us.

For the next three months, ten shuttlecraft made daily visits to the surface. Disease-control procedures remained in effect, yet during that time, biologists identified more than 100 species of edible plants and dissected hundreds of warm- and cold-blooded animals in search of potentially harmful diseases. As had been the case on other colonial worlds, no virus or bacteria that seemed harmful to Humans emerged.

At last, I held a meeting with Gayle and the other key members of the biological and medical staffs. We agreed that disease-control procedures could be dropped, at least until some problem developed, and that the scientific staff could move to the surface and set up a base. The base would allow the staff to make their observations at any time of day and on any day they needed. They would no longer be tied to the ship's shuttlecraft, which could carry only one-third of them to the surface at any given time. Living on the planet would provide important insight into the ways of nocturnal animals, night-time weather and other, possibly unknown, phenomena. From the surface base we would extend smaller temporary bases to observe, in detail, the areas of the planet that might serve as sites for future settlements. All these reasons seemed very good, but as far as I was concerned, the main reason for establishing the surface base was to get the scientific staff off the ship and hopefully improve morale. By this time, many of them were suffering from "spitchiness," the irritation that confinement in space brings to the inexperienced and undrugged space traveler.

Once we made our decision, it took another month to get the surface camp fully operational. Two spires of equipment, including shelters, laboratories, communications equipment, and a large laser fusion reactor had to be unpacked and set up. When this was done, the entire crew came down to the surface for a gigantic beach party. The cooks spent three days preparing the feast which included roasting a whole bush buffalo weighing some 650 kilograms and distilling 350 liters of rum for punch. The crew cast aside their inhibitions along with their uniforms as they spontaneously organized games and contests amidst the most picturesque setting I have ever seen. For the first time I truly appreciated the real significance of the pioneers' first Thanksgiving in New England. They gave thanks not only to God but to themselves for having worked hard and done a good job. After eight months of intense effort, the crew of the *Jennifer Freeman* had this coming.

The establishment of the base camp dramatically accelerated our studies. Before three more months passed, we had conclusively proven Athena's suitability for Humankind. I called a large meeting of all scientific group leaders to discuss our recommendations to GAILE and the data we would present to support them. When agreement was reached, we adjourned to prepare the data for loading aboard pilotless space probes that would carry it back to Earth. Because of the mission's importance, we launched duplicate probes and made the unprecedented request for a responding probe to confirm its safe arrival.

The *Jennifer Freeman* remained on Athena for three more years, gathering scientific data that would enable the future pioneers to plan their first colony. Work consumed every waking hour of the scientific staff, but they pursued it informally, without pressure. It seemed more like play. After a while they began to resemble wise old children, knowledgeable, logical, and cautious, yet filled with a child's curiosity about all the new, exciting things their world revealed to them. As work proceeded on the ground, the *Jenny* orbited overhead, weaving a complex pattern over the surface like string on a ball, surveying the land below with the viewpoint only a space ship can have.

My role changed too, during this phase of the mission. As before, the Ex-O ran the ship and my Research Director managed the scientific program on the ground. As Captain, I alternated between the ship and the surface trying to monitor all aspects of the mission. I consulted periodically with all my subordinates from the commissary officer and the ship's engineer to the field team leaders roaming the six continents. As the ultimate authority, I was often called upon to arbitrate disputes, and every other month I had to review and summarize the field data for a progress report to GAILE. During the course of all this, I saw much of the new planet's surface and developed great appreciation for its richness and beauty.

One day in the second year, I received a curious message from the ship's science officer. Ship's sensors had detected a substantial mass of pure metal on the surface, yet pure metals do not occur in nature. It couldn't have been one of our exploratory vehicles for they reported their locations to the ship, and the sensors were programmed not to recognize them. Furthermore, the metal exhibited none of the power radiation a living society would. I ordered an investigatory party at once, and we set off for the site near the northern coast of continent A2369.2. What we found made still more space history.

Arriving at the mysterious spot, we found ourselves hovering over dense jungle, lush, humid, and tropical. From the air we could see nothing, yet our instruments told us that below lay twenty to thirty metric tons of very pure titanalum/beryllium alloy whose metallurgy could not be traced to any GAIL member. The thick forest prevented our landing, so we receded half a kilometer from the site and burned a small landing patch in the forest. We advanced to the site on foot under the green canopy of foliage. The ground was almost clear and carpeted with a

THE MYSTERIOUS DOME OF ATHENA

A FULL SPECTRUM GAMMA SCAN OF THIS ALIEN STRUCTURE OF UNKNOWN ORIGIN TAKEN BY THE CREW OF THE JENNIFER FREEMAN IN 2368 ADTC REVEALS ITS TRUE EXTENT. THE GAMMA BEAM PASSES THROUGH ALL KNOWN MATERIAL OBJECTS, REVEALING THEIR MOLECULAR STRUCTURE AS WELL AS THEIR GROSS PHYSICAL SHAPE. THIS VISUAL READOUT, ENHANCED BY COMPUTER, ILLUSTRATES THE EXTENT TO WHICH NATIVE VEGETATION HAS SURROUNDED THE DOME. SOME VEGETATION AND FOREST DEBRIS HAS BEEN DELETED FOR CLARITY.

THE VERTICAL TUNNEL BENEATH THE DOME SERVES NO OBVIOUS PURPOSE AND ENDS ABRUPTLY AT A DEPTH OF 600 METERS.

FIG. 3.25

layer of composting leaves and flowers. When we reached the site, we came upon a geodesic dome approximately twenty-five meters in diameter. The dome appeared to have no doors or windows but was unmistakably hollow and the source of our metal readings. We detected no other signs of civilized life, and the vegetation grown up around the structure bespoke a long period of neglect. As we couldn't probe the dome with our sensors, I ordered a larger gamma scanner sent up from the base camp.

When it arrived, we set it up about 100 meters from the dome and took our first scan. Figure 3.25 shows a graphic readout of what we found. The dome was actually larger than it appeared, extending some five meters below grade to a foundation of fused and densely compacted ground. A long shaft, lined with alloy, reached down below the foundation. Scans at closer range revealed shuttered doors and windows so perfectly fitted that only a thin line masked by centuries of oxidation could be discerned from the outside.

We carefully drilled through the door with our powerful laser and found an empty house. It contained no furniture, save some oddly-shaped, built-in forms. Few pieces of equipment or implements gave any clue to who left them behind. Some tiny bits of apparatus lay tossed about, as if dropped carelessly by those departing: a section of fiberoptic, some circuit terminations, and fasteners of alien design and configuration. It looked much like our own base camp would look a year from then after we hauled up our gear, leaving only the shells of our buildings behind. The tunnel extending beneath the dome seemed to serve no purpose. It dead-ended abruptly at a depth of about 600 meters, as if it were part of some unfinished project.

The origin of the strange dome remains one of the great mysteries of Athena. Some leaves found inside indicate they were trapped more than 25,000 years ago! Explorers perhaps? And if so, what did they think of their find and when will they return?

I am sure the research staff and crew would have been content to remain on Athena forever, but supplies of irreplaceable consumables were running low, and an eager Earth awaited with thousands of questions for the scientists who had seen the new planet with their own eyes. Gayle summarized it best by saying, "We could stay here and screw around forever, Jimmy, but I think we've got what we came for."

Getting the crew to leave a new planet is the hardest job any exploratory commander has. As a crew member on an earlier mission, I remembered the thousands of justifications I had to remain on Romulus. The howls of protest from the researchers on Athena fell on sympathetic ears, yet outwardly I remained intransigent. The slightest sign of weakness would have lost all. Many begged to stay, saying that they would wait for the next ship. A few actually deserted and had to forcibly be rounded up. At last every one was accounted for and the equipment brought up and stowed. As the photon drives drove us from the Garden of Eden, we began to lose our childlike innocence again and fit ourselves into those well-worn molds that contained us on Earth.

I am not much of prognosticator or visionary, but if I had to predict one thing, I would predict that the future colony on Athena will be the most prosperous, creative, and beautiful Human society that has ever existed. Athena has all the beauty and diversity of Earth when it was new, yet the pioneers who settle her will bring with them the wisdom and experience of sixty centuries of civilization. Athenians need not grope and struggle and make the mistakes that Humanity has made on Earth. She will burst full-blown from the head of God, a mature, intelligent, compassionate world and so we named her, Athena.

* * * *

OPPORTUNITIES FOR IMMIGRANTS

Opportunities for immigrants on Athena are unlimited! Special skills and abilities are not as important to a new world as intelligence, resourcefulness, and the proven willingness to pitch in and work hard at whatever needs to be done. Spots remain open on the first pioneering vessel for anyone who wishes to build a world right, from the ground to the sky.

Interstellar Transportation and Communication

Almost everyone has ridden in levicars, hypergravs, or tube trains, but less than one percent of Earth's people have traveled in the airless void of interplanetary space. Consequently most pioneers ask the same questions about space travel. How is it done, how long will it take, how safe is it, and how comfortable will I be during the trip? Pioneers also want to know how they will communicate with their family and friends on Earth after arriving on their new planet. This section attempts to briefly answer each of these common questions. Readers who wish to learn more details about the workings of interstellar ships should refer to the special bulletin listed in Appendix B.

By any measure of our Earthly experience, the stars lie incomprehensibly distant. Comparing a trip from New York to Peking with a trip to our sun's *nearest* stellar neighbor, Alpha Centauri, is like comparing a distance of three centimeters to a trip halfway around the Earth! To travel such distances requires great speeds and large amounts of energy. Such speeds and energy levels alter concepts of space and time as we commonly perceive them on Earth, so calculating how long a trip between the stars will take becomes a very complex problem.

THE PHYSICS OF SPACE TRAVEL

We measure the distances to the stars in units called "light years." A light year is the distance that a beam of light travels during one Earth year. Though it makes a conveniently large measure of distance, approximately equal to 9.47 E12 kilometers, it has little relation to travel time between the stars.

For more than two centuries, the physics of Isaac Newton explained all the observable phenomena in our Solar system. Yet by the end of that period, physicists had observed a number of things that didn't fit Newton's theories. While trying to measure the speed of light, they noted that all observers measured the same speed, regardless of how they were moving relative to the light source. Newton's theories said velocities add linearly. The orbit of the planet Mercury also exhibited perturbations that Newtonian physics could not explain.

Albert Einstein resolved these dilemmas by rejecting two ideas implicit in Newton's physics; the existence of an absolute inertial frame and the notion that simultaneous events can be simultaneously observed. Einstein accepted the commonsense idea that the laws of physics can be formulated so they are the same for all observers, no matter how they are moving through space. After all, the Earth rotates on its axis and travels about the sun, and the sun orbits the galactic center, and the galaxy in turn moves through space along with thousands of other galaxies. Where then, is the absolute frame of reference that Newton postulated?

Starting with these new assumptions, Einstein rederived the laws of physics to arrive at some startling conclusions. His reasoning is too complex to describe here, but his conclusions are important to every space traveler. Einstein showed that an observer in motion *relative* to an object will observe that object to be shorter than he would if he were at rest relative to the object. Such differences can't be detected when speeds are low relative to the velocity of light, but at the speeds of interstellar spaceships they become significant.

Let's say that a space traveller gets into a space ship and heads for the planet Genesis about 44 light years away. If the ship accelerates to 95 percent the speed of light, Genesis now appears only 13.6 light years away. To travellers on board the ship, the trip appears to take just 14.3 years, but to observers watching from Earth or Genesis (all the planets have low velocities relative to each other), the trip appears to take three times as long. This simple example illustrates why travelers aboard a starship perceive that the trip takes less time than the people waiting for them on the planets. The effects of acceleration further complicate the computation of travel times, but the results, calculated by computers, are indicated in the data tables for each colony planet.

Unfortunately, Einstein's theories went on to predict that no material object could reach or exceed the speed of light relative to another object. Einstein calculated that as the relative speed of an object increased, its mass and therefore its resistance to acceleration would increase too. Experiments with atomic particle accelerators in the mid-20th century verified his calculations with stunning accuracy. Thus as the space age opened, and people began to send their first feeble probes to the other planets of the Solar system, the distances to the stars still seemed overwhelming. Even if technology could have developed new propulsion sources that would enable ships to approach the speed of light, the nearest stars would still have taken decades to reach.

Despite this dismal conclusion, Einstein's theories did much to advance the cause of space travel, for he predicted that matter could be converted into energy. The inefficient chemical rockets that powered early space vehicles could not possibly have reached the stars. Only direct conversion of large amounts of matter into energy could propel spaceships near the velocity of light. Before the next great breakthrough in physics, the theories of Einstein had formed the basis of the laser fusion drive that propelled Captain Jan De Wyze and the *Freedom 4* to Alpha Centauri and the discovery of Wyzdom. The round trip of just 8.6 light years required more than thirteen Earth years of travel time. Yet to the crew of the *Freedom 4,* the trip seemed to take five years each way. During that time they aged three years and four months less than the people waiting for them on Earth!

Thirteen years after the return of the *Freedom 4,* a young physicist named Raymond Krauchunas again rocked the scientific world with a new comprehensive theory of physics that unified the previously discordant concepts of matter, electricity and quantum mechanics. His theories encompassed all data that supported the theories of Newton, Einstein, Maxwell, and Planck, and went on to explain new data gathered during the voyage of the *Freedom 4* and other, early, near-light- velocity spacecraft. The singularities of mass and energy predicted by Einstein's theories bothered Krauchunas. Singularities are mathematical concepts of infinity or "infinitesimalness" that never in fact occur in nature. In the past, simple theories predicted the force of an elastic impact or the stress before a crack to be infinite. Yet more refined observation and analysis showed these predictions to be imprecise.

Krauchunas again reexamined the basic postulates of Einstein's physical theories and found in them subtle, hidden assumptions. Krauchunas rejected the notion that space-time need be described by any finite number of dimensions and postulated the existence of hyperspace. He went on to show how a space ship, or any self-propelling matter, could travel between points in Einstein's space-time continuum by taking a short cut through hyperspace. Since the object no longer transverses "real" space and time, no time need elapse during the course of the trip.

An analogy may be helpful in understanding hyper-light travel. Imagine two ants walking along a piece of cloth that has been dropped casually on the floor. The cloth is wrinkled and folded back on itself, and the ants are walking on the folded surface. As they walk, one ant makes a tiny hole in two overlapping pieces of the cloth and crawls through it. He emerges at another point on the cloth several inches from the other ant who does not pass through the hole but keeps walking along the surface of the cloth. The second ant doesn't reach the point on the cloth that the first ant has reached until some time has passed. Meanwhile the first ant has continued along the cloth and remains several inches ahead of the second ant.

In a sense, the first ant traveled faster than the second ant, but in another sense, the first ant did not travel through the same space as the second ant; so it becomes illogical to speak of his velocity relative to the second ant. Einstein envisioned space as a multidimensional analogy of the folded cloth. Krauchunas envisioned a way that people could make holes in the cloth to cross vast distances of "normal" space in no time at all. The path through hyperspace relates in a complex way to both the acceleration and the first derivative of acceleration, known as "jerk," at the instant of departure from real space-time.

Twenty-five years passed between the publication of Krauchunas's unified field theory and the launching of the first starship capable of hyperspace travel. The *Albert Einstein*, launched in 2112, made the round trip to Wyzdom in only thirty-eight months. Time aboard ship measured less than eighteen months. After the *Einstein's* return, the ICSE constructed a much larger vessel, the *Christopher Columbus*, to make a fifteen-year exploration of eight nearby star systems. Yet even though the concept of hyperspace flight was proven, the first pioneers used conventional sublight ships to transport the first permanent colony to Wyzdom.

Though the development of hyperspace travel (sometimes called "space warp") brought the stars within reach, reducing travel times to a matter of weeks required a second major scientific breakthrough. Prior to this, interstellar travel had been limited by the fact that Humans can tolerate no more than 1.3 g of acceleration for extended periods. Consequently, it took almost nine months for space ships to accelerate to the near light speeds needed to make the jump through hyperspace and another nine months to slow down again. Thus despite the advent of space warp, the minimum interstellar voyage took one and one-half years of Earth time (though the time seemed shorter to the passengers). Development of the artificial gravitational field allowed starships to attain much higher acceleration levels than those achieved in the early days of interstellar travel. These higher acceleration levels shortened trip times appreciably, and the g-field protected passengers and crew from the large acceleration forces undergone in the transition to hyperspace. The g-field also gave ship designers greater freedom to design their vessels, since the field could be used to resist acceleration forces that would tear the ship apart.

Within the confines of the ship, the artificial g-field creates a gravitational field identical in all respects to the gravitational field of a planet. It maintains this static field despite wide swings in the acceleration of the ship itself. Therefore, the g-field generators are not static devices but highly complex automatic equipment. Adjustments can be made so that the amount of gravity felt within the ship varies. Many novice space travellers believe it would be fun to make their journey in a weightless state, but although most people enjoy the sensation of weightlessness for rather brief periods, the weightless condition becomes rather annoying for the extended periods required by interstellar travel. In addition, people inexperienced with weightlessness can easily injure themselves by bumping into walls and ceilings. The crew adjusts the g-field of the ship during the voyage from the Earth value of 9.8 m/s^2 to the value of the destination planet. This allows pioneers' bodies to become gradually acclimated to their new planet's gravitation.

THE MODERN MODULAR STARSHIP

Figure 4.1 shows a modern starship, typical of the sort used to transport pioneers to the colonies today. Built from a set of standarized, modular components, this design offers great flexibility of mission and low cost. The three basic components of the ship are pods, spires, and photon drive engines. Pods and spires comprise the building blocks of the "mainframe," the section of the ship that carries people and cargo. The engines, linked to fuel storage modules, comprise the "power frame."

Cylindrical transfer spokes link the pods of the mainframe together into a hexagonal array. The spires that trail behind the pods contain most of the load-carrying volume, and a variety of special designs accommodate passengers or cargo to suit the mission. Pod access links connect the transfer spokes of the mainframe to a central cylinder called the "force field wand." The wand ties the mainframe and the power frame together. It contains both power and control circuits and affords the crew access to the ship's engines during flight. For missions requiring additional payload capacity, large mainframes can be constructed by adding additional pods and spires. Designers arrange the pods in hexagonal configuration like the cells in a bee's honeycomb because it offers compactness, symmetry and flexibility. Figure 4.2 shows a spatial projection of a much larger ship with four engines and a four hex mainframe. The scale on this drawing gives some idea of the enormous size of these vessels. The size of the pod, spire and engine modules does not change from one ship to the next; only the number of modules is altered.

Figure 4.3 illustrates some of the many configurations that can be built from the standardized modules. Higher thrust-to-payload ratios are desired for long-distance missions to allow longer jumps through hyperspace. Unfortunately high ratios can not be maintained in the larger vessels; they must sometimes make more than one hyperspace jump during a single crossing. The use of standardized modules greatly reduces the cost of ships by eliminating recurrent engineering costs and allowing shipbuilders to tool up for quantity production. Such standardization does not prevent manufacturers from upgrading engine designs or navigational equipment as improvements arise. On the contrary, the standard design actually facilitates such improvements because new subsystems fit easily into older vessels.

The modular starship appears to be a thin and fragile structure because designers no longer need to use large amounts of material to withstand acceleration-induced forces. Instead, a complex system of fields, energized by the ship's engines, binds the modules together. Today's starships can accelerate at up to 16 g continuously and can momentarily attain higher levels than this during hyperspace transition.

Figure 4.4 shows an enlargement of a typical single hex mainframe. The g-field wand at the center of the hexagon generates a uniform anti-acceleration field which grips all six spires and pods. As the ship accelerates forward the anti-acceleration field cancels the effect of all but one g of acceleration within the ship. During negative acceleration (slowing), the anti-acceleration field's polarity is reversed to prevent people inside from being thrown up to the ceiling.

Little lateral force ever acts on the ship, so it is not designed to withstand much lateral loading. During turns, however, some lateral force tends to deform the hexagon. To prevent this, magnetic fields of opposite polarity act between the central wand and the spires to maintain tensile force within the transfer spokes and tension cables. This design saves expensive material and weight, since tension members can be made much lighter than compression members.

Figures 4.5 and 4.6 illustrate typical spire and pod modules. The scale of these drawings indicates the tremendous carrying capacity of each of these structures. Comparing these two figures with hex array of Figure 4.4 shows that the decks of the spires and pods run perpendicular to the longitudinal axis of the ship. In other words, a person standing in the ship would look straight up to face the front of the ship and would look straight down to face the engines.

The citadel structure at the end of each force field wand houses three additional field generators that protect the ship and its passengers from collisions with foreign objects. At near light speeds, an impact between the ship and a particle no bigger than a grain of sand would melt a hole several inches in diameter in the strongest and most refractory of materials. Solid particles occur even less commonly in the interstellar void than they do in the space of our Solar system. Yet they do exist, ranging in size from fine dust to giant spheres several times the mass of the planet Jupiter! There is a slim but finite probability that a ship could collide with such a particle during the trillions of kilometers of an interstellar crossing, and just one such collision would be fatal.

Figure 4.7 is a schematic profile of the ship's defense fields. The familiar g-field offers the first line of defense. Physically identical to the anti-acceleration fields of the ship and the anti-gravity fields of levicars, the polarity of the citadel g-field is reversed so that it repels mass instead of attracting it. The field is focused into a beam which extends up to 100 million kilometers in front of the ship. When it encounters a particle at this range, it exerts just a slight force on both the ship and the particle. This gentle nudge is usually sufficient to deflect the particle from the ship's path, just as the bow wave of a boat keeps floating debris from striking the hull.

The g-field grows stronger as the ship and the obstacle get closer until, within a certain critical range of the ship, sensors automatically activate the "weak field" generator, the ship's second line of defense. The w-field neutralizes the weaker of the interatomic forces. This causes all matter within the field to break up into a cloud of free protons, neutrons, and electrons. A magnetic field, emanat-

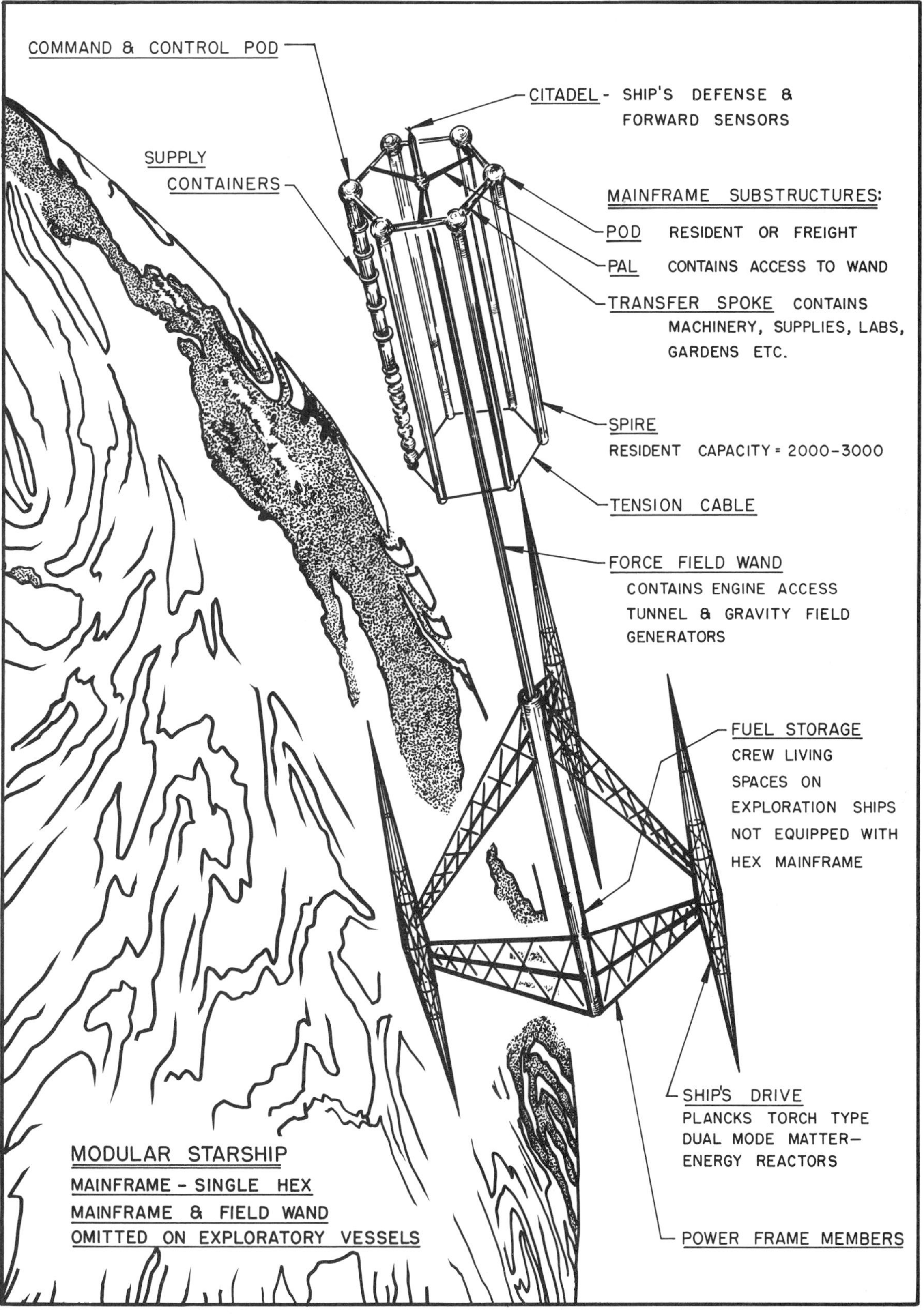
COMMAND & CONTROL POD
CITADEL - SHIP'S DEFENSE & FORWARD SENSORS
SUPPLY CONTAINERS
MAINFRAME SUBSTRUCTURES:
POD RESIDENT OR FREIGHT
PAL CONTAINS ACCESS TO WAND
TRANSFER SPOKE CONTAINS MACHINERY, SUPPLIES, LABS, GARDENS ETC.
SPIRE
RESIDENT CAPACITY = 2000-3000
TENSION CABLE
FORCE FIELD WAND
CONTAINS ENGINE ACCESS TUNNEL & GRAVITY FIELD GENERATORS
FUEL STORAGE
CREW LIVING SPACES ON EXPLORATION SHIPS NOT EQUIPPED WITH HEX MAINFRAME
SHIP'S DRIVE
PLANCKS TORCH TYPE DUAL MODE MATTER-ENERGY REACTORS
POWER FRAME MEMBERS
MODULAR STARSHIP
MAINFRAME - SINGLE HEX
MAINFRAME & FIELD WAND OMITTED ON EXPLORATORY VESSELS

FIG. 4.1

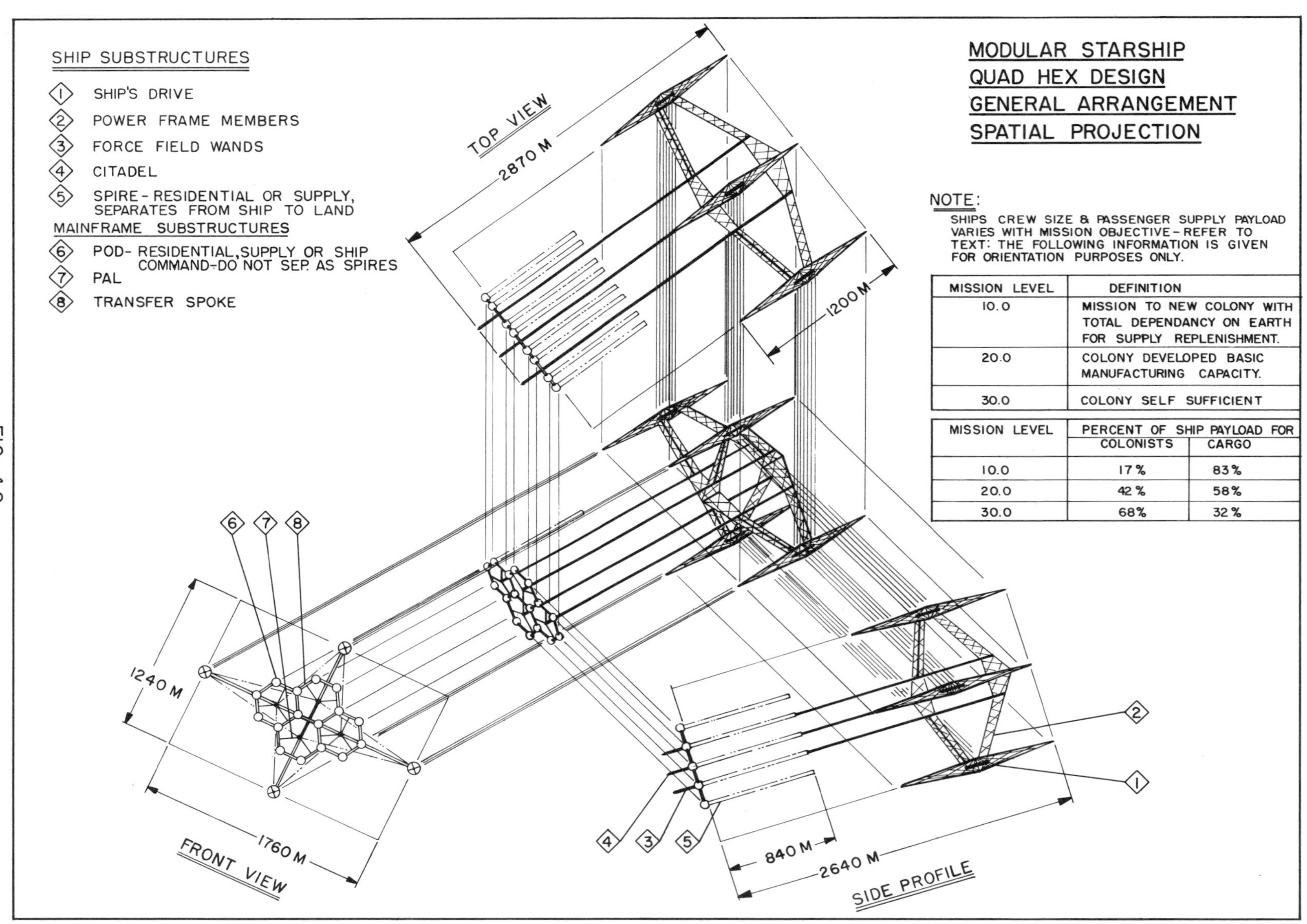

MISSION LEVEL	DEFINITION
10.0	MISSION TO NEW COLONY WITH TOTAL DEPENDANCY ON EARTH FOR SUPPLY REPLENISHMENT.
20.0	COLONY DEVELOPED BASIC MANUFACTURING CAPACITY.
30.0	COLONY SELF SUFFICIENT

MISSION LEVEL	PERCENT OF SHIP PAYLOAD FOR	
	COLONISTS	CARGO
10.0	17 %	83 %
20.0	42 %	58 %
30.0	68 %	32 %

FIG. 4.2

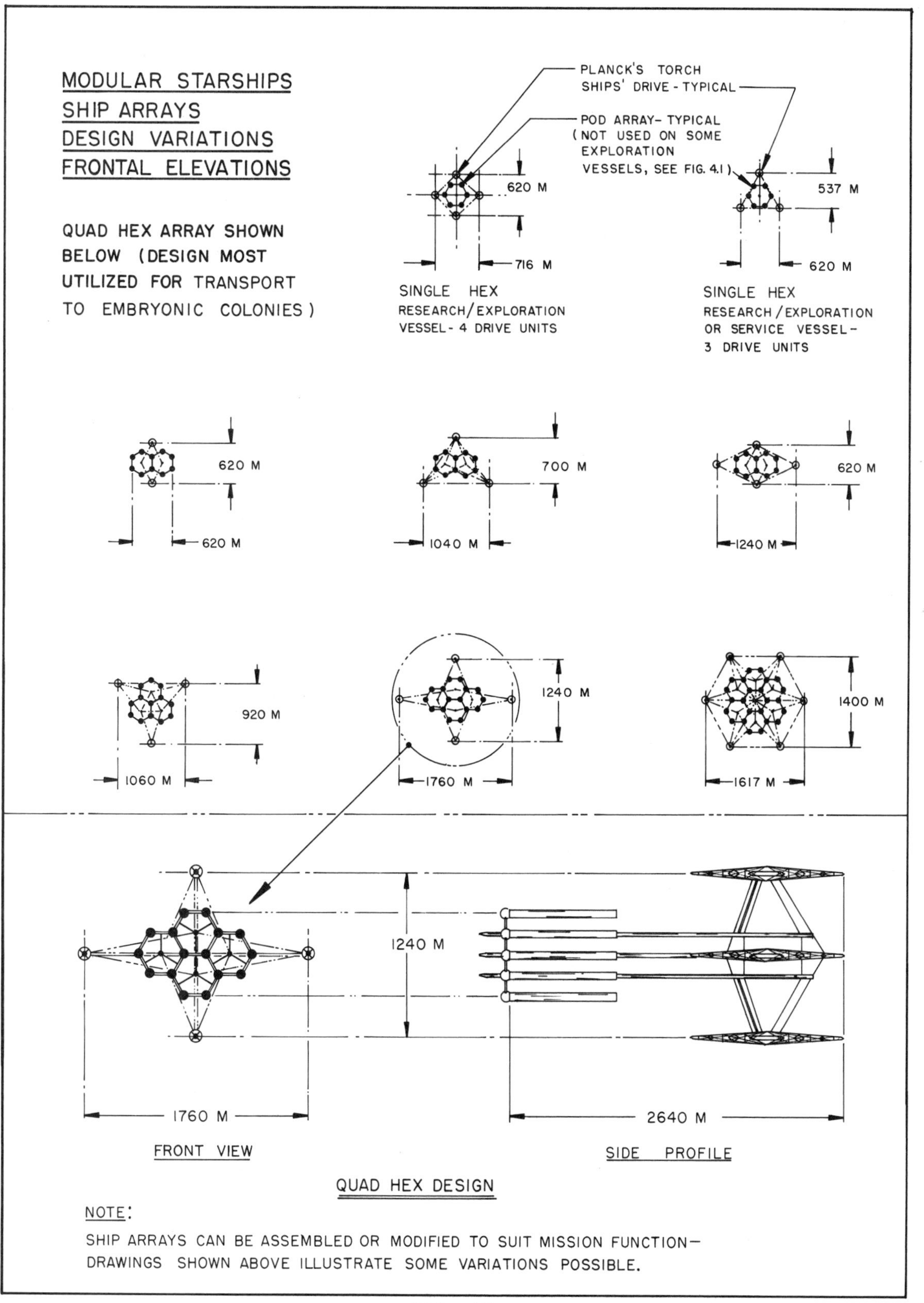
MODULAR STARSHIPS
SHIP ARRAYS
DESIGN VARIATIONS
FRONTAL ELEVATIONS
QUAD HEX ARRAY SHOWN BELOW (DESIGN MOST UTILIZED FOR TRANSPORT TO EMBRYONIC COLONIES)
PLANCK'S TORCH SHIPS' DRIVE - TYPICAL
POD ARRAY- TYPICAL (NOT USED ON SOME EXPLORATION VESSELS, SEE FIG. 4.1)
620 M
716 M
SINGLE HEX
RESEARCH/EXPLORATION VESSEL- 4 DRIVE UNITS
537 M
620 M
SINGLE HEX
RESEARCH/EXPLORATION OR SERVICE VESSEL- 3 DRIVE UNITS
620 M
620 M
700 M
1040 M
620 M
1240 M
920 M
1060 M
1240 M
1760 M
1400 M
1617 M
1240 M
1760 M
2640 M
FRONT VIEW
SIDE PROFILE
QUAD HEX DESIGN
NOTE:
SHIP ARRAYS CAN BE ASSEMBLED OR MODIFIED TO SUIT MISSION FUNCTION— DRAWINGS SHOWN ABOVE ILLUSTRATE SOME VARIATIONS POSSIBLE.

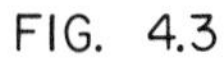
FIG. 4.3

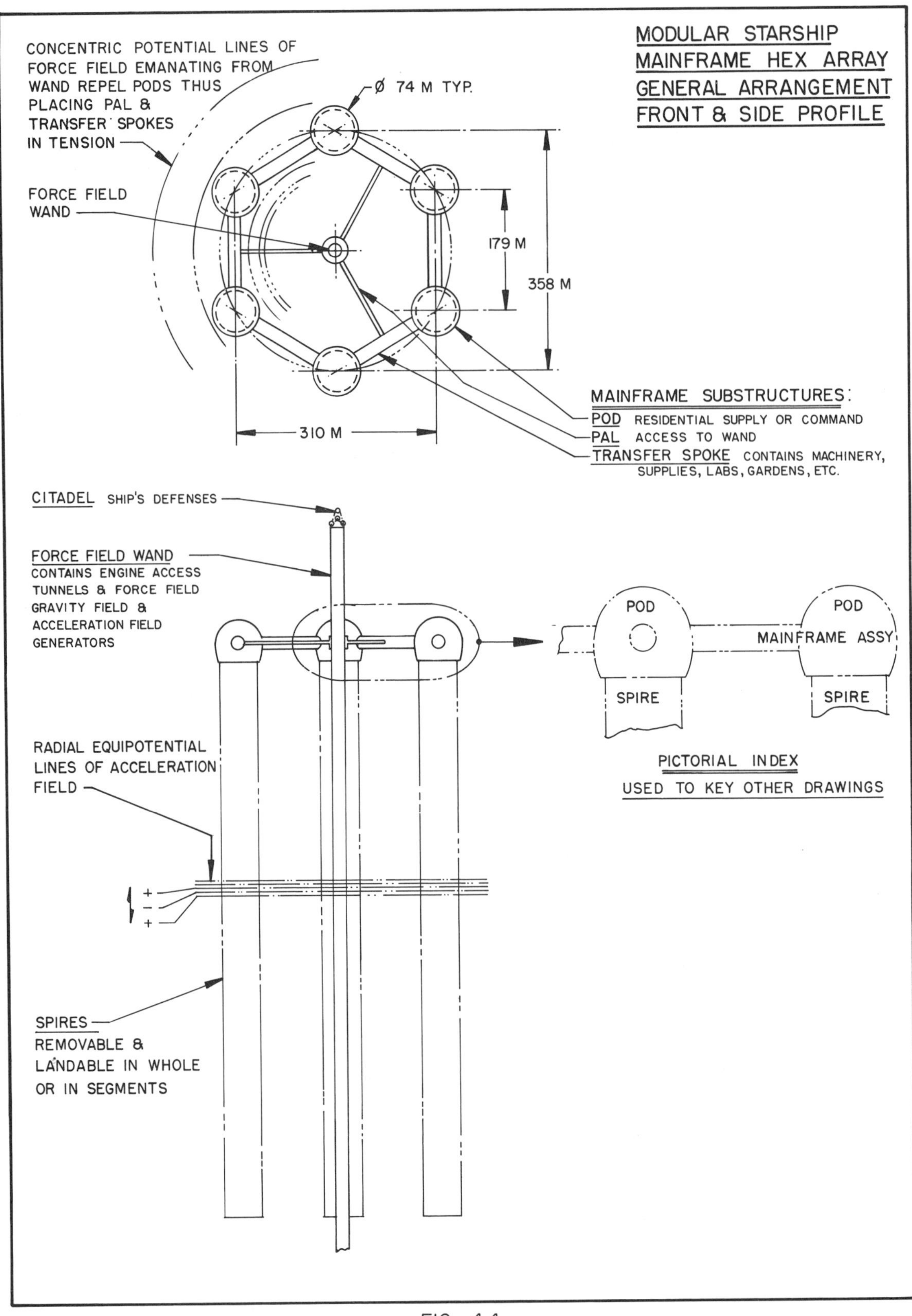
MODULAR STARSHIP
MAINFRAME HEX ARRAY
GENERAL ARRANGEMENT
FRONT & SIDE PROFILE
CONCENTRIC POTENTIAL LINES OF FORCE FIELD EMANATING FROM WAND REPEL PODS THUS PLACING PAL & TRANSFER SPOKES IN TENSION
Ø 74 M TYP.
FORCE FIELD WAND
179 M
358 M
310 M
MAINFRAME SUBSTRUCTURES:
POD RESIDENTIAL SUPPLY OR COMMAND
PAL ACCESS TO WAND
TRANSFER SPOKE CONTAINS MACHINERY, SUPPLIES, LABS, GARDENS, ETC.
CITADEL SHIP'S DEFENSES
FORCE FIELD WAND
CONTAINS ENGINE ACCESS TUNNELS & FORCE FIELD GRAVITY FIELD & ACCELERATION FIELD GENERATORS
POD
POD
MAINFRAME ASSY
SPIRE
SPIRE
PICTORIAL INDEX
USED TO KEY OTHER DRAWINGS
RADIAL EQUIPOTENTIAL LINES OF ACCELERATION FIELD
+
−
+
SPIRES
REMOVABLE & LANDABLE IN WHOLE OR IN SEGMENTS

FIG. 4.4

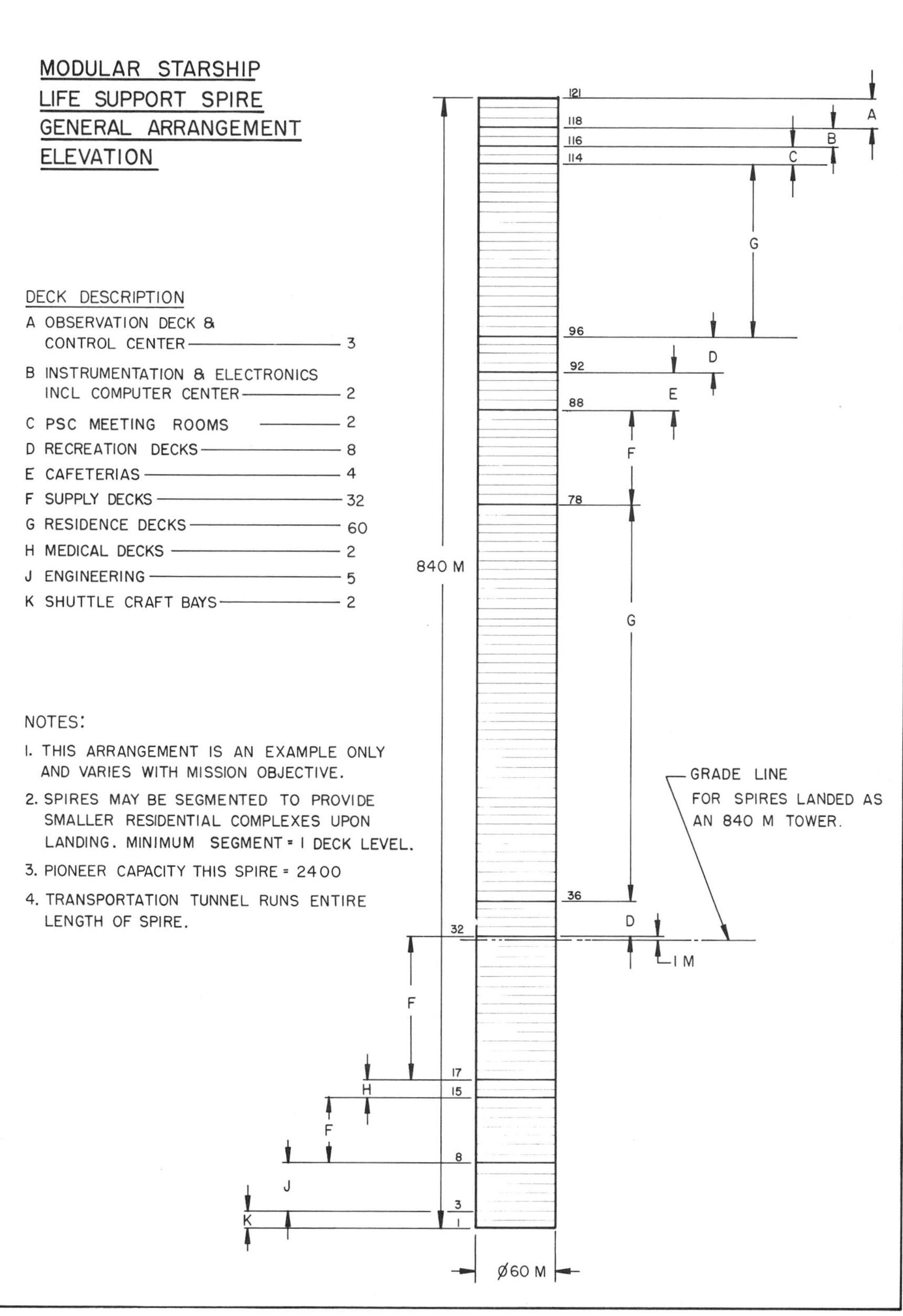

MODULAR STARSHIP
LIFE SUPPORT SPIRE
GENERAL ARRANGEMENT
ELEVATION
DECK DESCRIPTION
A OBSERVATION DECK & CONTROL CENTER 3
B INSTRUMENTATION & ELECTRONICS INCL COMPUTER CENTER 2
C PSC MEETING ROOMS 2
D RECREATION DECKS 8
E CAFETERIAS 4
F SUPPLY DECKS 32
G RESIDENCE DECKS 60
H MEDICAL DECKS 2
J ENGINEERING 5
K SHUTTLE CRAFT BAYS 2
NOTES:
1. THIS ARRANGEMENT IS AN EXAMPLE ONLY AND VARIES WITH MISSION OBJECTIVE.
2. SPIRES MAY BE SEGMENTED TO PROVIDE SMALLER RESIDENTIAL COMPLEXES UPON LANDING. MINIMUM SEGMENT = 1 DECK LEVEL.
3. PIONEER CAPACITY THIS SPIRE = 2400
4. TRANSPORTATION TUNNEL RUNS ENTIRE LENGTH OF SPIRE.
121
118
116
114
96
92
88
78
36
32
17
15
8
3
1
A
B
C
G
D
E
F
G
D
F
H
F
J
K
840 M
GRADE LINE FOR SPIRES LANDED AS AN 840 M TOWER.
1 M
Ø60 M

FIG. 4.5

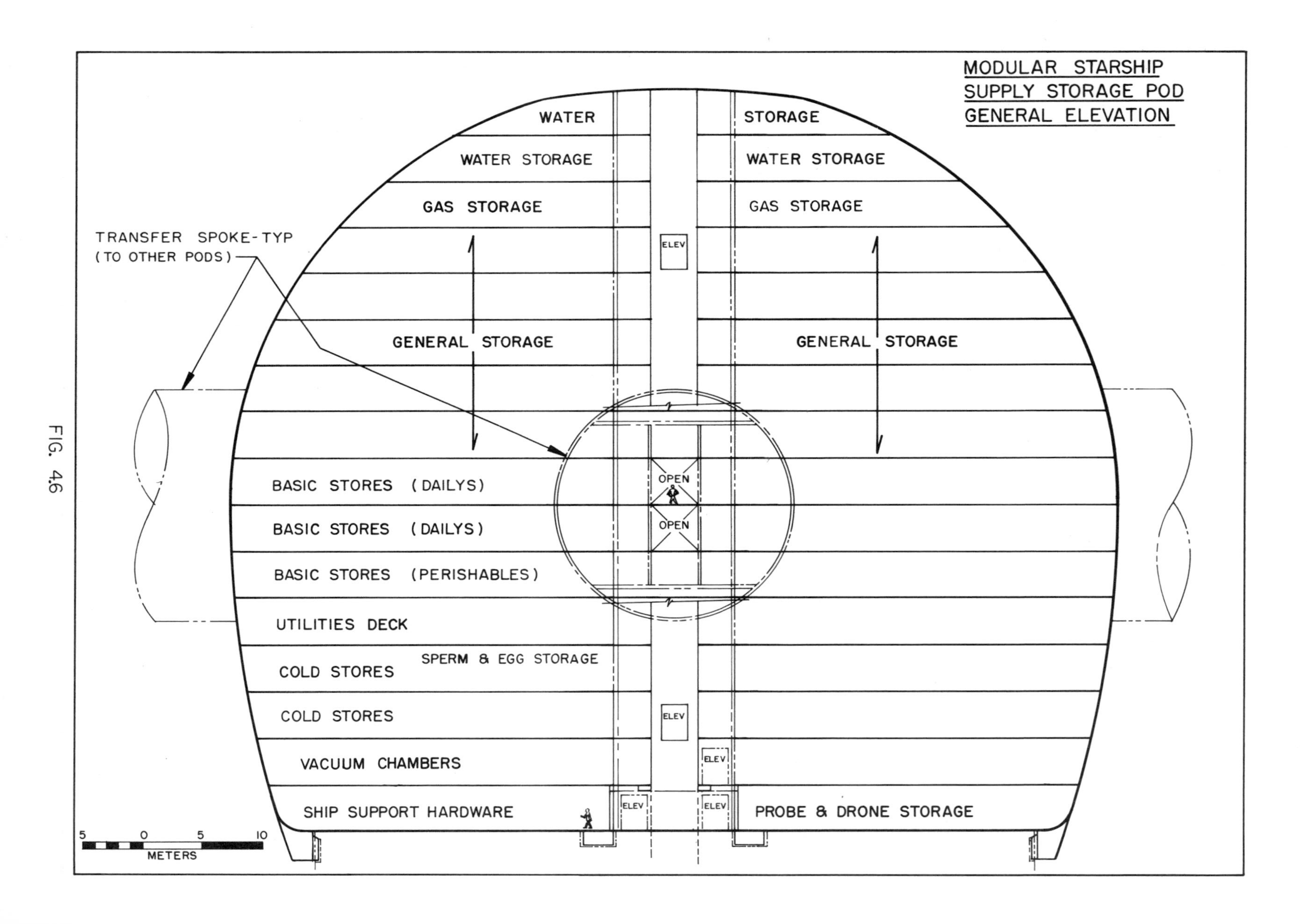
MODULAR STARSHIP
SUPPLY STORAGE POD
GENERAL ELEVATION
WATER
STORAGE
WATER STORAGE
WATER STORAGE
GAS STORAGE
GAS STORAGE
TRANSFER SPOKE-TYP
(TO OTHER PODS)
ELEV
GENERAL STORAGE
GENERAL STORAGE
OPEN
OPEN
BASIC STORES (DAILYS)
BASIC STORES (DAILYS)
BASIC STORES (PERISHABLES)
UTILITIES DECK
COLD STORES
SPERM & EGG STORAGE
COLD STORES
ELEV
VACUUM CHAMBERS
ELEV
SHIP SUPPORT HARDWARE
ELEV
ELEV
PROBE & DRONE STORAGE
5
0
5
10
METERS

FIG. 4.6

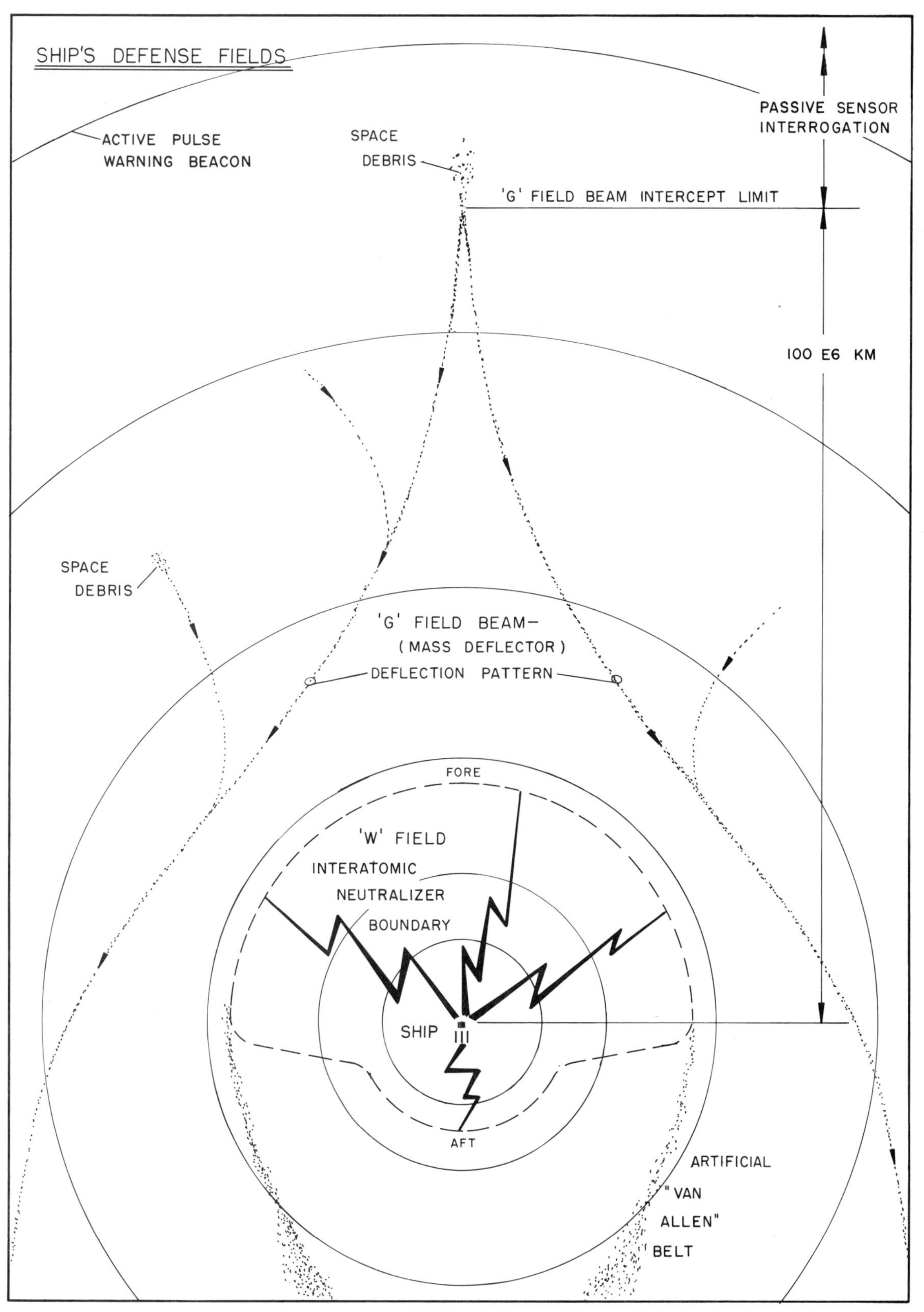
SHIP'S DEFENSE FIELDS
ACTIVE PULSE WARNING BEACON
SPACE DEBRIS
PASSIVE SENSOR INTERROGATION
'G' FIELD BEAM INTERCEPT LIMIT
100 E6 KM
SPACE DEBRIS
'G' FIELD BEAM— (MASS DEFLECTOR)
DEFLECTION PATTERN
FORE
'W' FIELD
INTERATOMIC NEUTRALIZER BOUNDARY
SHIP
AFT
ARTIFICIAL "VAN ALLEN" BELT

FIG. 4.7

ing from the citadel, captures the protons and electrons, forming an artificial "Van Allen belt" about the ship. Just before arrival, the crew de-energizes the magnet to release the charged particles to space. The g-field and conventional neutron shielding protect the ship's occupants from the w-field's neutrons.

GAIL starship's take special precautions to avoid energizing the w-field when another starship is in their path. All vessels carry wideband transponders which emit conventional EM radiation over a series of specific frequencies. Association ships also monitor an even wider range of frequencies and employ long-range sensors tuned to detect unnatural energy sources just in case alien ships of unknown origin should cross their path. Unprovoked firing of the w-field at such an alien might be misinterpreted as a hostile act.

Today's starships employ the "photon drive," sometimes referred to as the "Planck's Torch," to accelerate them to the near light speeds required for hyperspace transition. The photon drive uses a high energy beam to provide "massless" thrust. It offers the most efficient conversion possible of matter into ship's kinetic energy. Yet despite the photon drive's advantages the inefficiency of their power reactors severly limited the payloads of early Planck's torch ships.

Payloads increased dramatically following the development of the comprehensive unified field theory, a joint product of the Ardotian and Human civilizations which led to the conversion of matter into antimatter, in 2284. This provided the ultimate power source, since matter and antimatter, upon mixing, release 100 percent of the energy predicted by Einstein's well known formula, $E = mc^2$. This breakthrough greatly enhanced Humankind's ability to colonize distant worlds. For the first time, large amounts of people and cargo could be transported across distances of more than 25 light years. The colonization of Genesis and the mining of Mammon became practical.

Spaceship engines must annihilate an enormous amount of mass to reach near light velocities. Accelerating a payload to 98 percent light velocity and bringing it to rest again requires the destruction of about 100 times the mass of the payload! Payload-to-fuel ratios have changed little from the days of chemical rockets, only the speeds have increased.

Since the advent of nuclear-powered space travel, hydrogen has served as the basic fuel. Because interstellar starships need enormous amounts of hydrogen, they obtain it from large, uninhabited planets, such as our Solar system's giant Jupiter. Jupiter's atmosphere is composed mainly of hydrogen. Prior to taking on passengers, the starships fuel from a low orbit around the planet. Because hydrogen is of very low density, even in liquid form, it is converted to a dense plasma for storage by stripping the protons of their electrons. The plasma, about 1,000 times as dense as liquid hydrogen and nine times as dense as lead, is stored within a magnetogravetric field at more than 30,000° Kelvin. In this form it resembles the substance that makes up the white dwarf stars. The plasma conversion is not at all wasteful; only matter in the plasma state can be converted to antimatter.

Jupiter may seem like a long way to go for fuel, but the only major source of hydrogen on Earth is its oceans. The people of Earth have deemed the permanent loss of large quantities of water and the consequent ecological upset unacceptable. Besides, compared with interstellar distances, a trip to Jupiter is a short hop.

Figure 4.8 shows a typical photon drive engine. As the dimensions on the drawing indicate, it is a very large piece of equipment and a very expensive one too. Anywhere from two to six engines may be used to propel a starship, and the double-ended design allows the drive to provide reverse thrust without turning the ship around. The engines are mounted well aft of the inhabited sections of the ship and may be jettisoned individually if the reaction gets out of control (highly unlikely in view of the built-in safety devices).

Life Aboard Ship

Since interstellar travellers must spend from nine to eleven weeks aboard their vessel, designers have tried to make the ship as comfortable as possible within the severe constraints imposed by interstellar travel. The ship provides all forms of life support for its crew and its thousands of passengers, including breathable air, water, heat, light, and food. Wastes must be recycled and air must be scrubbed of carbon dioxide and other impurities. To the greatest extent possible, the ship is designed to be a closed ecological system. Hydroponic gardens using artifical light grow the bulk of the ship's food. The plants also serve to convert carbon dioxide into breathing oxygen. To

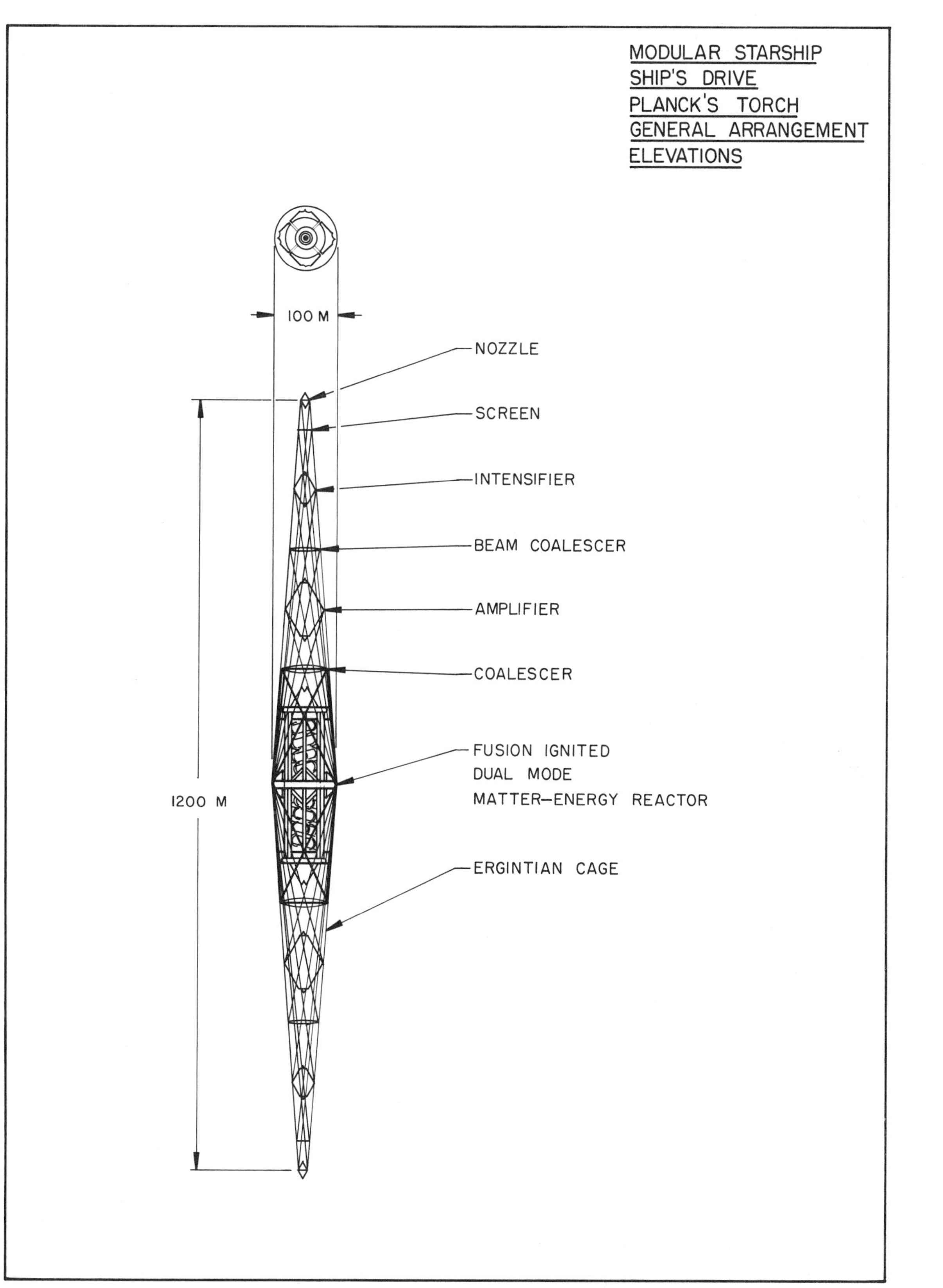
MODULAR STARSHIP
SHIP'S DRIVE
PLANCK'S TORCH
GENERAL ARRANGEMENT
ELEVATIONS
100 M
NOZZLE
SCREEN
INTENSIFIER
BEAM COALESCER
AMPLIFIER
COALESCER
FUSION IGNITED
DUAL MODE
MATTER–ENERGY REACTOR
1200 M
ERGINTIAN CAGE

FIG. 4.8

minimize the weight of water carried, all waste water is recycled and purified. Some waste waters provide a nutrient base for the gardens. Ship's water storage also serves as a compact supply of oxygen which can be electrolyzed to breathable form in emergencies.

The ship travels through an interstellar void whose temperature measures very near absolute zero. Though only a small fraction of the power of the photon drives need be tapped to provide heating for the passengers, a complex system of ducts and air movers is required to circulate air and heat throughout the ship. Walls are carefully insulated to prevent excessive thermal gradients and designers have minimized the machinery noise. In a closed environment, not only temperature, but humidity and odors also must be controlled to insure passenger health and comfort.

Human beings require more than the bare necessities of life when traveling in space. Privacy, companionship, recreation, and exercise all are needed for normal living, and it isn't easy to fulfill such needs within the confines of a spaceship. The earliest pioneers avoided such problems by making the interstellar journey in a state of "suspended animation." Most of the passengers and crew remained unconscious, at greatly reduced metabolic rates that actually slowed the aging process, but this technique posed many problems.

First, the systems required to maintain suspended animation were even more massive than conventional life support equipment. At lower metabolic rates, environmental temperature had to be controlled very precisely. Hot and cold spots on the body's surface also had to be prevented by a well-circulated cooling medium or tissue decay would set in. More than 1,200 liters of coolant per person had to be carried.

The complexity of life support for hibernating Humans made it both expensive and unreliable. Virtually all body functions had to be continuously monitored for each person, and automatic equipment was required to bring the hibernating traveller to consciousness if his individual life support failed. Such failures occasionally claimed lives, but more often the victim would be awakened to spend years of monotonous travel fully conscious.

Induced hibernation took time to initiate and terminate. Ship's crews spent months preparing pioneers on large colonial vessels for the voyage and still more months awakening them on arrival. Lastly, despite the advantages suspended animation offered, some of the crew had to remain awake throughout the entire trip to monitor the ship's functions and the health of the unconscious passengers. The monotony experienced by these crew members on year-long shifts was difficult for most to tolerate.

The greatly reduced trip times resulting from the development of space warp obsoleted suspended animation techniques. Conscious passengers can use the months of travel time to prepare for their life on a new world. Each colony planet has unique life forms, geography, and Human culture, and every new arrival should have at least a rudimentary knowledge of them. During the trip, all pioneers attend daily orientation classes covering almost every aspect of life, including flora and fauna, geography, economic, and political structure, types of housing, local cooking, and foods and colonial history (if any).

Pioneers also use the time aboard ship to plan the first months and years of their lives on their new world. The career they will pursue, the location where they will settle, and the type of home they will ultimately build all require careful consideration. During the course of the trip, the crew gradually adjusts the ship's gravity and atmospheric composition from Earth normal to that of the destination planet. The change, though barely perceptible to the passsegers, gently acclimates them to the new conditions.

Life support spires (figure 4.5) contain most of the pioneers' accommodations on board the starship. On missions to the embryonic (post-Brobdingnag) colonies, the spires are detached from the pods and lowered to the surface to serve as temporary housing for the new immigrants. The developed colonies do not need such supplemental housing, nor do they require imported equipment. Ships bound for these planets also carry passengers in the pod structures as shown in figure 4.9. Room size and design in both the pods and spires are similar.

GAILE apportions living space on starships to family units. Single people get single rooms with private baths, and double-room suites accommodate married couples without children. Larger families with children live in multi-level apartments. Figure 4.10 shows a typical deck plan in a residential spire. An elevator and utility shaft form the central core of the spire while individual living units and recreation rooms line the outside wall. The area in between, known as the "common," provides needed open space, a place for floor meetings, and a waiting area for the elevators.

Figures 4.11 and 4.12 illustrate a typical multi-level family living unit. Each unit is completely furnished and carpeted. It offers all the convenience and luxury appliances found in an Earth-bound apartment. An environmental control system regulates temperature, air circulation, humidity, and sonofield intensity to suit the occupants. An entertainment center plays music and shows holofilms retrievable from the library in the ship's computer. A data terminal allows access to written material in the library as well as the computer's run and programming facilities. Pioneers can access anything in the ship's library except the ship's technical manuals. Triple buffering protects the ship's navigational data and programs from computing errors by passenger users.

All multi-person living units come equipped with mini kitchen facilities consisting of a preservator, a food master, and micro-oven. Common dining facilities can also be found in each spire for those who don't wish to cook their own meals. Colonist-cooks direct meal preparation, and every pioneer who eats in the dining rooms must spend part of the trip assisting with meal preparation and serving. The ship's crew is too small to provide this service to the large number of passengers that each vessel carries, although the crew provides technical assistance in servicing the food machinery.

Passengers need more than just living quarters aboard a starship. They also need instruction, exercise, and a little fun during their long voyage, so each ship carries one or more pods devoted to these purposes as shown in figure 4.13. The activity pods serve as focal points for all passenger activities during the voyage, and as much as 30 percent of the ship's crew may serve as counselors and maintenance personnel in support of these activities.

All pioneers are urged to exercise daily because the confined and sedentary life aboard ship does not keep them in fit condition. Colonial living, in contrast with life on Earth, requires more physical strength and stamina, especially during the first months after arrival. Exercise facilities in the Activities Pods include a swimming pool, several automated exercise gyms, a dance floor, and courts for the most popular games on the destination colony.

Diversionary activities aboard ship emphasize things that pioneers can do for themselves. These include theaters for plays and concerts, recreational gardens, passenger-operated cafes and a holovision studio for producing shows and information broadcasts. No one has an excuse to be bored!

Upon arriving at the colony planet, the crew detaches the spires for landing on the planet. Figure 4.14 illustrates the landing sequence. A space tug, transported with the first pioneering vessel, meets the ship and lowers the spires one at a time to the surface. The spires land at prepared sites and may be guyed to withstand surface winds. On the surface, spires may be left as one great tower, or they may be cut into smaller segments and spread around the countryside. Spires generally serve the colony planet for many years. After the original pioneers move out, the structures are converted to even more spacious apartments, offices or laboratories.

Starship Crews

Professional crews command and operate all GAILE starships. All officers graduate from the GAILE space academy and their training is augmented by years of actual experience in space. A typical starship captain logs from twenty to forty years in space prior to attaining his command. Starship crews perform all critical operation and maintenance tasks aboard ship, but because the cost of providing crews is so high, GAILE leaves non-critical chores, such as cooking, cleaning, class instruction, and direction of social activities to specially trained and compensated pioneers.

Except when making minor repairs in the spires or when unloading cargo, the crew lives and works entirely in the pods and spokes of the mainframe (refer to figure 4.1). The pods contain not only crews quarters, but all of the critical equipment that must remain with the ship after detaching the spires.

The Command and Navigation pod shown in figure 4.15 forms the brain and nerve center of the ship. It houses the ship's bridge, main computer, general offices and the quarters of all officers and flight crew. Crew members responsible for non-flight operation, such as maintenance, scientific observation and passenger services live in other pods of the ship.

The captain's quarters and those of the other high level officers are located above the bridge. This affords them some measure of privacy, since the ship's computer controls access to all decks above the general offices. Cameras in the elevators enable the computer to recognize and admit only those crew members who live in the command pod. No crew member except the top officers may enter the bridge, engineering or computer decks except during their assigned watch hours, and only the captain or the first officer may alter watch assignments.

The ship's bridge, shown in figure 4.16, contains no windows or viewing ports. Such openings would have little purpose, since at near light speeds the unaided Human eye could not see an object in time to avoid collision. Even during orbit insertion and unloading operations, cameras provide a more complete view of the surroundings than windows could afford. Normally the flight crew controls the ship exclusively from the circular room at the center of the bridge deck, called the "inner bridge." The main screen, covering one-quarter of the inner bridge walls, employs a free-format display that allows the crew to project a single large image or a variety of smaller images. During spire undocking, for example, the captain can simultaneously display a picture of every spire in space, as well as a view of the planet below. Data as well as pictoral images can also be displayed in both digital and graphic form. A smaller "patch screen" located at the end of the command pedestal allows the captain or watch officer to display data he wishes to see, while leaving the main screen display uninterrupted.

A watch maintained on the inner bridge at all times consists of a watch officer, a navigator, a helmsman, and a communications officer. On the engineering control deck below, two flight engineers continuously monitor the ship's power and life support functions. The proximity of the bridge to the senior officers' quarters makes them available around the clock to cope with unusual situations.

The navigator and helmsman do not really plot the course or steer the ship. Steering of the ship is performed automatically by a section of the main computer, reserved exclusively for this purpose. Human reactions cannot cope with the high speed of interstellar travel or the complexities of relativistic inertial guidance. Yet the watch officers do perform important functions. During the early part of the trip, the navigator and helmsman work continuously to prepare for the hyperspace maneuvers, using the computers to calculate the length of the jump and cross-checking against the ship's actual velocity and position. After the hyperspace jump, the watch must update and correct optical sitings in order to place the ship on the exact course to its destination. The watch also must be present to cope with emergencies, alarms, and malfunctions. Often they must dispatch maintenance personnel to calibrate or repair sensors, or to reprogram the computer.

The "outer bridge" houses still more control and monitoring panels. These are staffed during the first shift only. Crew members perform routine tests of all ship's systems during these hours and schedule maintenance as required. The main communications panel on the outer bridge also handles interplanetary data transmission as described in the next subsection.

During the transition to and from hyperspace, called "warp maneuvers," crew members occupy every station on the bridge. Both the captain and first officer are always present, and two screen operators assist the communications officer in controlling the bridge displays. The chief navigator, assisted by two computer programmers, assists at the warp control panel. This contains special displays and programming functions used only at this time. There is still a measure of uncertainty when travelling in hyperspace. Even today, starship crews cannot calculate the precise point where the ship will emerge. Though the actual jump through hyperspace seems instantaneous, warp maneuvers generally last several hours. Crews must check and recheck calculations before programming the final trajectory into the navigation system, and after emerging, it may take additional hours to determine exactly where the ship is. A small miscalculation during this critical period could place the starship light years from its planned destination.

Figure 4.17 shows the most critical equipment area on the ship—the main computer deck. It lies immediately below the engineering deck, a logical choice because of the high degree of interconnection between these two areas. The main computer controls and monitors virtually every ship function from passenger life support and space conditioning to the photon drive engines. The computer appears very large by modern standards because a high degree of redundancy is designed into the entire system. Every computer function has been installed in triplicate and all interconnections have been arranged to minimize data loss in the event of fire or other damage. Total loss of the main computer would literally incapacitate the entire ship. Small microprocessors located throughout the vessel would maintain life support and emergency communications equipment would let the ship call for help, but both the ship's helm and engines could no longer be controlled.

GAILE launches probes on rigid schedules and along predetermined paths to avoid collision with starships. Occasionally a probe crosses the path of a starship and is destroyed by the ship's w-field beam. Sometimes probes veer off course and head directly toward an inhabited planet. To guard against this, each probe contains a self-destruct mechanism which senses the approach of a massive object. Should this malfunction the self-destruct mechanism may be activated by maser from the target planet, or the planet's w-field defense may be used to destroy the errant probe.

The exchange of information between planets forms the cornerstone of all GAILE's programs. Earth acts as a relay station for information between all of the colony planets, since the cost of exchanging probes between each colony would be prohibitively expensive. Earth also provides the common link to the other GAILE species, tying Earth's remotest colony to the remotest Chlorzi and Ardotian worlds.

Though most information exchanged between the planets is scientific or technical in nature, pioneers can also use this great communications system to send personal messages. Though one can not switch on the telescreen and talk directly to one's parents on Earth, pioneers can send written letters, recorded monologues and personal pictures in the probes. The round-trip time for a single "letter" generally ranges from one to four months, though many pioneers send several messages without waiting for a reply.

Though it would take a major technological breakthrough to enable people to converse with the colonies instantaneously, ship and probe builders are continuing to design faster vehicles that will bring our worlds closer together. Someday round-trip voyages between the stars may be affordable by everyone.

The ship's flight crew lives in the middle decks of the command pod and the rest of the crew has similar quarters in other pods of the mainframe. The crew on immigrant vessels typically numbers three to five percent of the passengers, and includes not only the flight crew, but maintenance personnel, medical staff, counselors and science staff as well. Figure 4.18 shows a typical crew's residence deck. GAILE provides crew members with very spacious and comfortable quarters, for they spend far longer amounts of time in space than the pioneers.

Figure 4.19 shows a typical apartment for a single crew member. It contains all the comforts and conveniences of a modern home on Earth and affords the crew member a private place to relax and read or watch holovision as well as a place to sleep. Though most crew members take their meals in the dining halls, a small kitchenette allows them to fix snacks or to cook meals if they enjoy doing it. This typical quarter is approximately the size allotted to married pioneers.

Ships bound for remote planets such as Romulus or Athena will carry a scientific study pod. Since the ship is approaching the limits of explored space, it provides a unique platform for astrophysical observations. Figure 4.20 illustrates a typical sciences pod. In addition to a variety of scientific and medical laboratories, it includes extra residential quarters for the scientists GAILE wishes to send on round trips to one of the col-

SYMBOLS USED ON SHIP DRAWINGS

SYMBOL	INTERPRETATION
C	Corridor
CE	Captain's elevator
DF	Drinking fountain
DN	Down
E	Elevator
F	Facilities
GR	Game room
K	Kitchen
L	Living room
PI	Personnel identification station (computer linked)
S	Stores
TYP	Typical (of other symbols shown on drawing)
V	Vending machine
WC	Watch captain

Table 4.1

onies. To be invited on such a trip is a great honor and privilege. Few Earthbound scientists who work with data from the other planets ever have the chance to visit them.

INTERPLANETARY COMMUNICATION

People have enjoyed instantaneous communication between all parts of the Earth for centuries, but unfortunately, communications between Earth and its planet colonies is not so convenient. All long-distance communication on Earth uses some form of electromagnetic radiation to carry the information. Such radiation can travel no faster than the speed of light. If conventional masers were used to transmit data from Earth to the planet colonies, it would take from 4.3 to 77 Earth years for it to get there.

By employing space warp, starships greatly reduce the apparent travel time between the planets and therefore offer the fastest medium for communication between them. Before departing from Earth for one of the colonies, every starship loads its computers with the latest news and technical information. During the weeks required to reach near light speeds, ships receive information updates from Earth by ordinary maser communication. After warp maneuvers, the ship begins to transmit this new information to the destination planet at once, and maser waves travelling at the speed of light reach the colony before the ship. When the ship returns from the colony, the process is reversed.

Since ships make voyages to each of the planets only once or twice a year, pioneers had to devise another way to communicate with Earth. Monthly and sometimes twice-monthly, Earth exchanges information probes with each of the colonies. The probes resemble tiny, pilotless spaceships. Each contains computer memory banks, guidance systems and self-focusing maser transmitters, and is propelled by a tiny neuron drive engine. Ultraminiaturization of electronics and engine parts has reduced the mass of each probe to less than 100 kilograms, but each must carry more than 20 times its mass in fuel. Though the probe's neuron engines are less efficient than mixed matter drives, they need not carry enough mass to bring them to a stop. Instead the probes are aimed on a trajectory that will miss the destination planet. After they emerge from hyperspace, they transmit their load of information via maser beam, then continue past the planet and into its sun.

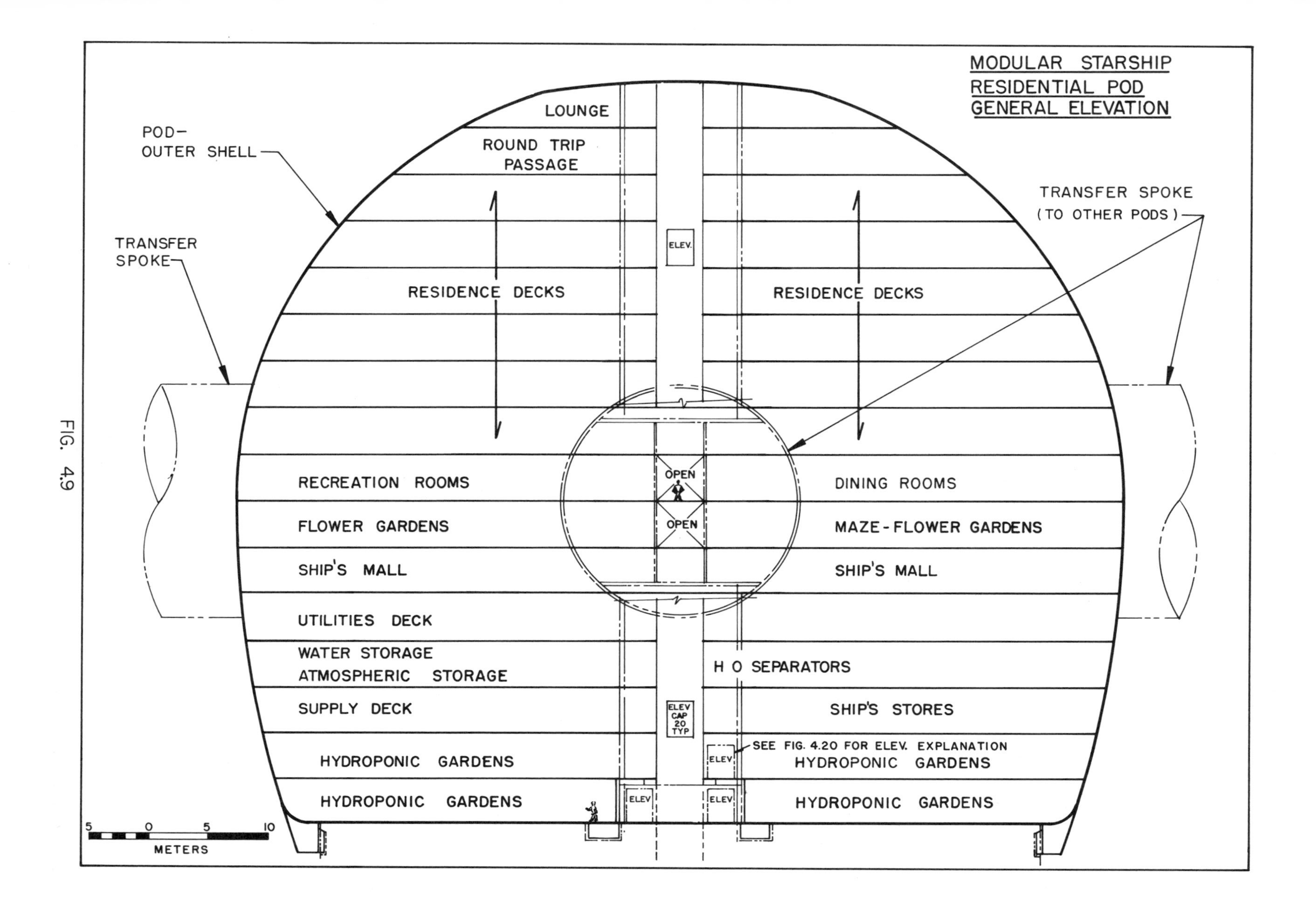
MODULAR STARSHIP
RESIDENTIAL POD
GENERAL ELEVATION
POD-
OUTER SHELL
TRANSFER
SPOKE
TRANSFER SPOKE
(TO OTHER PODS)
LOUNGE
ROUND TRIP
PASSAGE
ELEV.
RESIDENCE DECKS
RESIDENCE DECKS
OPEN
OPEN
RECREATION ROOMS
DINING ROOMS
FLOWER GARDENS
MAZE-FLOWER GARDENS
SHIP'S MALL
SHIP'S MALL
UTILITIES DECK
WATER STORAGE
ATMOSPHERIC STORAGE
H O SEPARATORS
SUPPLY DECK
ELEV
CAP
20
TYP
SHIP'S STORES
SEE FIG. 4.20 FOR ELEV. EXPLANATION
ELEV
HYDROPONIC GARDENS
HYDROPONIC GARDENS
ELEV
ELEV
HYDROPONIC GARDENS
HYDROPONIC GARDENS
5
0
5
10
METERS

FIG. 4.9

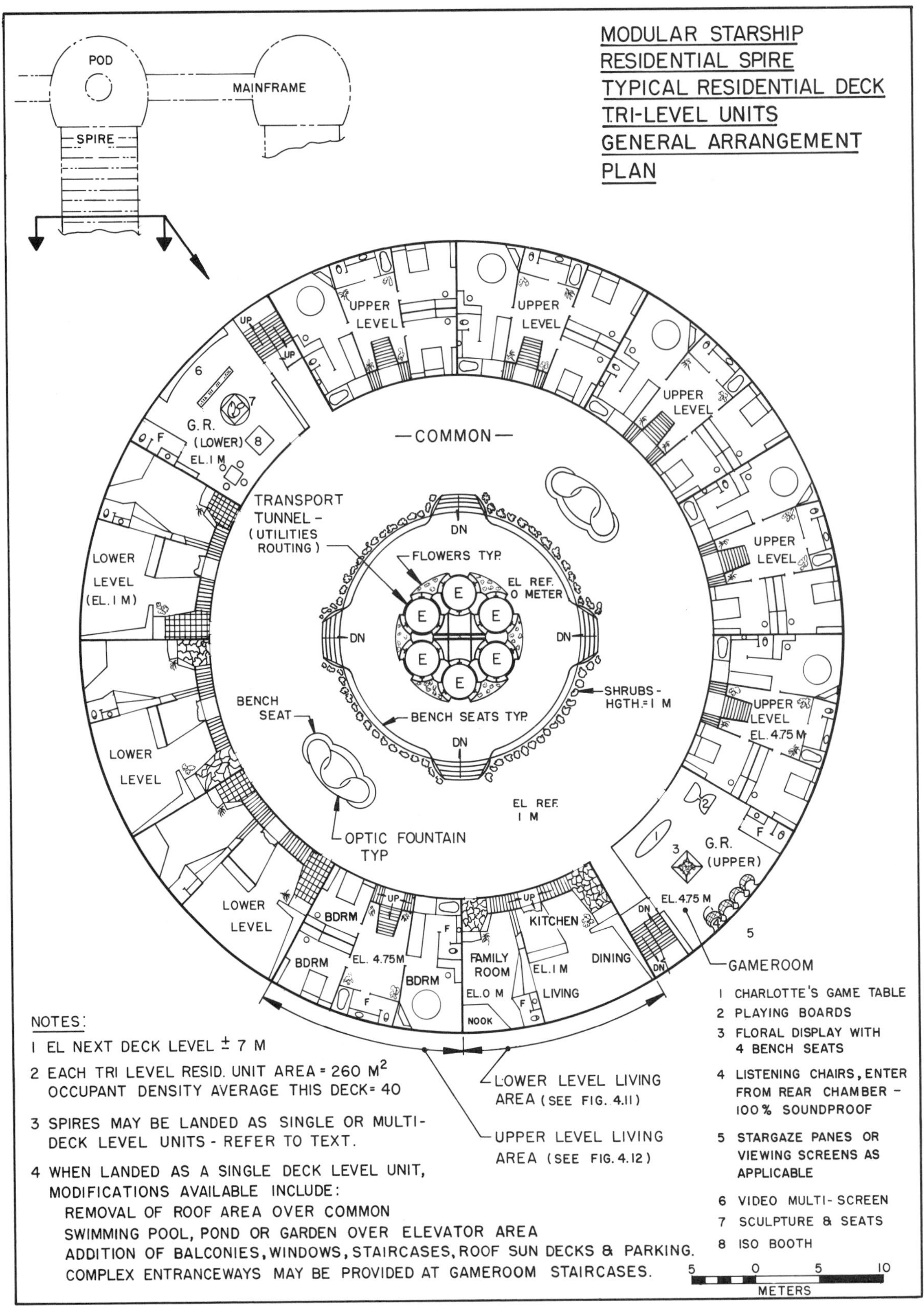
MODULAR STARSHIP
RESIDENTIAL SPIRE
TYPICAL RESIDENTIAL DECK
TRI-LEVEL UNITS
GENERAL ARRANGEMENT
PLAN
POD
MAINFRAME
SPIRE
UPPER LEVEL
UPPER LEVEL
UPPER LEVEL
UPPER LEVEL
UPPER LEVEL
EL. 4.75 M
UP
UP
G.R. (LOWER)
EL. 1 M
F
6
7
8
—COMMON—
TRANSPORT TUNNEL - (UTILITIES ROUTING)
DN
FLOWERS TYP.
EL REF. 0 METER
E
DN
DN
DN
SHRUBS - HGTH. = 1 M
BENCH SEAT
BENCH SEATS TYP.
LOWER LEVEL (EL. 1 M)
LOWER LEVEL
LOWER LEVEL
OPTIC FOUNTAIN TYP
EL REF. 1 M
G.R. (UPPER)
EL. 4.75 M
1
2
3
5
GAMEROOM
BDRM
BDRM
BDRM
EL. 4.75 M
FAMILY ROOM
EL. 0 M
NOOK
KITCHEN
EL. 1 M
LIVING
DINING
LOWER LEVEL LIVING AREA (SEE FIG. 4.11)
UPPER LEVEL LIVING AREA (SEE FIG. 4.12)
1 CHARLOTTE'S GAME TABLE
2 PLAYING BOARDS
3 FLORAL DISPLAY WITH 4 BENCH SEATS
4 LISTENING CHAIRS, ENTER FROM REAR CHAMBER - 100% SOUNDPROOF
5 STARGAZE PANES OR VIEWING SCREENS AS APPLICABLE
6 VIDEO MULTI-SCREEN
7 SCULPTURE & SEATS
8 ISO BOOTH
NOTES:
1 EL NEXT DECK LEVEL ± 7 M
2 EACH TRI LEVEL RESID. UNIT AREA = 260 M²
OCCUPANT DENSITY AVERAGE THIS DECK = 40
3 SPIRES MAY BE LANDED AS SINGLE OR MULTI-DECK LEVEL UNITS - REFER TO TEXT.
4 WHEN LANDED AS A SINGLE DECK LEVEL UNIT, MODIFICATIONS AVAILABLE INCLUDE:
REMOVAL OF ROOF AREA OVER COMMON
SWIMMING POOL, POND OR GARDEN OVER ELEVATOR AREA
ADDITION OF BALCONIES, WINDOWS, STAIRCASES, ROOF SUN DECKS & PARKING.
COMPLEX ENTRANCEWAYS MAY BE PROVIDED AT GAMEROOM STAIRCASES.
5 0 5 10
METERS

FIG. 4.10

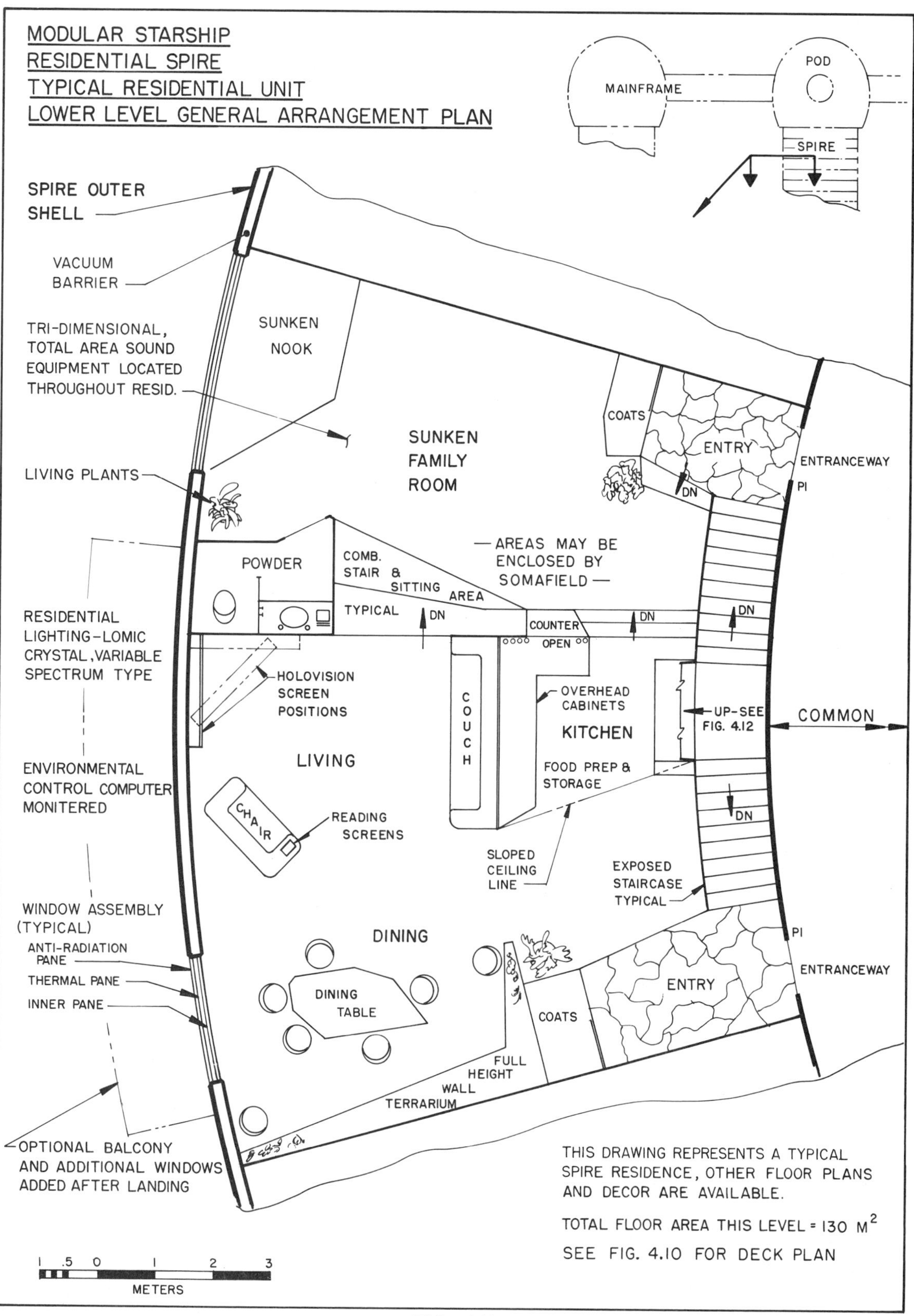
MODULAR STARSHIP
RESIDENTIAL SPIRE
TYPICAL RESIDENTIAL UNIT
LOWER LEVEL GENERAL ARRANGEMENT PLAN
MAINFRAME
POD
SPIRE
SPIRE OUTER SHELL
VACUUM BARRIER
TRI-DIMENSIONAL, TOTAL AREA SOUND EQUIPMENT LOCATED THROUGHOUT RESID.
LIVING PLANTS
RESIDENTIAL LIGHTING-LOMIC CRYSTAL, VARIABLE SPECTRUM TYPE
ENVIRONMENTAL CONTROL COMPUTER MONITERED
WINDOW ASSEMBLY (TYPICAL)
ANTI-RADIATION PANE
THERMAL PANE
INNER PANE
OPTIONAL BALCONY AND ADDITIONAL WINDOWS ADDED AFTER LANDING
SUNKEN NOOK
SUNKEN FAMILY ROOM
COATS
ENTRY
DN
ENTRANCEWAY
PI
POWDER
COMB. STAIR & SITTING AREA
TYPICAL
DN
— AREAS MAY BE ENCLOSED BY SOMAFIELD —
COUNTER
OPEN
DN
DN
HOLOVISION SCREEN POSITIONS
COUCH
OVERHEAD CABINETS
KITCHEN
UP-SEE FIG. 4.12
COMMON
LIVING
FOOD PREP & STORAGE
CHAIR
READING SCREENS
DN
SLOPED CEILING LINE
EXPOSED STAIRCASE TYPICAL
DINING
PI
ENTRANCEWAY
DINING TABLE
COATS
ENTRY
FULL HEIGHT WALL TERRARIUM
THIS DRAWING REPRESENTS A TYPICAL SPIRE RESIDENCE, OTHER FLOOR PLANS AND DECOR ARE AVAILABLE.
TOTAL FLOOR AREA THIS LEVEL = 130 M²
SEE FIG. 4.10 FOR DECK PLAN
1 .5 0 1 2 3
METERS

FIG. 4.11

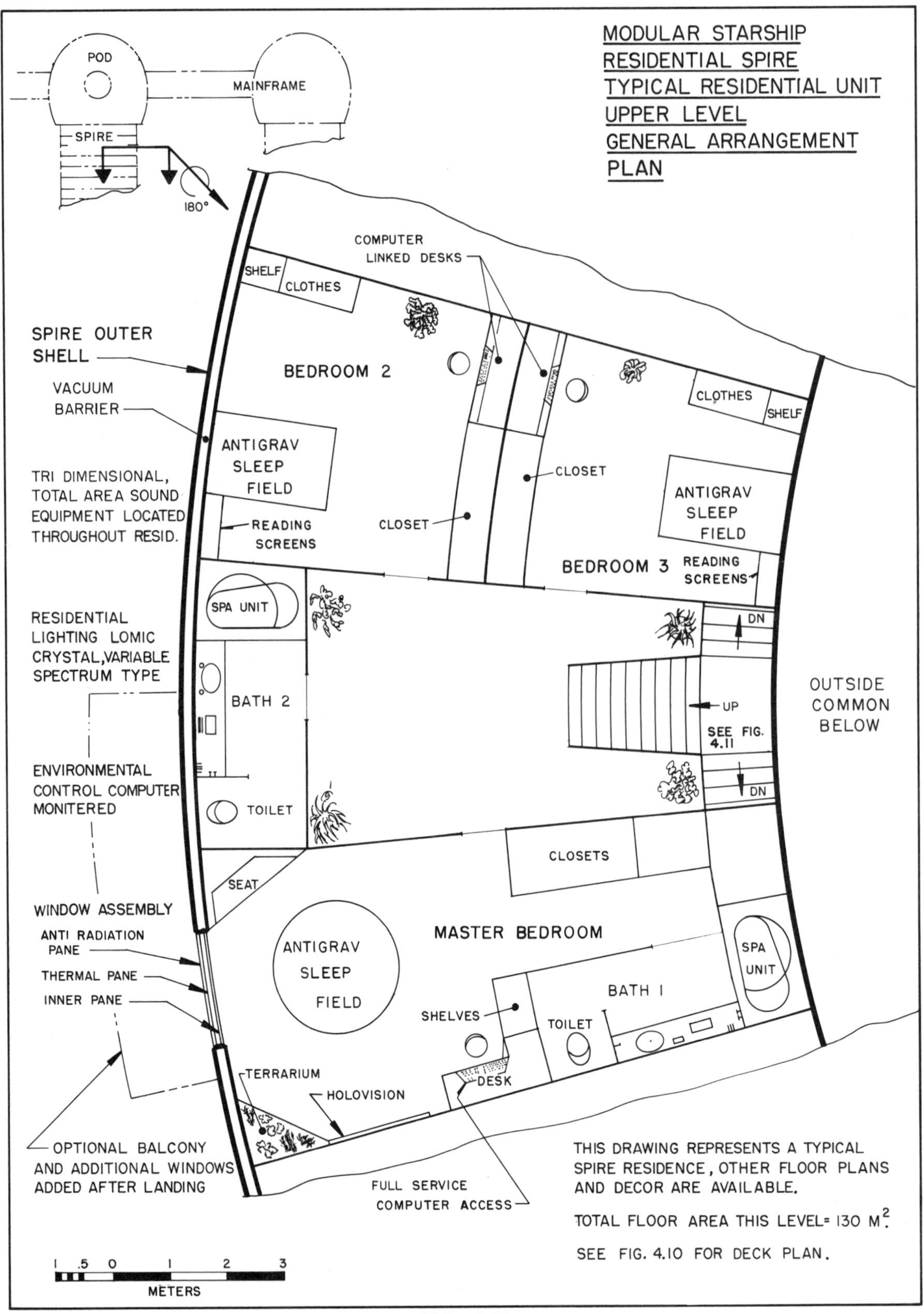
MODULAR STARSHIP
RESIDENTIAL SPIRE
TYPICAL RESIDENTIAL UNIT
UPPER LEVEL
GENERAL ARRANGEMENT
PLAN
POD
MAINFRAME
SPIRE
180°
COMPUTER
LINKED DESKS
SHELF
CLOTHES
SPIRE OUTER
SHELL
BEDROOM 2
VACUUM
BARRIER
CLOTHES
SHELF
ANTIGRAV
SLEEP
FIELD
CLOSET
TRI DIMENSIONAL,
TOTAL AREA SOUND
EQUIPMENT LOCATED
THROUGHOUT RESID.
ANTIGRAV
SLEEP
FIELD
READING
SCREENS
CLOSET
BEDROOM 3
READING
SCREENS
SPA UNIT
DN
RESIDENTIAL
LIGHTING LOMIC
CRYSTAL,VARIABLE
SPECTRUM TYPE
OUTSIDE
COMMON
BELOW
BATH 2
UP
SEE FIG.
4.11
ENVIRONMENTAL
CONTROL COMPUTER
MONITERED
TOILET
DN
CLOSETS
SEAT
WINDOW ASSEMBLY
MASTER BEDROOM
ANTI RADIATION
PANE
ANTIGRAV
SLEEP
FIELD
SPA
UNIT
THERMAL PANE
BATH 1
INNER PANE
SHELVES
TOILET
DESK
TERRARIUM
HOLOVISION
OPTIONAL BALCONY
AND ADDITIONAL WINDOWS
ADDED AFTER LANDING
FULL SERVICE
COMPUTER ACCESS
THIS DRAWING REPRESENTS A TYPICAL
SPIRE RESIDENCE, OTHER FLOOR PLANS
AND DECOR ARE AVAILABLE.
TOTAL FLOOR AREA THIS LEVEL= 130 M².
SEE FIG. 4.10 FOR DECK PLAN.
1 .5 0 1 2 3
METERS

FIG. 4.12

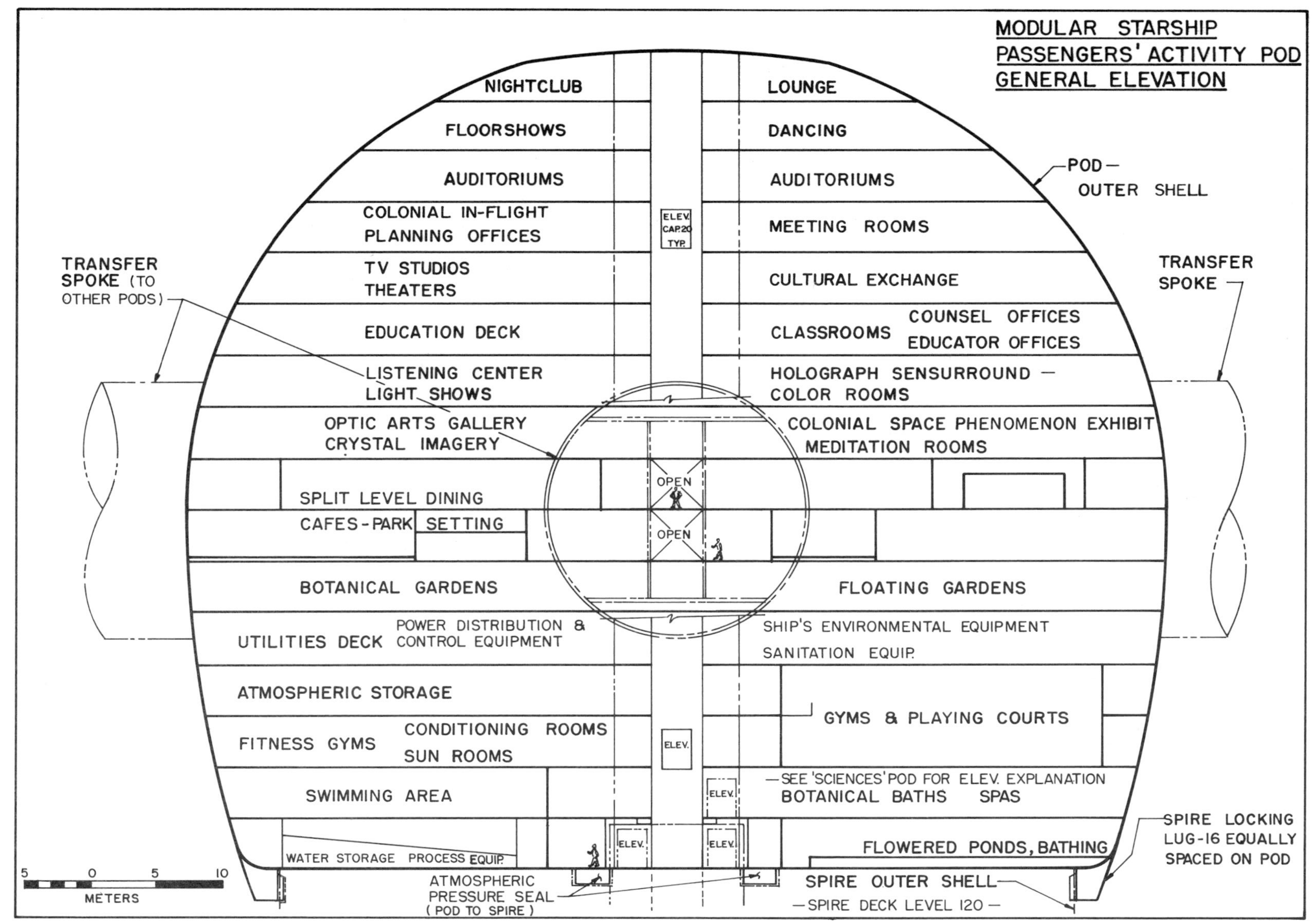
MODULAR STARSHIP
PASSENGERS' ACTIVITY POD
GENERAL ELEVATION
NIGHTCLUB
LOUNGE
FLOORSHOWS
DANCING
AUDITORIUMS
AUDITORIUMS
POD —
OUTER SHELL
COLONIAL IN-FLIGHT
PLANNING OFFICES
ELEV. CAP.20 TYP.
MEETING ROOMS
TRANSFER
SPOKE (TO
OTHER PODS)
TV STUDIOS
THEATERS
CULTURAL EXCHANGE
TRANSFER
SPOKE
EDUCATION DECK
CLASSROOMS
COUNSEL OFFICES
EDUCATOR OFFICES
LISTENING CENTER
LIGHT SHOWS
HOLOGRAPH SENSURROUND —
COLOR ROOMS
OPTIC ARTS GALLERY
CRYSTAL IMAGERY
COLONIAL SPACE PHENOMENON EXHIBIT
MEDITATION ROOMS
OPEN
SPLIT LEVEL DINING
CAFES-PARK SETTING
OPEN
BOTANICAL GARDENS
FLOATING GARDENS
UTILITIES DECK
POWER DISTRIBUTION &
CONTROL EQUIPMENT
SHIP'S ENVIRONMENTAL EQUIPMENT
SANITATION EQUIP.
ATMOSPHERIC STORAGE
GYMS & PLAYING COURTS
FITNESS GYMS
CONDITIONING ROOMS
SUN ROOMS
ELEV.
— SEE 'SCIENCES' POD FOR ELEV. EXPLANATION
SWIMMING AREA
ELEV.
BOTANICAL BATHS SPAS
SPIRE LOCKING
LUG-16 EQUALLY
SPACED ON POD
ELEV.
ELEV.
FLOWERED PONDS, BATHING
WATER STORAGE PROCESS EQUIP.
5 0 5 10
METERS
ATMOSPHERIC
PRESSURE SEAL
(POD TO SPIRE)
SPIRE OUTER SHELL
— SPIRE DECK LEVEL 120 —

FIG. 4.13

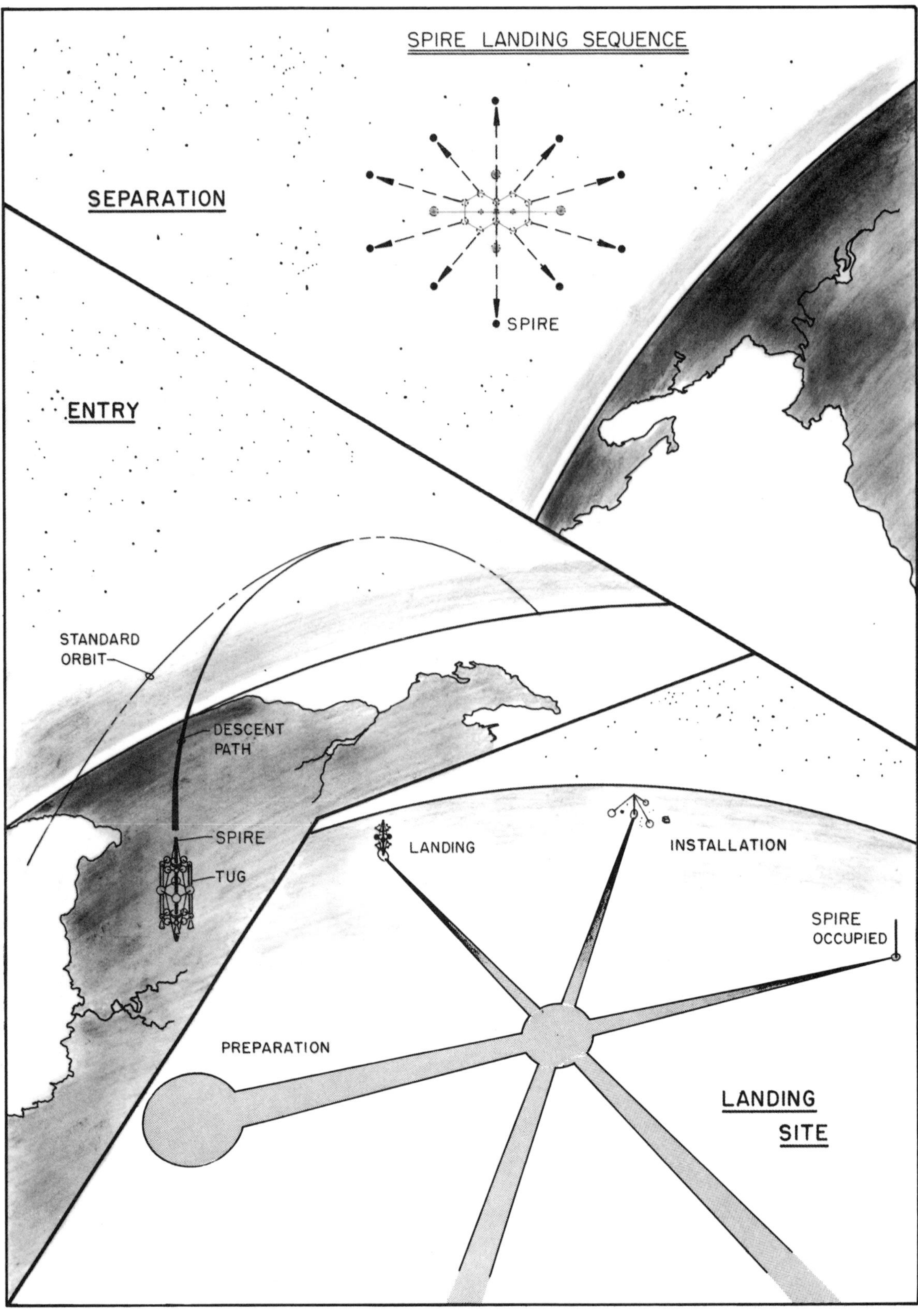
SPIRE LANDING SEQUENCE
SEPARATION
SPIRE
ENTRY
STANDARD ORBIT
DESCENT PATH
SPIRE
TUG
LANDING
INSTALLATION
SPIRE OCCUPIED
PREPARATION
LANDING SITE

FIG. 4.14

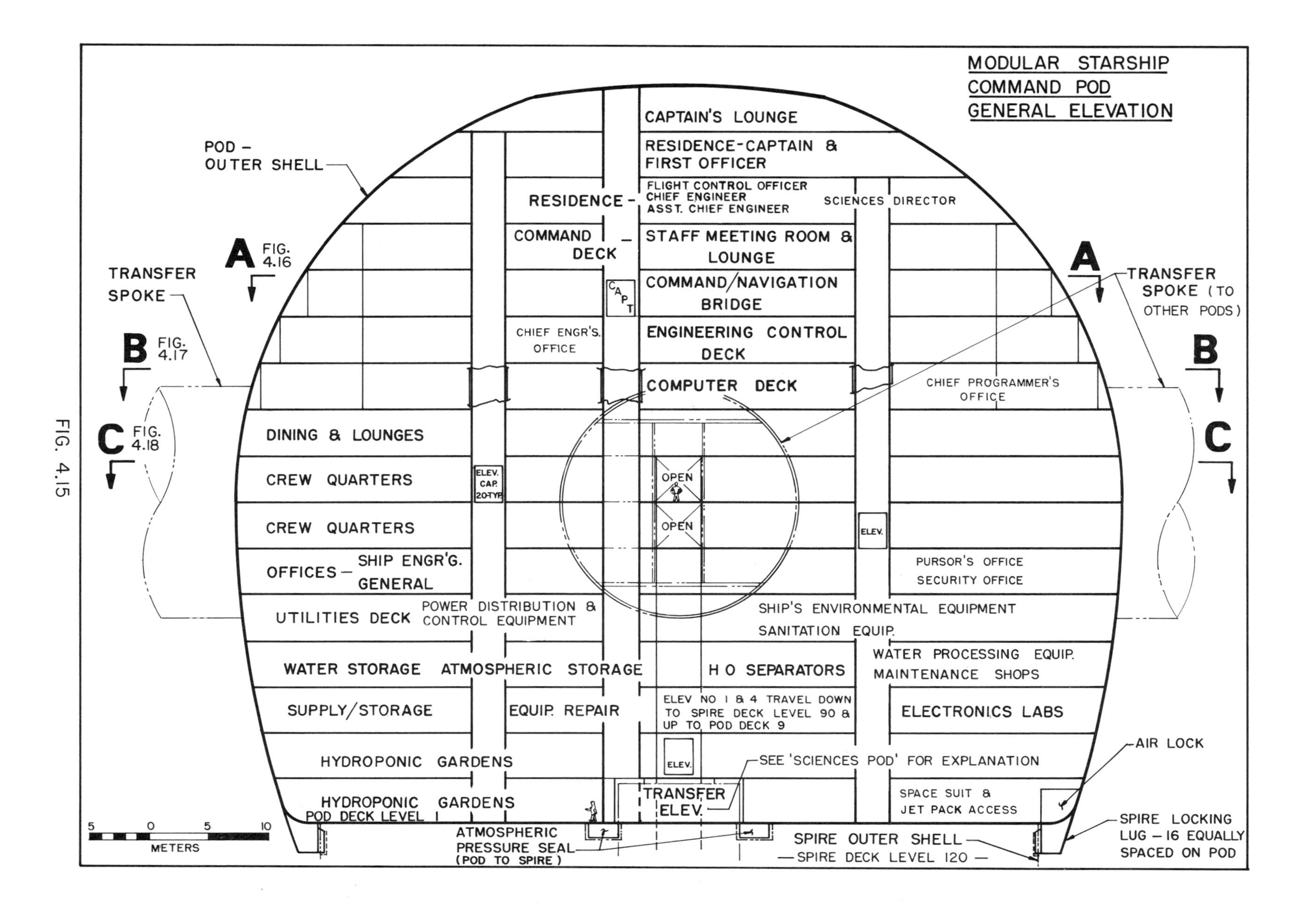
MODULAR STARSHIP
COMMAND POD
GENERAL ELEVATION
POD – OUTER SHELL
TRANSFER SPOKE
A FIG. 4.16
B FIG. 4.17
C FIG. 4.18
TRANSFER SPOKE (TO OTHER PODS)
A
B
C
CAPTAIN'S LOUNGE
RESIDENCE-CAPTAIN & FIRST OFFICER
RESIDENCE -
FLIGHT CONTROL OFFICER
CHIEF ENGINEER
ASST. CHIEF ENGINEER
SCIENCES DIRECTOR
COMMAND DECK -
STAFF MEETING ROOM & LOUNGE
CAPT
COMMAND/NAVIGATION BRIDGE
CHIEF ENGR'S. OFFICE
ENGINEERING CONTROL DECK
COMPUTER DECK
CHIEF PROGRAMMER'S OFFICE
DINING & LOUNGES
CREW QUARTERS
ELEV. CAP. 20-TYP.
OPEN
OPEN
CREW QUARTERS
ELEV.
OFFICES – SHIP ENGR'G. GENERAL
PURSOR'S OFFICE
SECURITY OFFICE
UTILITIES DECK
POWER DISTRIBUTION & CONTROL EQUIPMENT
SHIP'S ENVIRONMENTAL EQUIPMENT
SANITATION EQUIP.
WATER STORAGE
ATMOSPHERIC STORAGE
H O SEPARATORS
WATER PROCESSING EQUIP.
MAINTENANCE SHOPS
SUPPLY/STORAGE
EQUIP. REPAIR
ELEV NO 1 & 4 TRAVEL DOWN TO SPIRE DECK LEVEL 90 & UP TO POD DECK 9
ELECTRONICS LABS
HYDROPONIC GARDENS
ELEV.
SEE 'SCIENCES POD' FOR EXPLANATION
AIR LOCK
HYDROPONIC GARDENS
POD DECK LEVEL 1
TRANSFER ELEV.
SPACE SUIT & JET PACK ACCESS
5 0 5 10
METERS
ATMOSPHERIC PRESSURE SEAL (POD TO SPIRE)
SPIRE OUTER SHELL
— SPIRE DECK LEVEL 120 —
SPIRE LOCKING LUG – 16 EQUALLY SPACED ON POD

FIG. 4.15

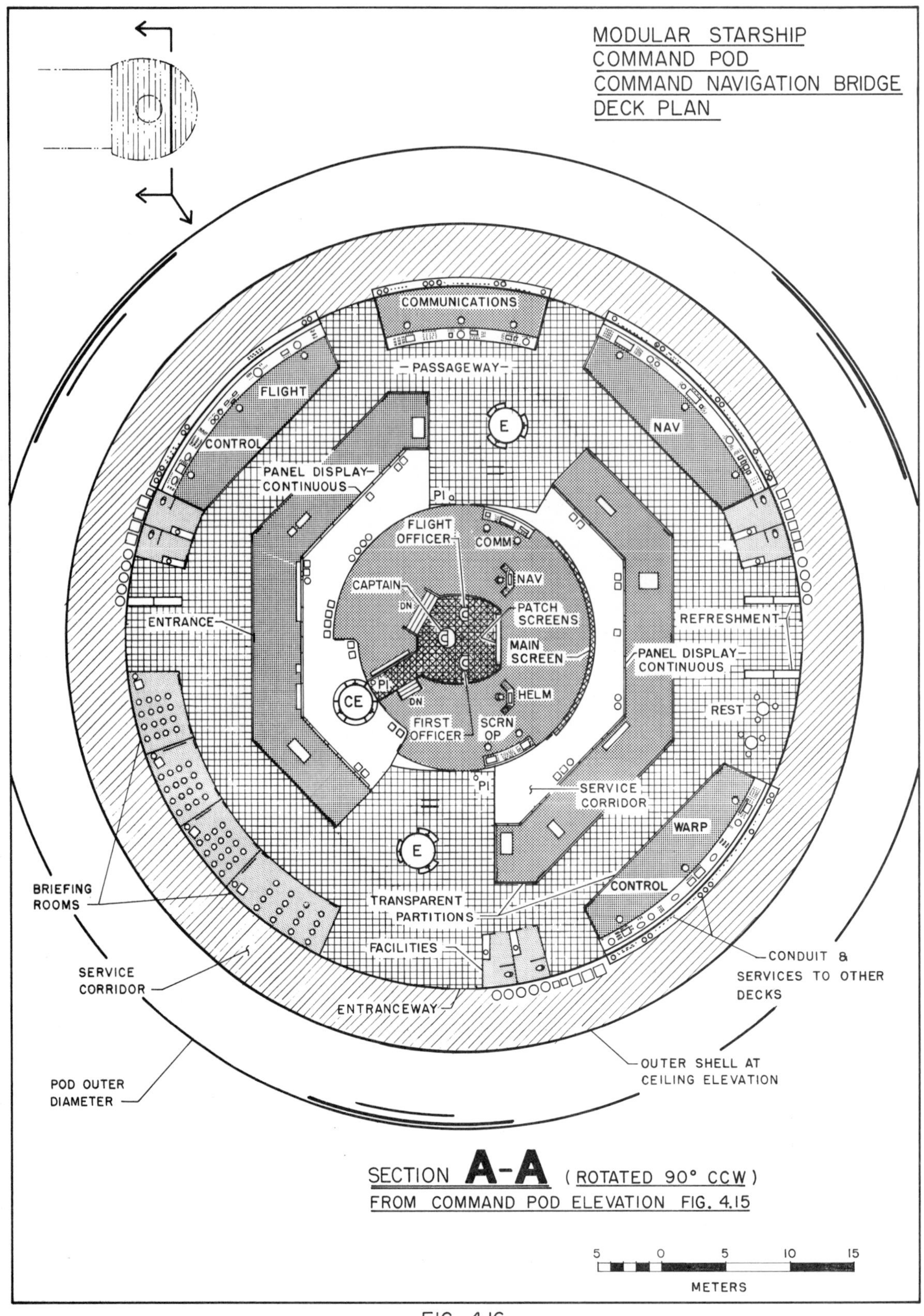
MODULAR STARSHIP
COMMAND POD
COMMAND NAVIGATION BRIDGE
DECK PLAN
COMMUNICATIONS
PASSAGEWAY
FLIGHT
CONTROL
NAV
E
PANEL DISPLAY CONTINUOUS
PI
FLIGHT OFFICER
COMM
NAV
CAPTAIN
DN
PATCH SCREENS
ENTRANCE
MAIN SCREEN
REFRESHMENT
PANEL DISPLAY CONTINUOUS
CE
HELM
REST
FIRST OFFICER
SCRN OP
SERVICE CORRIDOR
WARP
CONTROL
BRIEFING ROOMS
TRANSPARENT PARTITIONS
FACILITIES
CONDUIT & SERVICES TO OTHER DECKS
SERVICE CORRIDOR
ENTRANCEWAY
OUTER SHELL AT CEILING ELEVATION
POD OUTER DIAMETER
SECTION A-A (ROTATED 90° CCW)
FROM COMMAND POD ELEVATION FIG. 4.15
5 0 5 10 15
METERS

FIG. 4.16

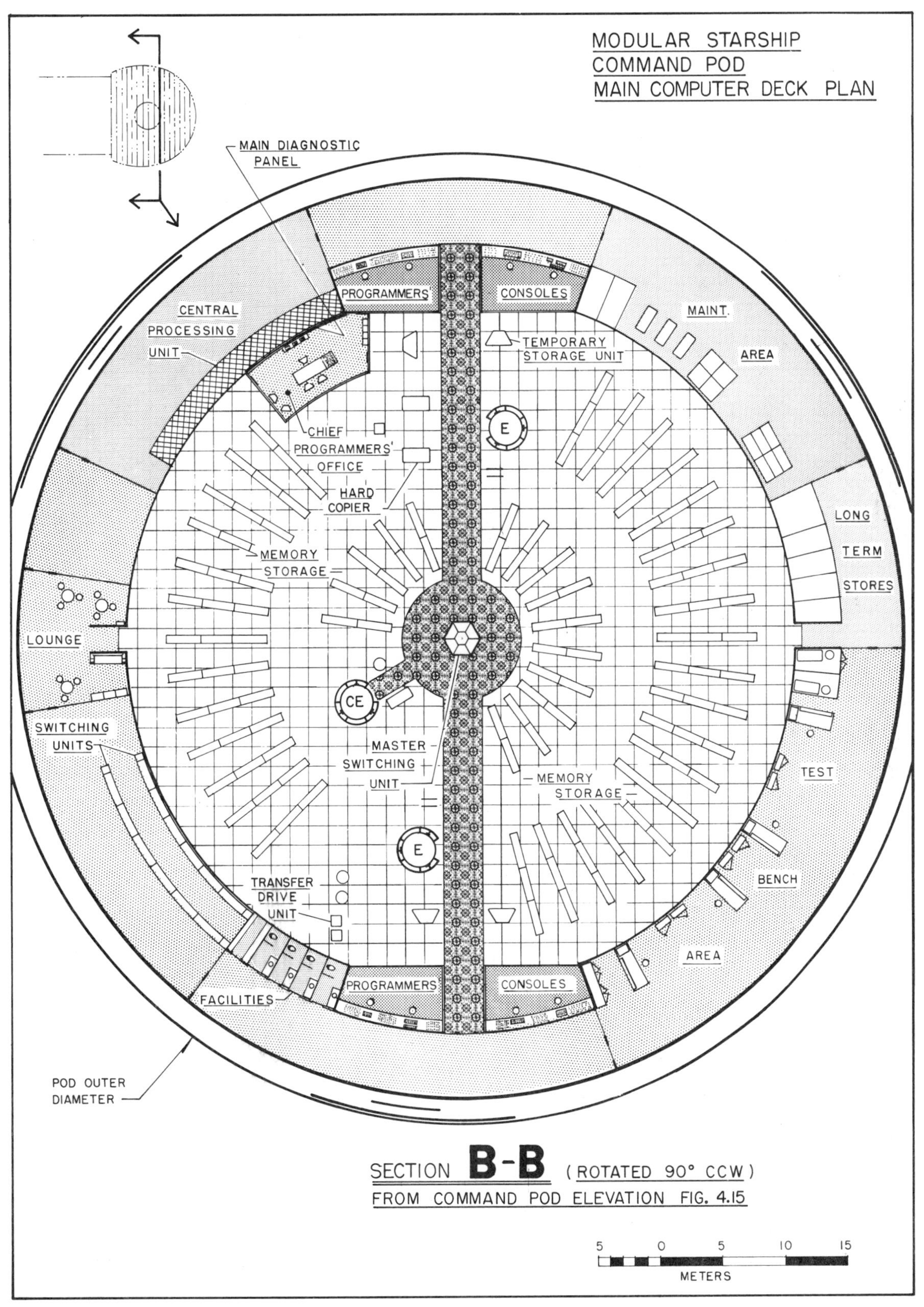
MODULAR STARSHIP
COMMAND POD
MAIN COMPUTER DECK PLAN
MAIN DIAGNOSTIC PANEL
PROGRAMMERS
CONSOLES
CENTRAL PROCESSING UNIT
MAINT. AREA
TEMPORARY STORAGE UNIT
CHIEF PROGRAMMERS' OFFICE
HARD COPIER
E
LONG TERM STORES
MEMORY STORAGE
LOUNGE
CE
SWITCHING UNITS
MASTER SWITCHING UNIT
MEMORY STORAGE
TEST
E
BENCH
TRANSFER DRIVE UNIT
AREA
PROGRAMMERS
CONSOLES
FACILITIES
POD OUTER DIAMETER
SECTION B-B (ROTATED 90° CCW)
FROM COMMAND POD ELEVATION FIG. 4.15
5 0 5 10 15
METERS

FIG. 4.17

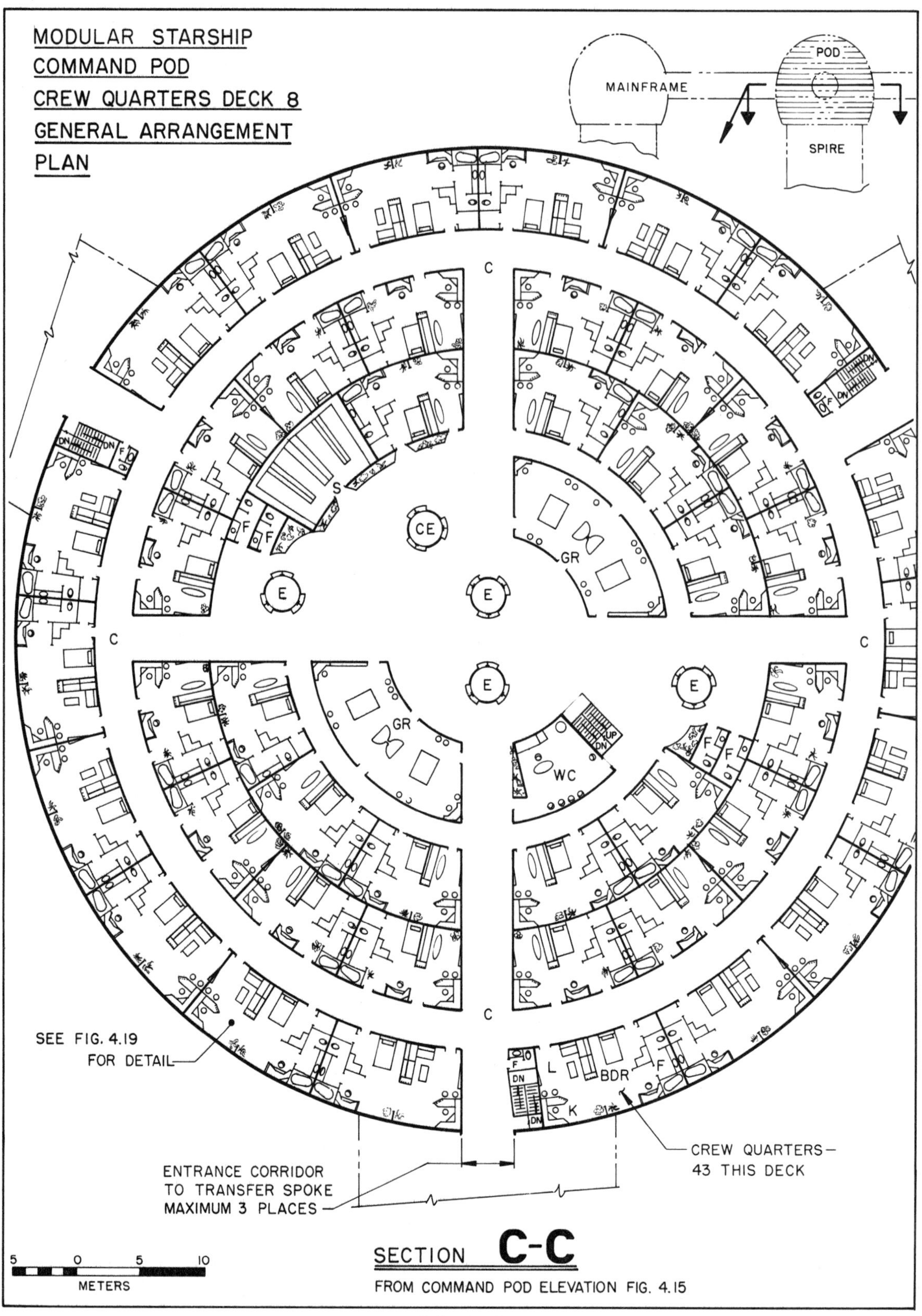
MODULAR STARSHIP
COMMAND POD
CREW QUARTERS DECK 8
GENERAL ARRANGEMENT
PLAN
MAINFRAME
POD
SPIRE
C
CE
E
GR
S
F
DN
UP
WC
L
BDR
K
SEE FIG. 4.19
FOR DETAIL
CREW QUARTERS—
43 THIS DECK
ENTRANCE CORRIDOR
TO TRANSFER SPOKE
MAXIMUM 3 PLACES
5 0 5 10
METERS
SECTION C-C
FROM COMMAND POD ELEVATION FIG. 4.15

FIG. 4.18

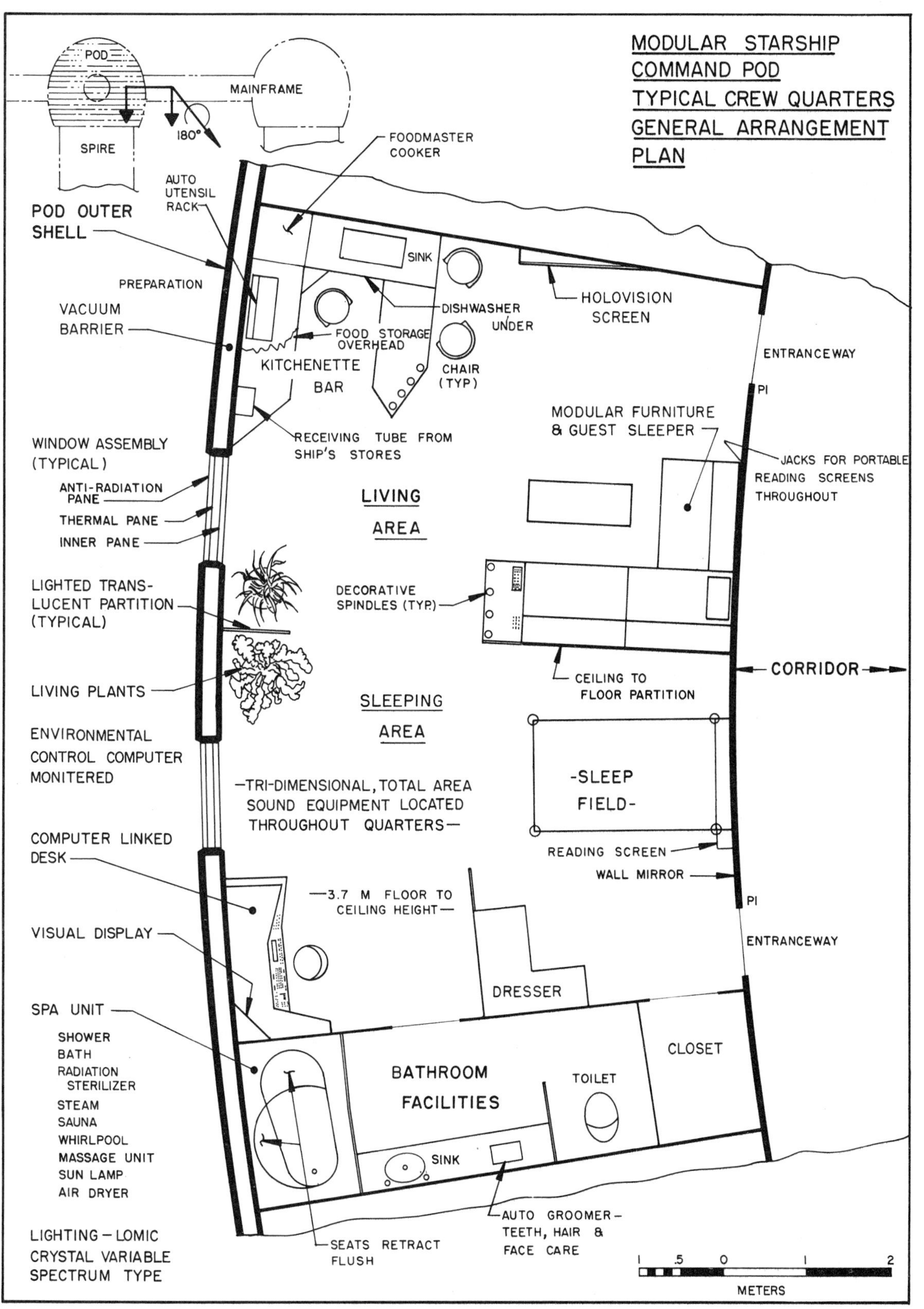
MODULAR STARSHIP
COMMAND POD
TYPICAL CREW QUARTERS
GENERAL ARRANGEMENT
PLAN
POD
MAINFRAME
SPIRE
180°
FOODMASTER
COOKER
AUTO
UTENSIL
RACK
POD OUTER
SHELL
PREPARATION
SINK
VACUUM
BARRIER
DISHWASHER
UNDER
HOLOVISION
SCREEN
FOOD STORAGE
OVERHEAD
KITCHENETTE
BAR
CHAIR
(TYP)
ENTRANCEWAY
PI
MODULAR FURNITURE
& GUEST SLEEPER
RECEIVING TUBE FROM
SHIP'S STORES
WINDOW ASSEMBLY
(TYPICAL)
JACKS FOR PORTABLE
READING SCREENS
THROUGHOUT
ANTI-RADIATION
PANE
THERMAL PANE
INNER PANE
LIVING
AREA
LIGHTED TRANS-
LUCENT PARTITION
(TYPICAL)
DECORATIVE
SPINDLES (TYP.)
CORRIDOR
CEILING TO
FLOOR PARTITION
LIVING PLANTS
SLEEPING
AREA
ENVIRONMENTAL
CONTROL COMPUTER
MONITERED
-SLEEP
FIELD-
-TRI-DIMENSIONAL, TOTAL AREA
SOUND EQUIPMENT LOCATED
THROUGHOUT QUARTERS-
COMPUTER LINKED
DESK
READING SCREEN
WALL MIRROR
PI
—3.7 M FLOOR TO
CEILING HEIGHT—
VISUAL DISPLAY
ENTRANCEWAY
DRESSER
SPA UNIT
SHOWER
BATH
RADIATION
STERILIZER
STEAM
SAUNA
WHIRLPOOL
MASSAGE UNIT
SUN LAMP
AIR DRYER
CLOSET
BATHROOM
FACILITIES
TOILET
SINK
LIGHTING – LOMIC
CRYSTAL VARIABLE
SPECTRUM TYPE
SEATS RETRACT
FLUSH
AUTO GROOMER–
TEETH, HAIR &
FACE CARE
1 .5 0 1 2
METERS

FIG. 4.19

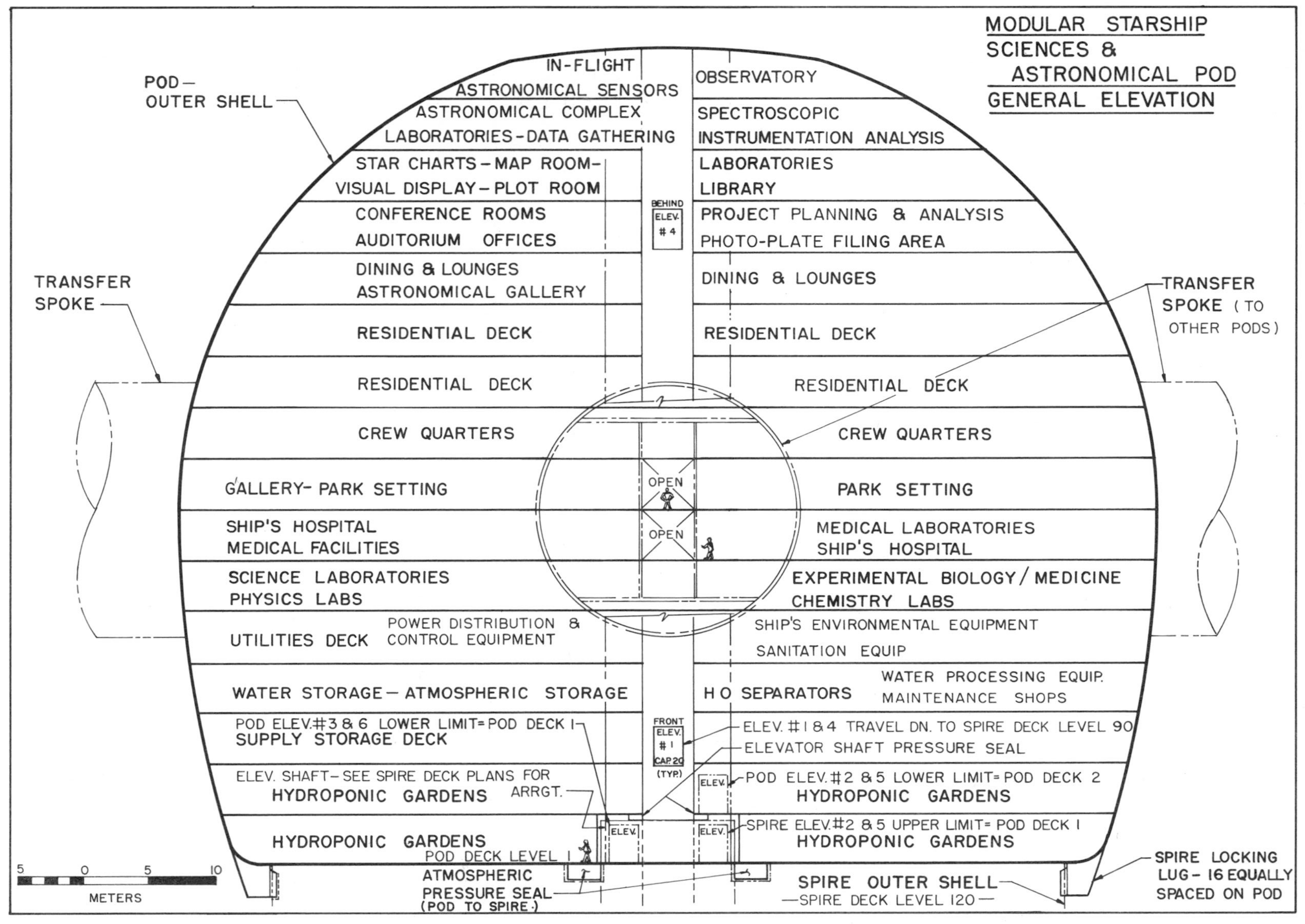
MODULAR STARSHIP
SCIENCES &
ASTRONOMICAL POD
GENERAL ELEVATION
POD—
OUTER SHELL
IN-FLIGHT
ASTRONOMICAL SENSORS
OBSERVATORY
ASTRONOMICAL COMPLEX
LABORATORIES-DATA GATHERING
SPECTROSCOPIC
INSTRUMENTATION ANALYSIS
STAR CHARTS-MAP ROOM-
VISUAL DISPLAY-PLOT ROOM
LABORATORIES
LIBRARY
BEHIND
ELEV.
4
CONFERENCE ROOMS
AUDITORIUM OFFICES
PROJECT PLANNING & ANALYSIS
PHOTO-PLATE FILING AREA
DINING & LOUNGES
ASTRONOMICAL GALLERY
DINING & LOUNGES
TRANSFER
SPOKE
TRANSFER
SPOKE (TO
OTHER PODS)
RESIDENTIAL DECK
RESIDENTIAL DECK
RESIDENTIAL DECK
RESIDENTIAL DECK
CREW QUARTERS
CREW QUARTERS
GALLERY- PARK SETTING
OPEN
PARK SETTING
SHIP'S HOSPITAL
MEDICAL FACILITIES
OPEN
MEDICAL LABORATORIES
SHIP'S HOSPITAL
SCIENCE LABORATORIES
PHYSICS LABS
EXPERIMENTAL BIOLOGY/MEDICINE
CHEMISTRY LABS
UTILITIES DECK
POWER DISTRIBUTION &
CONTROL EQUIPMENT
SHIP'S ENVIRONMENTAL EQUIPMENT
SANITATION EQUIP
WATER STORAGE — ATMOSPHERIC STORAGE
H O SEPARATORS
WATER PROCESSING EQUIP.
MAINTENANCE SHOPS
POD ELEV.#3 & 6 LOWER LIMIT=POD DECK 1
SUPPLY STORAGE DECK
FRONT
ELEV.
1
CAP. 20
(TYP.)
ELEV. #1 & 4 TRAVEL DN. TO SPIRE DECK LEVEL 90
ELEVATOR SHAFT PRESSURE SEAL
ELEV. SHAFT— SEE SPIRE DECK PLANS FOR
ARRGT.
HYDROPONIC GARDENS
ELEV.
POD ELEV. #2 & 5 LOWER LIMIT= POD DECK 2
HYDROPONIC GARDENS
ELEV.
ELEV.
SPIRE ELEV.#2 & 5 UPPER LIMIT= POD DECK 1
HYDROPONIC GARDENS
HYDROPONIC GARDENS
POD DECK LEVEL 1
5 0 5 10
METERS
ATMOSPHERIC
PRESSURE SEAL
(POD TO SPIRE.)
SPIRE OUTER SHELL
—SPIRE DECK LEVEL 120—
SPIRE LOCKING
LUG - 16 EQUALLY
SPACED ON POD

FIG. 4.20

Appendix A

Glossary of Scientific Terms

adtc—"anno domini terrestrial calendar," placed after dates expressed in the Gregorian calendar on Earth. All dates in this bulletin are expressed adtc, unless otherwise noted. Other planets use different calendars depending on their location and the length of their year.

albedo—the fraction of light reflected by a planet after striking it

angiosperm—the class of flowering plants characterized by developing seeds in an ovary

astronomical unit—a distance of astronomical measurement approximately equal to the distance between the Earth and its sun

biome—a general term referring to one of several classifications of biological regions, such as tundra, chaparral, deciduous forest, desert grassland or rain forest. Biologists define unique classifications of biomes for each planet.

Earth standard year—a measurement of time made by atomic clocks. It is equal to the sidereal year 2154 adtc on Earth

Earth standard hours—a measurement of time computed by atomic clocks and equal to exactly 1/24th of the sidereal day on 1 January 2154

eccentricity—a mathematical measurement of the "egg shape" or oblateness of an ellipse. Precisely, the eccentricity of an ellipse whose semi-major axis equals 'a' and whose focci lie a distance 'd' apart is equal to d/a. The more elongated the ellipse, the larger the value of its eccentricity.

ecliptic plane—the plane of orbit of a planet about its sun

ecosphere—the region of space surrounding each star bounded by two imaginary, concentric spheres, in which a habitable planet could theoretically exist.

equanox—the first day of spring or fall; more precisely, that moment at which the intersection of the planet's equator and its ecliptic point directly toward its sun

extra terrestrial—an adjective referring to things outside of the Earth

GAIL—the Galactic Association of Intelligent Life, an association of scientifically advanced, technologically sophisticated life forms from several planets, organized to promote the exploration and development of space and the advancement of their common knowledge.

GAILE—the Earth's administrative branch of GAIL, which engages in exploration and colonization of space on behalf of the people of Earth.

genus—the biological classification immediately above species; the scheme of classification in descending order is kingdom, phylum, class, order, *genus* and species.

hyperspace—all spaces outside of the "normal" space-time continuum in which the Earth and all known stars and planets exist. Starships travel through hyperspace in order to shorten travel time between the stars.

light year—an astronomical measure of *distance*

equal to the distance traveled by a beam of light in one Earth standard year.

obliquity—the angle between a planet's equator and the plane of its orbit about its sun (ecliptic)

parsec—a measure of distance approximately equal to 3.26 light years, now rarely used. A star lying one parsec from Earth would display an annual parrallax of one second (travel in an elliptic path spanning one second of arc in the sky).

photon drive—a type of modern starship engine, sometimes called a Planck's torch

phylum—the biological classification immediately below kingdom (see genus). Human beings belong to the phylum chordata.

primary—in astronomy, the object about which a smaller object orbits. The sun is the primary of Earth and Earth is the primary of the moon.

protists—in the biological scheme of classification, a third kingdom consisting of life forms, mostly microscopic, which are neither plant nor animal

Sol—the proper name of the Earth's sun

standard Earth time—time measured relative to an atomic clock standard on Earth. Each hour of standard Earth time equals 1/24 of the Earth standard day. All times expressed in this bulletin have been converted to Earth standard times, though each planet has its own time standard.

sun—the star about which any planet orbits

star type—a system of classifying stars according to their temperature and size. Temperature classifications from hottest to coolest are O,B,A,F,G,K,M. The number following the letter further divides the scale into subclasses. For example a G2 star is hotter than a G3. A roman numeral designation following the temperature class indicates the size of the star, from I for giants to VI for subdwarfs. All habitable planets orbit dwarf stars of size V (five). Example; Earth's sun is a type G2V star.

terrestrial—on or of the Earth. Sometimes used in a more general sense to describe Earth-like planets. The eight colonies may be classified by planetologists as terrestrial.

Von Roenstadt's habitability factor—an index which considers a variety of physical and biological factors to gauge a planet's suitability for Human life. Though now superceded by more complex analysis, the Roenstadt factor provides a simple comparison for non-scientists. Earth is defined as 1.0. Planets habitable only in artificial environmental structures have Roenstadts less than 0.5. Totally uninhabitable worlds have Roenstadts of 0. GAILE will not consider colonizing planets with factors less than about 0.75.

warp—the deformation of space caused by gravity or by high rates of acceleration. Starships "warp" space prior to making the transition to hyperspace, hence this transition is sometimes called "space warp" or "warp maneuvers."

Appendix B

OTHER GAILE BULLETINS ABOUT THE PIONEERING PROGRAM

DATA FILE CODE	TITLE
GAI-SB-4-123	WYZDOM—The First New World in Space
GAI-SB-4-231	POSEIDOUS—Wondrous Water World
GAI-SB-4-303	BROBDINGNAG—A Guide to the Land of Giants
GAI-SB-4-330	GENESIS—Human Life on a Lifeless World
GAI-SB-4-427	MAMMON—World of Wealth and Contrast
GAI-SB-4-560	YOM—A Guide to the Triangle World Planet
GAI-SB-4-611	ROMULUS—Life on the Twin Worlds
GAI-SB-4-671	The Pioneer's Guide to Athena (with a Synopsis of the Precise Plan for Development)
GAI-GB-2-61	Starships of the Galactic Association
GAI-GB-2-16	The Galactic Association of Intelligent Life—A Short History for Humans

Appendix C

GENERAL INFORMATION BULLETINS FOR NON-HUMANS

Data Access Code	Applicable GAIL Member	Subject
717-GAI-43-4-732	Ardotians	Colonial Program
362-GAI-51-1-135	Chlorzi	Colonial Program
471-GAI-23-5-341	Ergints	GAIL Master Expansion Plan
534-GAI-13-6-542	Minutae	General Information about GAIL and text of proposed agreement

Appendix D

PRELIMINARY APPLICATION FOR THE GAILE PIONEERING PROGRAM

Instructions

Supply all of the following information using the data form in the back of this bulletin, or facsimile:

1. Print your full name and the address to which you want your computer response sent in the spaces indicated.
2. Print the following physical data in the spaces indicated:

   Your age in Earth years
   Your height in centimeters
   Your mass (Earth weight) in kilograms

   Indicate your sex by placing the appropriate number in the box.
   (1) M (2) F (3) TM (4) TF
3. Answer all of the following questions. Be sure to read the instructions for each section and each question carefully before answering. Answer as accurately and as honestly as you can. There are no "right" or "wrong" answers. Failure to answer honestly will simply alter the computer's fair assessment of your chances for selection.

Descending Order Preference:

Read each question and the responses that follow it carefully. Print the number of the response that indicates your *first* and *second preference* in the appropriate box.

A. Indicate your *first* and *second* choice for the planet where you wish to emigrate.
   (1) Wyzdom
   (2) Poseidous
   (3) Brobdingnag
   (4) Genesis
   (5) Mammon
   (6) Yom
   (7) Romulus
   (8) Athena
   (0) No second choice

B. Of the choices listed below, indicate the *most* important and the *second most* important reasons why you wish to apply for the pioneering program.
   (1) to seek adventure
   (2) to live amidst surroundings more beautiful and natural than Earth's
   (3) because I dislike the restrictive laws and political structure of Earth
   (4) because I relish the challenge of designing and building a new world
   (5) to ride aboard a starship
   (6) to have more room and open space, free from people
   (7) because new planets offer opportunities to become rich
   (0) none of the above (may be used twice)

C. Of the personal traits listed below, indicate those you have that will help you *most* and *second to most* in becoming a successful pioneer.
   (1) physical strength or athletic ability
   (2) intelligence or analytical ability
   (3) good health
   (4) resourcefulness, ability to adapt well to new situations
   (5) creativity

(6) sense of humor
(7) integrity, honesty
(0) none of the above (may be used twice)

D. Indicate which of the following responses describes *best* and *second best* your feelings of anxiety about becoming a pioneer.
(1) the thought of leaving Earth forever
(2) the thought of leaving family and friends forever
(3) fear of disease or dangerous alien life forms
(4) fear of space travel
(5) looking for a job in an unfamiliar society
(6) leaving all Earthly wealth and possessions behind
(7) fear of the unknown
(0) none of the above (may be used twice)

E. Indicate your *first* and *second* choice fields of employment.
(1) life sciences—biology, medicine, farming
(2) physical sciences—physics, engineering, geology, manufacturing, mining, energy or materials transformation
(3) Human relations—social work, business, sales, law, theology, teaching, child care
(4) information sciences—programming, writing, accounting, communications, speech therapy
(5) entertainment—art, music, sports
(0) other (may be used twice)

F. Some pioneers request to work aboard the starship in order to earn additional credits to spend on their new planets. Of the areas listed below, indicate the two in which you feel *most* qualified and *second most* qualified to work.
(1) social direction/counseling
(2) kitchen/serving/domestic work
(3) power and life support systems
(4) colonial planning/administration
(5) recreation direction/entertainment
(6) ship's security force
(7) child care
(0) don't know/none of the above (may be used twice)

G. Indicate which of the following shipboard facilities you think you will use on your journey *most* and *second* to most.
(1) exercise rooms and playing courts
(2) reference library
(3) theaters, holovision, sound and light shows
(4) scientific laboratories
(5) training classes, programmed instruction
(6) pioneers steering committee planning offices
(0) none of the above/no opinion (may be used twice)

H. Indicate which of the following you think are the *most* and *second most* important functions of government.
(1) police and legal protection, courts of law
(2) community social services and unemployment insurance
(3) centralized planning for a free enterprise economy
(4) control of all jobs and ownership of all means of production
(5) ownership and control of all land
(6) population control
(7) there should be no government
(0) none of the above/no opinion (may be used twice)

Multiple Choice:

Read each question and all responses carefully. Print the *one best answer* that best describes your feelings. Pick only *one* answer to avoid distorting your results.

I. If you are not selected for your first or second choice planets, would you go to any planet GAILE chooses for you?
(1) yes (0) no

J. How many family members do you wish to accompany you to your new planet?
(1) spouse only
(2) two or more spouses
(3) one child
(4) spouse and one child
(5) two or three children
(6) spouse and two children
(7) more than three others
(0) no others

K. If you are accepted to the pioneering program but your spouse or loved ones(s) are not, will you leave without them?
(1) yes (0) no

L. Have you ever been convicted of a felony or served a prison sentence?
yes (0) no

M. Are you a practicing professional, certified technician or experienced worker in one of the following fields?
(1) medicine
(2) electrokinetics

(3) spaceship operations or space construction work
(4) geology, or minerals extraction or processing
(5) information systems
(6) agriculture
(7) power generation
(0) no

N. On your new planet, would you consider entering a field other than the one in which you have specialized training and experience?
(1) no (2) yes (0) I have no specific field.

Personal Preference:

For each pair of statements listed below, print the number of the one that best describes your feelings. Even if you have no strong feelings about either statement, be sure to choose one from each pair.

O. (1) I can't wait to leave Earth.
(2) I can't wait to land on a new planet.

P. (1) I would like to see a live trup.
(2) I would like to tour a starship.

Q. (1) I would like to talk to a starship captain.
(2) I would like to talk to a colonial governor.

R (1) I enjoy changing jobs and learning to do new tasks.
(2) I enjoy working at a task until I become really proficient at it.

S. (1) I am concerned about my safety aboard the starship.
(2) I am concerned about my safety on a new planet.

T. (1) I appreciate the comforts and conveniences of modern technology.
(2) I long for a simpler, less complicated lifestyle.

U. (1) I think wealth and power are the best measures of success.
(2) I think happiness and freedom to do what I enjoy are the best measures of my success.

V. (1) I like to work at what I do best.
(2) I like to work wherever people need me.

W. (1) I like city life.
(2) I like rural life.

X. (1) When I get up in the morning, I like to wonder what the new day will bring me.
(2) I don't like things that upset my daily routine.

Y. (1) I like meeting new people.
(2) I like being with my friends.

Z. (1) The success of the colonies will depend on individuals' efforts.
(2) Fate or divine guidance will determine the future success of the colonies.

HAVE YOU ANSWERED ALL OF THE QUESTIONS?

4. If you are responding via data terminal, address your file to GAILEPRELIM4 and include your uniform billing code.

If you are responding via hardcopy, mail the data form or facsimile along with the processing fee of $3 in check or money order to:

GAILE Pioneering Program
P.O. Box 178387
San Diego, California
North American Region 92117
Earth, Sol

LEAVE BLANK:

GAILE – PIONEERING PROGRAM

PRELIMINARY APPLICATION – DATA FORM

APPLICANT START HERE:

NAME: ______________________
LAST

______________ ______________
FIRST MIDDLE

ADDRESS: ______________________
NUMBER & STREET

______________________ ________
CITY & STATE ZIP CODE

PHYSICAL DATA: ____ ____ ____ ____
AGE HEIGHT – CM WEIGHT – KG SEX

— DESCENDING ORDER PREFERENCE —

QUESTION →	A	B	C	D	E	F	G	H
FIRST CHOICE	☐	☐	☐	☐	☐	☐	☐	☐
SECOND CHOICE	☐	☐	☐	☐	☐	☐	☐	☐

— MULTIPLE CHOICE —

QUESTION →	I	J	K	L	M	N
	☐	☐	☐	☐	☐	☐

— PERSONAL PREFERENCE —

QUESTION →	O	P	Q	R	S	T	U	V	W	X	Y	Z
	☐	☐	☐	☐	☐	☐	☐	☐	☐	☐	☐	☐

SEND COMPLETED DATA FORM WITH $3 PROCESSING FEE TO:

GAILE PIONEERING PROGRAM
P.O. BOX 178387
SAN DIEGO, CALIFORNIA 92117
NORTH AMERICAN REGION
EARTH, SOL